Pauper & Prince

PAUPER & PRINCE

RITCHEY, HALE, & BIG AMERICAN TELESCOPES

Donald E. Osterbrock

THE UNIVERSITY OF ARIZONA PRESS
Tucson & London

The University of Arizona Press

∞ This book is printed on acid-free, archival-quality paper.
Manufactured in the United States of America.

98 97 96 95 94 93 6 5 4 3 2 1

Library of Congress Cataloging-in-Publication Data
Osterbrock, Donald E.
Pauper and prince : Ritchey, Hale and big American telescopes / Donald E. Osterbrock.
p. cm.
Includes bibliographical references and index.
ISBN 0-8165-1199-3
1. Ritchey, G. W. (George Willis), b. 1854. 2. Hale, George Ellery, 1868–1938. 3. Astronomy—United States—History. 4. Telescopes—Design and construction—History. 5. Astronomers—United States—Biography. I. Title.
QB36.R39088 1993
522'.1973'09—dc20 92-42704
CIP

British Library Cataloguing-in-Publication Data
A catalogue record for this book is available from the British Library.

Dedicated to the memory of

John S. Hall

(1908–1991)

Friend, fellow astronomer, and historian,
student of George Willis Ritchey's
life and career

Contents

Illustrations

Preface

GEORGE WILLIS RITCHEY was a famous name in American astronomy in the first quarter of the twentieth century. He was the prophet and builder of the first large, successful American reflecting telescopes. An optician without equal, he was also a creative telescope designer and a master of astronomical photography. Ritchey worked closely with George Ellery Hale, the phenomenal fund-raiser and organizer of observatories. Yet their very success led to tension, estrangement, and finally to Ritchey's dismissal and banishment by Hale. Most astronomers, scientists and historians today know of Ritchey only through the reports of Hale and his close associate and successor as director of Mount Wilson Observatory, Walter S. Adams. We think of Ritchey in connection with the Ritchey-Chrétien system, used in almost all large reflecting telescopes today, and perhaps with the 100-inch Mount Wilson mirror, but know little about his life and career. Both ended in apparent failure, and only scanty records of his achievements remain. Hale and Adams clashed with Ritchey, and won out, at least in the short run. Their version of Ritchey's story has survived. Yet in the end, his visions have proved true. Our present very large telescopes have demonstrated the validity of many of Ritchey's predictions, which seemed fantastic if not impossible to most of the astronomers of his day. This book tells his story.

His career was so intertwined with Hale's that it is impossible to treat either one's career without often mentioning the other. Hence this book focuses on Ritchey, but includes the telescope-building side of Hale's life, and his interactions with Ritchey. It is the story of Ritchey, the optical expert, sure of his own ideas, impractical, visionary, but with all the right ideas, far ahead of his time; of Hale, the promoter, wonderful at conceiving new projects, observatories and observing programs, skillful at explaining his ideas in simple terms and persuading very wealthy men to provide the funds for them, expert at collecting power into his own hands and using it to bring his dreams to fruition; and of Adams, the research scientist, stable, sane, with a fierce drive to use telescopes to study the universe. He made himself indispensable to Hale, and undermined Ritchey. In the end he survived them both. History does repeat itself, and these same three types endure in each generation, though few today have the near mythic qualities of Ritchey, Hale and Adams.

My own interest in Ritchey was enhanced by the fact that he lived for many years of his youth and young manhood in Cincinnati, where I was born and grew up. His brother, J. Warren Ritchey, was a teacher whom I saw often at Hughes High School, where I was a student. A generation earlier both of them had been students at the same school, at its original, downtown location. I have visited the old Ohio River towns where Ritchey was a boy, Tuppers Plains and Pomeroy, Ohio and Evansville, Indiana, have lived in Chicago, where he lived, and have worked at Yerkes and Mount Wilson observatories, where he worked years before. These associations all helped me understand his life.

Many people have aided me in my work on this book. First and foremost I thank the late John S. Hall, my friend, fellow astronomer, historian, and student of Ritchey's life and career. He shared his insights and source materials freely with me, and read drafts of all the chapters in this book. His comments have been invaluable to me. To his memory the book is dedicated.

Another astronomer who helped me very greatly is James Felten, an expert genealogical researcher. He found many facts for me, especially about Ritchey's family, and their life in southeastern Ohio. My brother,

Carl H. Osterbrock, and my friend, John E. Ventre, also helped greatly with local research in Cincinnati, and at its university. Ritchey's niece, Catherine Ritchey Miller, very kindly provided me with many important and written records of their family's history, as well as with her own keen insights.

My main source of information was written material in archives, and I am especially grateful to Dorothy Schaumberg at the Shane Archives, Judith Lola Bausch at Yerkes Observatory, Janice Goldblum, at the National Academy of Sciences, Ronald S. Brashear, at the Huntington Library, Brenda G. Corbin, at the U. S. Naval Observatory, and Ray Bowers and Susan Vasquez, both at the Carnegie Institution of Washington, for their intelligent, dedicated assistance in tracking down important new leads. I also thank James D. Ebert and Maxine F. Singer, at the Carnegie Institution, Lewis M. Hobbs, D. A. Harper and Richard G. Kron, at Yerkes Observatory, George W. Preston and Ray J. Weymann, at the Observatories of the Carnegie Institution of Washington, and Irwin I. Shapiro at Harvard College Observatory for opening their institutions' archives to me for this historical research. Horace W. Babcock and Allan Sandage were especially helpful in providing written records from the early days of Mount Wilson Observatory, as well as many acute insights into its history.

I am grateful to Malcolm S. Longair, of the Royal Observatory, Edinburgh, David W. Dewhirst, of the Institute of Astronomy, Cambridge, and Peter D. Hingley, of the Royal Astronomical Society for the letters and other material they sent me concerning Ritchey's correspondents and activities in Great Britain. Lubos Perek provided important information from Czechoslovakia on Vincent Nechvíle.

I am particularly grateful to Françoise Le Guet Tully, who searched the Cercle Henri Chrétien archives in Nice, France for material on Ritchey, and sent copies of all of it to me. She located additional material at the Academy of Sciences archives in Paris, and at the St. Gobain Archives, in Blois. She, Sylvain Veilleux, and Stephane Courteau, the latter both then graduate students at Lick Observatory, helped me greatly in understanding fully some of the letters and memoranda in French.

Leon Panetta and his Congressional office staff were very helpful in providing references to bills and committee hearings involving Ritchey and the telescope he built for the U. S. Naval Observatory. Richard B. Burgess, of West Covina, California, allowed me to consult freely and use for my research his magnificent collection of Ritchey material. It includes very many photographs and manuscripts, as well as two experimental cellular mirror disks which Ritchey made. Ritchey left this material to his son Willis, who, before his death, gave it to Richard's late father Stanley Burgess, a neighbor and fellow manual training teacher in the Los Angeles school system. Some day I hope a wise museum director will exhibit these surviving physical examples of the work of a pioneering astronomical optician.

W. Malcolm Browne, who worked closely with Ritchey in Washington, helped me in several long interviews, which he followed up with written answers to my later questions. Clinton B. Ford, the late Ralph Haupt, Minoru "Paul" Nakada, and Chester Watts, Jr. also freely contributed their memories of Ritchey, through interviews or in writing. Daniel J. Schroeder, Harland W. Epps and Daniel H. Schulte provided me with explanations and calculations of many optical concepts, especially those connected with the Ritchey-Chrétien and related optical systems. Aden Meinel and Arthur A. Hoag added their memories of putting these systems into astronomical use. Owen Gingerich, John Lankford and Steven J. Dick, fellow historians of astronomy, sharpened many of my ideas and interpretations through discussions of my work in progress with me. To all of these friends I am extremely grateful.

My wife, Irene H. Osterbrock, counseled me throughout this work. Her first-hand knowledge of astronomy, astronomers and observatories, from a different perspective from my own, made it possible for her to see much that I otherwise would have missed.

Finally I wish to thank most sincerely Margaret Best, at the Institute for Advanced Study, and Gerri McLellan, at Lick Observatory, who entered the entire manuscript in word processors, and went through revision after revision of each chapter with me, and Pat Shand, who made the final corrections and revision at Lick. I greatly appreciate the skill, accuracy, dedication and interest with which they carried out this work.

I did the historical research for this book over many years, going back to 1982. I wrote the first eight chapters during 1989–90, when I was at the Institute for Advanced Study on sabbatical leave, partly supported by an Ambrose Monnell Foundation fellowship grant for "distinguished visitors," while also holding the Otto Neugebauer Fellowship in the History of Science. I am very grateful to the Institute, its then director, Marvin L. Goldberger, and its professor of astronomy, John N. Bahcall, for their generous support of my work, and interest in it. I completed the book at Lick Observatory, and am also most grateful to it, its director Robert P. Kraft, and the University of California for their longtime support of my astronomical and historical efforts.

The appendix contains a brief, simple explanation of reflecting telescopes and the terms used to describe their mirrors and optical arrangements. Readers who are not familiar with these types of reflectors may wish to consult the appendix, and its Figure A.1, from time to time as they read the descriptions of Ritchey's telescopes.

Pauper & Prince

Ohio and Indiana

1864 - 1891

GEORGE WILLIS RITCHEY first looked through a telescope at the age of three, according to family legend. His father, a fine craftsman, had made it from two small glass lenses and a roll of paper.[1] A decade and a half later, a year or two after he finished high school, Ritchey bought an 8½-inch (diameter) telescope mirror from John A. Brashear,[2] the noted telescope and optical instrument maker, who had then only recently opened his first small shop.[3] It was a high-quality mirror, which probably cost about $60, approximately the amount Ritchey earned at his job each month.[4] This purchase indicates his very real interest in astronomy and also the difficulties he had in financing it. Over and over in the years to come Ritchey was to build telescopes for others that he could not afford himself, and to spend or lose all his own money several times over in the process.

Ritchey's work and ideas would completely change the science of astronomy. When he came into the world on December 31, 1864, the largest telescope in America was the 18½-inch refractor built by Alvan Clark & Sons for the University of Mississippi, which had gone instead to the Dearborn Observatory in Chicago. When he died on November 4, 1945, the largest telescope in the world was the 200-inch reflector, built by the

California Institute of Technology and approaching completion on Palomar Mountain, California. George Willis Ritchey did not build either of these telescopes, but he was the leader of the revolution in concept and scale that lay between them.

Ritchey was born on the last day of the last full year of the Civil War, the year of Ulysses S. Grant's costly victories in Virginia and William T. Sherman's burning of Atlanta and march to the sea. Ritchey died in the year World War II ended, the year of Adolf Hitler's death, of the surrender of the German Wehrmacht, of Hiroshima and Nagasaki. Ritchey was born in a little settlement in the Ohio River Valley; he died on an orange ranch in the outskirts of the megalopolis of Los Angeles. America had developed from the barely industrialized eastward-facing republic of Abraham Lincoln to the technological, transcontinental nation of Harry Truman. Ritchey and American astronomy grew and developed with the country; their telescopes grew and developed with them. Much of what we know about the nature of the universe, nearly everything that we knew about it before the coming of the space age, was learned through the giant reflecting telescopes that Ritchey dreamed of, built, and foretold.

Ritchey was a product of the melting pot of his time, a mixture of immigrant and old American stock. His grandfather, George Ritchey, had brought his young family to the United States from Ireland in 1841. George Ritchey was born about 1809 in County Down, Ireland. It is one of the Six Counties, the traditionally Protestant region of Ulster that is now Northern Ireland. George Ritchey's background, like those of most of his friends and neighbors, was Scottish and Presbyterian; his parents had come over the sea from Scotland after the final defeat of the Irish rebellion in 1798. He married Sarah Clifford, another Scotch-Irish Presbyterian, in Bangor, County Down, about 1833. They both spoke with strong Scottish accents throughout their lives and worshipped a mighty, avenging Presbyterian God.[5]

Although we know practically nothing about his life in Ireland, George Ritchey must have been mechanically inclined, as all his sons and grandsons were after him. Probably he worked at a small mill and learned to maintain and operate steam engines, just then coming into widespread use. George and Sarah Ritchey lived in Bangor, a few miles northeast of Belfast, the largest city and industrial center of Ulster. In Bangor the Ritcheys' first child, James, who became George Willis

Ritchey's father, was born on May 28, 1834. Four years later came Samuel, and then, after another two years, John.[6] As their family grew, the Ritcheys saw the situation in Ireland deteriorate. In the south, essentially an occupied nation, the English rack-rent policies impoverished the Catholic peasantry. In the north the British divide-and-conquer tactics favored the Scotch-Irish Protestants but held them back industrially in comparison with England. Clearly, economic bad times were coming.

George Ritchey, like many of his compatriots, decided to take his family to the New World while he still could. In the early summer of 1841 he and his wife sold most of their possessions and with their three sons sailed across the Irish Sea from Belfast to Liverpool, the main port of embarkation for America. On July 18, with ninety-four other passengers, nearly all of them mechanics, farmers, or laborers, and their families, they embarked in the brig *Congress*, bound for the United States.[7] In fifty-two days under sail they crossed the North Atlantic Ocean and landed in New York on September 7, 1841. They had left the United Kingdom of Great Britain and Ireland in the fourth year of young Queen Victoria's reign, to become citizens of the United States of John C. Calhoun, Henry Clay, and Daniel Webster.

The Ritcheys came at the beginning of the great wave of immigration from Ireland; it crested later in the decade as the combination of absentee rule and the potato blight culminated in the great famine of 1846 and 1847. The only hope for the poverty-stricken Catholic peasants of the south was to escape to New York, Boston, and other large American cities on the East Coast, where they landed. George Ritchey was not poor. He was not rich either, but he could afford to take his family directly to Meigs County, Ohio, traveling inland by lake, river, and canal boat, wagon, and perhaps railroad. There on October 25, 1841, he bought a farm at the crossroads village of Tuppers Plains.

Southeastern Ohio, where Meigs County is located, was first settled only half a century earlier, in the years just after 1794, when Mad Anthony Wayne and his little army defeated the Indians at the Battle of Fallen Timbers, far to the northwest. That victory broke the power of the tribes, leaving the whole state open to unchallenged white settlement. Pomeroy, the county seat of Meigs County, was on the Ohio River forty miles south of Marietta, the first American settlement in Ohio that survived, and the first capital of the Northwest Territory. Ohio became a state in 1803, and by 1841, when George Ritchey and his family arrived, the population of Meigs County had grown to twelve thousand. Tuppers Plains, where he bought his farm, is 5 miles west of the Ohio River, between Marietta and Pomeroy, almost 150 miles due east of Cincinnati, at that time the largest city in Ohio.

The very first settlers in Meigs County had come from nearby Virginia (now West Virginia), but most of the early pioneers were New Englanders. Meigs County is hilly country, rich in coal and salt deposits, the basis of the early industry in the region. However, the area around Tuppers Plains is flat, rich agricultural land that was heavily farmed for wheat, corn, and hay in 1841 when the Ritchey family settled there. By then there were many other Scotch-Irish Presbyterians just like the Ritcheys living in Meigs County.[8] The Ritcheys' farm was in Olive Township, along its boundary with Orange Township, the name and color symbols for Protestant domination of Northern Ireland.

George Ritchey was hard working, intelligent, and successful. He prospered in Ohio. He bought his first farm for $146 at a sheriff's sale; less than a year later, in July 1842, he bought a larger, adjoining farm for $1,000, partly financed by a mortgage. Included in this purchase was a mill, the only one in the area. Probably, it was a grain mill, powered by horses walking on a treadmill.[9] George Ritchey, with his mechanical knowledge and experience, converted it to steam and added power saws, making it over into a combined grist- (grain) and sawmill. The surrounding hills were heavily wooded, and Ritchey's sawmill converted the trees into lumber for the barns and farmhouses that went up as the population of Orange and Olive townships grew.

As the Ritcheys prospered, the family grew, too. The Ritcheys' fourth child and first daughter, Sarah, was born around 1842, shortly after their arrival in America. In 1844 came George, and then in succession three daughters, Jane, Mary, and Frances, the last born about 1854.[10] By then, George Ritchey was a solid pillar of the community; in 1846 he and his wife Sarah had deeded to the Trustees of the Orthodox Presbyterian Church of Tuppers Plains, for the nominal price of one dollar, a parcel of land six rods square "to erect a house thereon for the worship of Almighty God." A year later he was elected one of the Trustees himself.[11]

All the children attended school in Tuppers Plains, and in 1851 the Ritcheys sent their oldest son James to nearby Marietta College, a Presbyterian institution that was one of the first colleges in Ohio. He was one of sixteen freshmen, all male, and he registered in the Scientific Department. He only stayed at Marietta for one year, but he studied algebra, geometry, trigonometry, mensuration, surveying, navigation, analytic geometry, and drawing, as well as English, French, and the Bible.[12] James Ritchey was very musically inclined. He was an excellent singer and learned to play the organ. He had worked with his father in the mill since childhood and had learned how to make fine wooden furniture. After he returned from college he went to work full-time at the farm and mill. James combined his woodworking skills and musical talents, making himself a melodeon so that he could play at home. Before long he became the church organist, then its choir director.[13]

On January 1, 1863, when he was twenty-eight years old, James Ritchey married Eliza Arvilla Gould. She was of old American stock, a direct descendant of John and Priscilla Alden of the Mayflower and Plymouth Plantation. On her father's side Eliza was also descended from Zaccheus Gould, who had emigrated to America only eighteen years after the Mayflower's first voyage. Many generations later their descendants Benjamin Gould and Lydia Alden were married, and in 1808 they moved on from western Massachusetts to Marietta. Some years later they settled a few miles north in Salem Township, Ohio (near the present Lower Salem). Its Presbyterian church was the home of one of the first temperance societies in Ohio and did not admit anyone "who will not abstain . . . from ardent spirits." In this church the Goulds' son Daniel married Anna L. Sharp in 1831. She was a daughter of John Sharp, "a man of large stature and inclined to corpulency" who was also "a man of force and influence in the affairs of Washington County in his time." He had been a state representative, a state senator, and a county judge, and during the War of 1812 had recruited a company of the Ohio Volunteer Infantry and served as its captain. Sharp had set out the first apple orchard on the Little Muskingum River and was called by an admiring biographer "the Solon of his community."[14]

Daniel and Anna Gould had ten children, of whom Eliza was the eighth. Her mother died when Eliza was only six years old, and her father died just three years later.[15] Eliza was raised by her stepmother and relatives and had to do housework to earn her keep, but she managed to attend various country schools and then a small-town academy. She taught school for a year or two before she married James Ritchey on January 1, 1863, just a month before her twenty-first birthday.[16] Later that same year James's brother Sam married his bride Anna. Both sons continued to work with their father at the family mill.

Their two younger brothers were fighting with the Union Army in the Civil War by then. They had enlisted in the 116th Ohio Volunteer Infantry in August 1862. Almost immediately, after only a few days of drilling in their camp at Marietta, the regiment was rushed across the Ohio River to the defense of Parkersburg, threatened by a Confederate raid. Within a few weeks the regiment joined the Federal service, and after one month's training, they were sent to the field in Virginia. Their first battle was in January 1863, a few days after their brother James's wedding. In March, the 116th marched into the Shenandoah Valley, where they spent the best part of two years fighting for the Union. They ended the war in Grant's army before Richmond, John by then the regimental sergeant-major, and George, who had been named the best soldier in the regiment in February 1865, a corporal.[17]

Meanwhile, in Tuppers Plains, James and Eliza Ritchey's first child, Luther Gould Ritchey, was born on October 1, 1863. He was named for Eliza's brother, William Luther Gould, who had been killed in 1862 while serving in the Union Army in Virginia. Their second son, George Willis Ritchey, always known in the family as Willis, was born December 31, 1864. He was named for his grandfather and for his mother's brother James Willis Gould, who had died as a baby. Another son, Samuel Edward (Ned) Ritchey, came a year and a half later, and then, at the end of 1869, a fourth son, James Warren Ritchey.[18]

In the fall of 1867 James and Eliza Ritchey and their family moved from Tuppers Plains to Pomeroy, as did George and Sarah Ritchey and their younger daughters Jane, Mary, and Frances.[19] Samuel and Anna Ritchey stayed at Tuppers Plains and took over the farm.

Meigs County's population had grown to 31,000 and Pomeroy's to 5,000. It was a long, narrow town, crowded along one main street in the narrow strip of flat land between the Ohio River and the hills just to its west. Pomeroy was a prosperous little industrial center. Twenty-seven salt "furnaces" produced five thousand barrels of salt each day. Their coal fires belched clouds of black smoke as they evaporated the water from the brine, drawn up by pumps from deep wells. In addition, there were steel mills, machine shops, tanneries, a brewery, and an organ factory.

George Ritchey had become a wealthy entrepreneur. His mill was successful and with its profits he started a furniture business in Pome-

(*Left*) Eliza Gould Ritchey, George Willis Ritchey's mother. Courtesy of Catherine Ritchey Miller.

(*Right*) James Ritchey, George Willis Ritchey's father. Courtesy of Catherine Ritchey Miller.

roy. He and his son James were skilled cabinetmakers, and Ritchey & Son was soon a profitable enterprise. They specialized in high-quality, custom-made furniture.[20] In 1870 George Ritchey reported the value of the real estate he owned, chiefly his furniture factory, warehouse, and store, as $12,000, and his personal fortune as $4,000.[21] Those figures should be multiplied by a factor of fifteen to translate them into modern terms. The Scotch-Irish immigrant had become a wealthy American. His daughter Sarah had married T. Curtis Smith, who had been the regimental surgeon for the 116th Ohio Volunteer Infantry and was now a leading light of the Meigs County Medical Society.

Then on January 8, 1871, disaster struck. A burning ember carried by the wind struck the roof of Ritchey & Son's building and set it on fire. The whole establishment, the furniture it contained, and the lumber piled high around it burned to ashes. It was a total loss, and it was uninsured.[22] George and James Ritchey had to begin over. For a time they thought of moving to Charleston, West Virginia, where George, Jr., was living, but in the end they decided to rebuild in Middleport, the thriving town at the landing just south of Pomeroy.[23] In 1871, with Smith and other investors, they incorporated as the Middleport Furniture Company, borrowing heavily to get started again.[24] They built an imposing brick factory.[25] As in Pomeroy, they specialized in high-quality, made-to-order furniture pieces.[26] But it was a bad time to be in debt. Before long the Panic of 1873, a great depression that actually lasted many years, overtook them. Business fell off, they could not meet their obligations, and the business failed. Although family legend blamed it on the "old man's [George's] pigheadedness," economic forces over which he had no control were undoubtedly the real culprits.[27] On September 25, 1875, an auctioneer sold the Middleport Furniture Company at a public sale. It had cost $20,000 three years before; it went for $8,258 at the auction.[28] The Ritcheys were poor again.

A few months later James Ritchey took his family west to make a new start in Evansville, Indiana.[29] It is another old Ohio River town, just a few miles east of the mouth of the Wabash River, the border there between Illinois and Indiana. James and his brother George had gone to Evansville immediately after the auction and had found jobs as cabinetmakers at the Armstrong Furniture Company. Once they were established in Evansville, James brought out his wife and children in April 1876. Willis was then just eleven years old.

Within a year James had risen to foreman in the Armstrong Furniture Factory. He was a superb craftsman and an effective leader. Within a few months he superintended the building of a large paneled bank counter, "the finest wooden counter in the State and probably not excelled in this country," as a special order for the First National Bank of Middleport.[30] It was shipped back to Meigs County by riverboat. George, Sr., soon joined them in Evansville and before long was also working in the factory as a cabinetmaker, under his son. Theirs had always been a large, mutually supportive family, and now James, just over forty years old, had become the actual head of the clan. They all worked hard and were paying off their debts in Pomeroy. James and Eliza Ritchey's last child, John Paul Ritchey, was born in 1878; their oldest son, Luther, started work in the furniture factory during his summer vacation from school when he was fifteen years old.

In November of that year, James Ritchey moved his family again, to Cincinnati, the Queen City of the West, as it was then often called. Situated on the Ohio River roughly halfway between Pomeroy and Evansville, it was a metropolis of 250,000 people. James Ritchey became the foreman at the Meader Furniture Company's factory. His brother George went back to farming, and their father returned to Middleport. The old man's wife, Sarah, had died sometime between 1870 and 1880, but his three unmarried daughters were still living with him.[31]

Less than a year after James and Eliza Ritchey moved to Cincinnati, their oldest son Luther died at the age of sixteen. No record of the cause of death exists today, but it was probably a fever or disease, many of which could be lethal in those days before antibiotics.[32] Willis was left the oldest of four brothers. All of them had attended school from the age of five or six. Willis had started in Pomeroy, finished grade school in Evansville, and started high school there. In Cincinnati he attended Hughes High School, then located at Fifth and Mound streets, only a few blocks from his family's home at Mound and Chestnut streets. His younger brother Warren, who later taught physics and mathematics at Hughes, remembered that Willis had been a student there for a year and a half, and presumably he graduated in the summer of 1881, at the age of sixteen.[33]

Young Willis went to work at the Meader Furniture Factory in 1881, as his brother Luther had before him, and as his brother Ned would soon afterward. Everyone in the Ritchey family helped out.

In 1882 Willis entered the University of Cincinnati, a small municipal institution then only twelve years old. Tuition was free for all residents of Cincinnati. Ritchey enrolled in the School of Design (alternatively called at various times the Art Department).[34] It was located on Walnut Street between Fourth and Fifth streets, in the downtown area less than a mile from Ritchey's home. In the first-year class in drawing and design he had seventy-seven fellow students, the majority of them women. Ritchey's interest was probably largely in the "design," or mechanical drawing, aspects of the course. It was a four-year program, but after only one year he transferred to the special scientific course for 1883–84. It was taught in the Academic Department, which was located at a different site between Clifton and McMicken avenues, just at the edge of the central Cincinnati basin. It was little over a mile north of Ritchey's home. Being a student in the "special" scientific course meant that he could concentrate entirely on science but would not earn a degree.[35]

After only one year in this course, Ritchey dropped out of the university and went back to work full-time at the furniture factory. He was planning marriage to Lillie May Gray, and he had to earn some money. She was the daughter of Jones Gray, a carpenter and cabinetmaker who lived and worked in Middleport for many years, undoubtedly at least some of them in the Ritcheys' furniture company. Gray's wife Margaret died when Lillie was very young, and the girl had to do housework to earn her board from an early age. Lillie's adopted brother, James Gray, another cabinetmaker, had gone to Evansville to work with the Ritcheys. Their families must have been close and may have been distantly related. Lillie Gray and Willis Ritchey were married in Cincinnati on April 8, 1886. He was then twenty-one years old and she was nineteen, both far too young for marriage in his parents' opinion, but "contrary to all [their] expectations [Lillie] made him a fine and helpful wife."[36] Willis and Lillie moved to a separate flat at Ninth and Mound streets, close to his parents' home.

By now Willis was certainly deeply interested in astronomy. According to his later recollection, he had decided to become a professional astronomer before he finished high school.[37] He had visited the Cincinnati Observatory for the first time on February 22, 1883, while still a student in the School of Design, and he returned frequently, often with groups of his friends.[38] The observatory was, according to its partisans, the oldest professional observatory in the United States; it had been completed in January 1845, a few months before Harvard College Observatory. The Cincinnati Observatory had been the project of Ormsby MacKnight Mitchel, a graduate of West Point and a popular lecturer. He had convinced the citizens of Cincinnati that they needed an observatory, organized the Cincinnati Astronomical Society, and sold enough memberships in it at twenty-five dollars a share to buy a 12-inch refractor and erect a building to house it. The site was on Mount Ida, one of the seven hills on which Cincinnatians liked to say their city had been built. It overlooks the central basin of the city. John Quincy Adams, the former president of the United States, who had lobbied unsuccessfully for a national observatory, came west from Massachusetts to give the dedication speech on November 10, 1843. The grateful citizens of Cincinnati changed the name of Mount Ida to Mount Adams in his honor.

In 1860 Mitchel, who received no salary and had to support himself by lecturing and writing about astronomy, departed for greener pastures.[39] The observatory, with no program or reason for existence except so that its members might look at stars through the 12-inch, gradually decayed. Then in 1870 the University of Cincinnati came into existence as a municipal institution. Its leaders naturally felt that the observatory should become part of their organization. The Mount Adams site was by then unsuitable because of smoke pollution, the great bane of observatories located in the cities of those days. The soft coal used for heating and for running factory machinery poured out tremendous quantities of thick, black smoke, through chimneys innocent of any suppression devices. Whether the sky was cloudy or not, it was often covered by a dark pall of smoke. Ash and soot were deposited on dome and telescope lens alike.

Thus it was not difficult for all concerned with the Cincinnati Observatory to work out a mutually beneficial arrangement. The university acquired the telescope and moved it to a new site on Mount Lookout in the eastern part of the city, far from the smoke of the central industrial basin. The land on Mount Adams was released and sold to the Catholic Church, allowing the Astronomical Society to pay off its debts. (The Holy Cross Church was built on this site; later the Monastery, a modern office building, was erected in its place.) On August 28, 1873, the cornerstone that Adams had laid was ceremoniously transferred to the new site, and by 1874 the observatory was once again in business.[40]

In September 1886 Ritchey re-entered the university, now as a regular first-year student and candidate for a B.S. degree. He was given a part-time job as student assistant. No doubt this was largely because of his strong interest in astronomy, but also it reflected his skill and ability with any kind of tools. To make ends meet he had to continue to work part-time at his regular cabinetmaking job.

The Cincinnati Observatory was then in the doldrums. After the university had taken it over, the first director had been Ormond Stone, but he left for a better job at the University of Virginia in 1882, before Ritchey ever visited the observatory. After Stone's departure Herbert C. Wilson, the one and only graduate student in astronomy, was put in charge as "Astronomer Pro-tem." In 1884 Jermain G. Porter came in as the new director, but Ritchey apparently had little contact with him. The young student's sponsor was Dr. Henry T. Eddy, whose title was professor of mathematics, civil engineering, and astronomy. He also was dean and executive officer of the tiny university from 1884 through 1888. Eddy taught in the Academic Department; Ritchey took his courses there but went to the observatory (by streetcar, or on foot if he came home after observing early in the morning before the streetcars began running) to work. He apparently received no inspiration from anyone, and his main memory afterwards was of studying on his own the papers of the English pioneers of reflecting telescopes, Andrew A. Common and Isaac Roberts, in the observatory library.

Ritchey, a craftsman like his father and grandfather, set up in his home his first little "astronomical laboratory," or shop, as anyone else would have called it. There at Ninth and Mound streets, while he was still a student at the University of Cincinnati, he made, according to his later memory, his first model of a telescope mounting on his "flotation" principle, which he was first to use on the 60-inch Mount Wilson reflector twenty years later.[41] And there, or previously in his parents' home, he certainly did build his own first reflecting telescope, using the mirror he had bought from Brashear but making the mounting himself.[42]

Though he had resolved to become an astronomer, Ritchey decided within two years that he could remain a student no longer. He had to support himself and his wife, and his job would leave too little time for his studies. No doubt he also wanted to get out into a more exciting part of the astronomical world than the Cincinnati Observatory. He wrote Brashear in Pittsburgh to ask for a job in his optical shop. Brashear had no opening but encouraged Ritchey by telling him that he might find a position as an assistant to Samuel P. Langley, the professor of astronomy at the Western University of Pennsylvania (now the University of Pittsburgh) and director of its Allegheny Observatory. In December 1887 Ritchey applied there. He wrote that he was experienced in astronomical computations and at observing with the transit circle, used for determining time. He knew analytic geometry and trigonometry, and had begun calculus, then a second-year university subject. He was eager to work and he needed money.[43]

Unfortunately for his hopes, Langley had no job for him. Ritchey begged for even a part-time job at Allegheny but to no avail.[44] Shortly afterward, Ritchey dropped out of the University of Cincinnati for the second and last time and with his wife moved to Chicago.[45] He had managed to land a full-time job as a teacher of woodwork at the Chicago Manual Training School. Located at Michigan Avenue and 12th Street, a little south of the center of the city, it was an independent high school devoted to educating boys through shopwork as well as academic instruction. It was sponsored by the Commercial Club of Chicago, a businessmen's group, and its board of directors included Marshall Field, George M. Pullman, and several other wealthy Chicago civic leaders. Among its students were poor boys who were training for their lifework as mechanics and artisans, and rich boys whose fathers believed that the discipline of working with tools would prepare them for their roles as future company presidents.[46] Ritchey's skills as a cabinetmaker had brought him his new job as a woodwork teacher, even though he had less than two full years of college, without a single education course.

Ritchey's brother Ned later became a woodwork teacher also. After finishing Hughes High School and working full-time for a few years, he began teaching at the Cincinnati Technical School and then in 1892 followed his brother to Chicago. There he taught shop for many years at Lane Manual Training High School, part of the Chicago public school system. Ned wrote a textbook on woodwork that clearly expresses the shop teacher's philosophy, which he, his brother, and their father all shared. His book starts with a very good description of trees and their wood, including drawings and descriptions of leaves and seeds. The qualities of different types of wood, their uses, and their properties are given in great detail. Throughout the book there is great emphasis on exactness, care in work, checking, inspection, pride in doing a job well and in the finished product. "Perfectly" is one of the most frequent descriptions of how a piece of work is to be done. There are many questions to test the student's knowledge of what has been learned. The how and why of each operation, such as the use of a rip saw, the use of a hack saw, and so forth, are described in detail. The importance of understanding shop machines, such as lathes, how they work, and how to use them with maximum efficiency are carefully explained. The beauty as well as the utility of the final product is always emphasized.[47]

These principles guided Willis Ritchey throughout his career as a mirror and telescope maker. He was a genius at applying them, and they enabled him to produce outstanding new research telescopes. But they also infuriated many of the workmen in his shops who did not want to spend their whole lives listening to lectures, just as they infuriated many astronomers who were more anxious to start working with a new telescope as soon as it might be considered operational, rather than wait years for a perfect final instrument.

That was all far in the future though, when Willis and Lillie Ritchey moved to Chicago in 1888. They lived in the suburban South Side, first near 35th Street, later in Englewood near 59th Street.[48] From there Ritchey could take the streetcar, or later the El, to the Manual Training School. Both the Ritcheys' children were born in Chicago, their daughter Elfleda on April 3, 1889, and their son Willis Gray on November 11, 1890.[49] Ritchey again had his own "private laboratory" in the basement. He also had access to the tools and machines at the school where he worked. He spent as much time as he could working and experimenting busily on his astronomical ideas. Then in 1891 he met George Ellery Hale, whose life was to be entwined with his own, for better or for worse, for the next forty-seven years.[50]

Kenwood Observatory

1868 - 1896

GEORGE ELLERY HALE came from a completely different world from that of George Willis Ritchey. Hale's parents were both born in New England of old American stock, stretching back for generation after generation to the early Pilgrim settlers. His father, William Ellery Hale, was the son of a Congregational minister who had preached temperance for many years, and then gave up the pulpit and served God in the world by starting an insurance company which sold policies only to teetotalers. Later he became an officer in a paper company in Beloit, Wisconsin, and there William E. Hale began his business career. George Ellery Hale's mother was the daughter of another Congregational minister, who divorced his wife and became a homeopathic physician. William E. Hale and Mary Browne met through their New England connections and were married in 1862. They set up housekeeping in Chicago, where he had moved as the agent of the Rock River Paper Company.[1]

William E. and Mary B. Hale had two children, both of whom died in the first year of their lives, before George Ellery Hale was born on June 29, 1868. At that time the Hales lived in a house on North La Salle Street, in what is now the center of the Chicago financial district. In 1870 they moved to Kenwood, a prosperous suburb six miles south of their former home. The next year the great Chicago fire destroyed the whole downtown area, including North La Salle Street, but they were safe on the South Side.

William E. Hale had become a highly successful businessman, but much of his money was tied up in the Hale Building at State Street and Washington Boulevard, one block from the official center of Chicago. The building burned to the ground in the fire, but Hale's credit was good and he rebuilt and roared to success with Chicago. With his brother George W. Hale he founded an elevator company. It was the right business to be in at the right place at the right time. Chicago sprang up like a phoenix from the ashes, in the form of the skyscrapers that made its skyline famous, and William E. Hale & Co. supplied the elevators that made those skyscrapers possible. William E. Hale was soon a very wealthy man.[2]

George Ellery Hale's parents doted on him. He was infinitely precious to them, for they feared that, like their first two children, he would die as an infant. He was a sickly boy, but he survived. Soon he had a sister, Martha, two years younger, and then, after another five years, a brother, William B. ("Will"), but George was always the favorite. He started his education at Oakland Public School, near his home, but he did not like it and was frequently sick. Finally, he came home with typhoid fever when he was twelve. When he recovered, his parents took him out of Oakland and put him in the Allen Academy, a private school for boys at Michigan Avenue and 23rd Street, halfway between his home and downtown Chicago. The academy's headmaster, Ira W. Allen, a Harvard graduate who had studied in Germany, was a stern disciplinarian but a supportive mentor. He encouraged young George in his growing interest in science. During several of the years he was at Oakland and the Allen Academy, Hale and his brother and sister also had a private tutor, Dr. Edward F. Williams, who lived in their parents' house with them and helped them with their studies.[3] He was another Congregational minister.

As he was approaching his fourteenth birthday, young George met his first real astronomer. He was Sherburne W. Burnham, then forty-four years old, who lived in Chicago, between the Hale's home and the Allen Academy. Burnham, originally a New Englander, had served as a civilian stenographer with the Union Army in the Civil War and after it in the occupation of New Orleans. He had become interested in astronomy, and, after he settled in Chicago, had bought an excellent 6-inch telescope from Alvan Clark & Sons, the great American telescope making firm. With this instrument Burnham had become a skilled observer and discoverer of double stars, pairs of stars that revolve in orbits about one another, much as the moon revolves about the earth. Double-star orbital periods are much longer than a month, but by measuring the star's positions with respect to one another over many years it is possible eventually to plot out their orbits and deduce their dimensions. Burnham, with his keen eye and unrivalled powers of concentration, had made himself a world famous expert in this field. He had been hired by the James Lick Board of Trust to take his telescope to test as an observing site Mount Hamilton, California, where they planned to erect a telescope "superior to and more powerful than any telescope yet made" with the $700,000 Lick had left for the purpose. Burnham had spent six weeks at Mount Hamilton in the autumn of 1879 and discovered forty-two "new" (previously unknown) double stars; he pronounced it an excellent site and the Lick Trustees had started construction. In 1881 Burnham had left Chicago again for a job as an astronomer at Washburn Observatory of the University of Wisconsin, but that had lasted less than a year, and now he was back again, working by day as a court reporter in the United States District Court and observing double stars in his backyard observatory by night.[4]

Hale had gone to see Burnham to get help with a telescope. The young student had tried to make one with a simple lens, but it did not work, and he was seeking expert advice. Burnham told Hale he needed something more sophisticated than that and said that he knew just where he could buy an excellent, second-hand Clark refractor. The previous year, in the first letter of his that has been preserved, Hale had asked Santa Claus for a "small tin tackle-box, like they have at Wilkinson's, a bait-box, some tackle, a good Strong jack-knife, a bob-sled, books 'ad infinitum'."[5] Now he rushed home and asked his father for the telescope. He needed it. He had a good reason and a deadline. There was to be a transit of Venus across the face of the sun, a rare astronomical event, on December 6, 1882, and he *had* to observe it. He argued so persuasively that his father went to see Burnham and asked where he could buy the telescope for his son. George had it before December 6, and he observed the transit. This scenario was to be reenacted many times over the course of the next sixteen years, until William E. Hale died in 1898.

The young student became more and more interested in astronomy. He set up the telescope, a first-class instrument, in his own little observatory on the roof of his parents' house. Martha and Will became his adoring assistants. Through Burnham, Hale met George W. Hough, the dignified, elderly professor of astronomy and director of Dearborn Observatory, located in Douglas Park, at 35th Street near the Lake Michigan shore. Hough showed Hale its telescope, the 18½-inch Clark refractor that had been the largest telescope in America when it was completed during the Civil War. Hough's speciality was observing the planet Jupiter; he and Burnham were practitioners of traditional positional and solar-system astronomy.

This old astronomy was boring to Hale, but he became interested in spectroscopy, the method of analyzing light, to study directly the atoms in laboratory light sources, their composition and nature. With a telescope, these same methods could be applied to the sun, the stars, and other astronomical light sources. Hale read avidly the books and magazines his parents brought home for him on these subjects. He made a little improvised spectroscope, but it was not good enough to show him what he wanted to see. He needed a better spectroscope. He told his father why he needed it. His father bought it for him. With this instrument Hale could see the bright spectral lines in kerosene and gas flames, electric arcs, sparks, and other laboratory sources, and some of those same lines as dark absorption features in the spectrum of the sun. A spectroscope analyzes, or spreads out the light in wavelength, or color, and each element has its own characteristic lines, which identify it wherever it is. Hale's first spectroscopes used glass prisms to break up the light into its individual wavelengths.

Soon he wanted a better spectroscope with higher dispersion—the ability to spread out the light more—and thus make it possible to see more lines. He decided he needed a diffraction grating, a reflecting mirror with thousands of parallel, closely spaced lines ruled on it, which is better than a prism for producing high dispersion. Henry A. Rowland, the great American physicist, had developed a method for ruling these gratings, which were sold only by John A. Brashear, the telescope and optical instrument maker in Allegheny, Pennsylvania. Hale wrote to

Brashear and ordered a small grating from him; when he received the grating, a letter from the kindly, outgoing optician and salesman came with it. A few months later Hale decided he needed a larger grating. Traveling alone by train, he went to Allegheny, a suburb of Pittsburgh, to get it. Brashear was amazed to find that George E. Hale of Chicago, who had been corresponding so learnedly with him about the lenses, prisms, and gratings he needed, was not a middle-aged scientist as he had imagined, but a seventeen-year-old boy. But Hale knew what he was talking about, and his father's money and the air of confidence that it gave him made Brashear take him seriously. Brashear introduced young Hale to Samuel P. Langley, the professor of astronomy and director of Allegheny Observatory. He was one of the great pioneers of American astrophysics, the study of the nature of the universe by spectroscopy. He had named this subject the "New Astronomy" and had written a book with this title; Hale became his devoted reader and followed his example in concentrating his research efforts on the sun.

In 1886 Hale graduated from the Allen Academy. His parents took him on a tour of Europe. In France he was introduced to Jules Janssen, the Director of Meudon Observatory, another pioneer astrophysicist and student of the sun. In London, Hale found an even better prism spectroscope than any he had at home. He decided he needed it. It cost £40, or $200 in American money, equivalent to about $3,000 in our present currency. His father gave him the money, and Hale brought this Browning spectroscope back to America with him.

In the fall Hale entered the Massachusetts Institute of Technology, then commonly known as "Boston Tech." By this time he had definitely decided on a research career in astrophysics. His father had sent him to MIT on the recommendation of Daniel H. Burnham, the noted Chicago architect whose firm, Burnham and Root, had designed many of the buildings for which William E. Hale & Co. had provided elevators. In his classes Hale was a good student, but not outstanding; he was more interested in his own research projects than in spending all his time studying for his formal courses. During Hale's freshman year at Tech, his family back in Chicago moved into a magnificent new mansion at 4545 Drexel Boulevard in Kenwood, designed to their specifications by Burnham and Root. There was an elevator in the house, and the attic had been designed for Hale's laboratory. The day after he arrived home for his summer vacation, he was at work in the lab. He had soon set up his grating instrument, now equipped for photography and hence a spectrograph, and mounted his heliostat, a flat mirror driven by clockwork to track the sun and bring its light into the laboratory. He put up his telescope outside the house on a specially poured concrete pier at ground level. This was science as Hale enjoyed it.[6]

In his second year at Boston Tech, Hale wrote to Edward C. Pickering, director of the nearby Harvard College Observatory, and offered his services as a volunteer assistant. Pickering, a former physics professor at MIT, gladly accepted, and soon Hale was spending every Saturday afternoon and evening at the observatory, with its 15-inch refractor. Pickering and his staff were doing astrophysical research, not the old-time astronomy, and Hale reveled in actual photography and spectroscopy. During that year his father had built for him, on a lot next to their home on Drexel Boulevard, a separate brick laboratory building, again designed by Burnham and Root.[7] It was Hale's own Kenwood Physical Laboratory. It had a spectroscopy room big enough for a longer-focus Rowland concave grating, a photographic darkroom, and an office for Hale.[8] Soon after he arrived at home he had moved his equipment to the new building, and was busy again with his experiments.[9] Before the summer was over, he was able to get good photographic spectrograms that confirmed the presence of carbon "flutings" (molecular bands) in the spectrum of the sun, a subject of lively controversy at the time.[10] Hale was doing real research with the instruments and laboratory his father had bought for him.

In September 1888 he returned to Boston Tech for his third year, packed with physics courses and German, which Hale disliked, as well. Still he found time for his volunteer work at the Harvard College Observatory each Saturday and not only enjoyed it but passed all his MIT courses satisfactorily.[11] After the exams were over, Hale went down to Baltimore and met the renowned Rowland in his laboratory at Johns Hopkins. Hale had first written to him years earlier, before he ever entered Boston Tech, but had received no answer. Now the great physicist was initially gruff to the young MIT junior, but Hale won him over, stayed three hours, and got a complete tour of the laboratory. In the end Rowland gave Hale permission to buy one of his largest gratings from Brashear. Just a few days later Hale visited Charles A. Young, the pioneer American astrophysicist, at his Princeton University Observatory. Young, a genial old fellow, was happy to show Hale the telescope and spectroscope, and together they looked at the spectrum of the chromosphere of the sun.[12]

As soon as Hale got back to Chicago for his summer vacation he went to work again in his own observatory. He remounted his gratings following the suggestions Rowland had given him and was soon obtaining spectra of the sun and of various laboratory sources, most of them metals like magnesium, manganese, tin, and calcium, in flames, arcs, and sparks.[13] For Hale's twenty-first birthday that summer, his father gave him a sizable block of stock in a family corporation that was building a new office building in downtown Chicago. His relatives elected young George one of the three directors of this company, along with his father and his Uncle George.[14]

Early in June, Hale's father had ordered, from Brashear, a large professional-quality spectrograph for him to use on the telescope at Harvard. Hale intended to do his MIT senior thesis on the sun using the spectrograph, though he had not yet decided what his specific topic would be. This spectrograph, built to specifications provided by Hale, was "going to be a beauty, considerably larger than the one [Brashear] made for the Lick [Observatory]," whose 36-inch telescope, which had gone into operation the previous year, was the largest in America.[15]

Late that summer, Hale had his brilliant idea, the invention of the spectroheliograph. He had long been studying prominences, the large, transitory structures that can be seen at the edge of the sun. They have a characteristic spectrum like the chromosphere, indicating that they consist of hot, transparent gas. Prominences were first discovered at eclipses, when the much brighter light of the sun is blocked out by the moon and they become visible. Young had found, twenty years before, that large prominences are bright enough so that their strongest emission lines can be seen with a spectroscope mounted on a telescope, even when the sun is not eclipsed. Similarly, their spectra can be photographed with a spectrograph without an eclipse. These methods show whether prominences are present around the edge of the sun (or limb of the sun, as astronomers say), and where they are located, but do not allow their forms to be studied. Hale's idea was to scan the slit of the spectrograph slowly across the sun and at the same time scan a photographic plate at exactly the same rate across a slit fixed in the focal plane of the spectrograph at the position of one of the bright emission lines of the prominences. This process would build up an image of the edge of the sun in the light of this emission line and would thus form a picture of the sun in this monochromatic light, showing the prominences all around the edge of the sun. It was an extremely clever instrumental idea, and Hale quickly wrote letters describing it to his college chum, Harry M. Goodwin, and to Young at Princeton to establish his priority to the idea.[16] Young wrote back that he thought that Hale's "photographing device" would succeed if he could get photographic plates sensitive at Hα, the red hydrogen line that is one of the strongest features in the spectra of prominences. Young was more dubious about using D3, the yellow helium line that is somewhat weaker, but the enthusiastic Hale had already decided that the older man thought his "prominence scheme" would "work perfectly."[17]

Hale decided to choose the demonstration of his new method of photographing solar prominences as the topic for his senior thesis. Pickering was glad to let him bring the spectrograph that Brashear was building to the Harvard College Observatory and do the observational work there. Brashear completed the instrument by early October. It cost Hale's father $1,000. The spectrograph was too heavy for the rickety old Harvard 15-inch refractor, as Brashear had warned Hale it might be, and Hale had to abandon his plan to use that telescope for his thesis.[18] Instead he used a 12-inch telescope, mounted horizontally in a fixed position on two heavy brick piers and equipped with a large heliostat to feed the sunlight into it. Hale's plan was a good one; the trouble, as with many of his future observing programs, was that he did not have enough time to carry it out. There were all kinds of adjustments to make, experiments to try, problems with the heliostat mirror and the telescope lens becoming distorted in the heat of the sun, each of which required days to solve. At the same time Hale was taking his courses. By the end of the school year, as he himself wrote in his thesis, he had not been able to obtain any "photographs of any intrinsic value" of prominences, but he had demonstrated his spectroheliograph as a working device.[19]

Hale was graduated from Boston Tech on June 1, 1890, and two days later married Evelina Conklin, to whom he had been engaged from the beginning of his freshman year. He had met her in Madison, Connecticut, where his grandmother lived and where both their families went regularly on summer vacations. After their wedding and two days in New York, the newlyweds went to Niagara Falls and then "home" to Hale's parents' house in Chicago.[20] After a week there, they continued west by train, with stops in Colorado, Yosemite, San Francisco, and Lick Observatory, the high point of the honeymoon trip in Hale's estimation. It had gone into operation only two years before with its giant 36-inch refracting telescope, built with James Lick's money. Hale's mentor Burnham was now one of the five astronomers on its staff; he welcomed the young couple to Mount Hamilton, the site of the observatory, and introduced them to Director Edward S. Holden and his other colleagues.

That night Hale observed with spectroscopist James E. Keeler at the 36-inch. More than forty years later, as he composed his autobiographical notes, Hale could still not forget his first sight of that long tube, the largest telescope in the world, pointing up in the darkness of the great round dome toward the slit that seemed to him to be an opening into heaven.[21] The next day Holden offered him the chance to come to Lick as a volunteer observer, bringing his large spectrograph with him, to use it on the 36-inch refractor for solar research. Hale briefly considered this offer but declined it with thanks on the grounds that the "seeing," or steadiness of the magnified images of stars in a telescope, was poor in the daytime at Mount Hamilton, although it is ordinarily very good at night.[22]

Holden's offer pointed up Hale's problem of what he was to do next. At one time his father, who was a trustee of Beloit College, a Congregational institution, had dreamt that George might become a professor of astronomy there. However, when the budding young scientist visited Beloit at the end of his junior year, he found that the person who actually had the job, Charles A. Bacon, was not a doddering old fogy, as he had imagined, but a bright young graduate of Dartmouth who was "very much better fitted for the place than I ever expect to be." Bacon had practically no time for research and had to teach mathematics and surveying as well as classical astronomy. All these subjects bored Hale. He decided he did not want the job.[23]

Hale had also been wooed by Edward A. Tanner, the president of Illinois College, another Congregational school in Jacksonville, Illinois, to become its professor of astronomy.[24] This position did not interest Hale either. Upon his graduation, he received the offer of a job more in line with his own interests, as an assistant in physics at the University of Kansas, but the place and the salary, $1,000, held little attraction for him.[25] He had the vague idea of going to Germany, still the world center of physics in 1890, to do graduate work leading to earning his Ph.D. there. However, because Hale could hardly understand a word of German and dreaded studying it, there were obvious difficulties with this plan. He considered entering Johns Hopkins as a graduate student under Rowland and wrote him to explore this possibility.[26] Hale also considered entering Princeton as a graduate student, and Young assured him that he would be glad to have him bring his large spectrograph and use it on the 23-inch refractor there.[27]

In the end, however, Hale's father solved the problem of what the young astrophysicist should do next and where he should do it. His son needed an observatory for research. What better place for it than in Chicago, right next to the family mansion? William E. Hale decided to build a professional observatory for him there. He ordered a lens for a 12-inch refracting telescope from Brashear. It was made to George's specifications, with an especially long focal length, ideal for his solar research. From Warner and Swasey, the makers of the mounting of the Lick telescope, his father ordered a particularly sturdy telescope mounting so that it could carry his son's large, heavy spectrograph with its attachments that converted it into a spectroheliograph. The building was a two-story tower, designed by Burnham and Root, faced in stone to match the house, and topped by a rotating dome.[28]

While waiting for his observatory to be completed, George traveled to the East and consulted with Young about the spectroscopic research he planned to do on stars, in addition to his solar work.[29] Hale gave a week-long series of lectures on astrophysics at Beloit in December and repeated them at Illinois College in January. He again turned down a professorship at the latter institution, whose President Tanner wanted him on the permanent faculty in "the worst way" and offered him $1,500 a year to come.[30] In his Kenwood Physical Laboratory Hale did some very good spectroscopic research that had been suggested to him by Keeler. Hale's laboratory spectra helped confirm Keeler's observational result that the strongest emission line in the spectra of nebulae was not due to magnesium, as the eminent British astronomer J. Norman Lockyer had proposed. This was the subject of Hale's first published research paper, completed before his twenty-third birthday.[31]

His new telescope was completed and erected in its dome by spring, and with it Hale finally succeeded in getting good photographs of solar prominences in May.[32] He arranged a dedication ceremony for the new

Kenwood Observatory, 1891. From the *Sidereal Messenger.*

Kenwood Physical Observatory in June; his father brought Young out to Chicago from Princeton as the main speaker. The Princeton astronomer said he had never enjoyed two days more in his life than those at Kenwood and predicted a bright future for young Hale. Brashear, Hough, President Edward D. Eaton of Beloit College, Hale's tutor Williams, and two other ministers also made brief remarks. A few years later Hale's father estimated the value of the Kenwood Physical Observatory as $25,000, the equivalent of perhaps $400,000 in our present currency. It was certainly far superior to anything Beloit College or Illinois College had to offer, or to the Allegheny Observatory with its antique 13-inch refractor where Keeler, the outstanding young astrophysicist in America, had just gone as director. Yet Hale, after using his telescope for only two months, had already decided that it was inadequate and was planning to add a second one, especially designed for the blue spectral region instead of the visual or yellow region, on the same mounting.[33]

Soon after the dedication Hale left with his wife on a summer trip to Europe. He had asked Holden to nominate him for membership in the Royal Astronomical Society, the chief astronomical group in Great Britain, which he knew would be very helpful for making scientific contacts.[34] Though he did not want a job at Beloit College, the young astrophysicist had no hesitation in asking Eaton to appoint him to an unpaid, honorary professorship there so that he might legitimately refer to himself as Professor Hale during his travels abroad. The president, who still hoped to add the wealthy trustee's son to his faculty, was glad to oblige.[35]

In London Hale made his headquarters at the offices of the American Elevator Company, whose president was a friend and business associate of his father.[36] Hale traveled around England visiting the Royal Greenwich Observatory and the Oxford University Observatory as well as several physics laboratories. He met Lockyer and William Huggins, the pioneer English observational astrophysicists, who were bitter scientific rivals. Hale gave a paper on his research on prominences at the meeting of the British Association, the largest scientific society in the British Isles, held that summer at Cardiff in Wales. He went to the Continent and met Henri Deslandres, the outstanding French astrophysicist, at Meudon Observatory, and then made a lightning tour of observatories in Germany, Switzerland, and Italy.[37] At the end of September he and Evelina headed back to Chicago.

Soon Hale was immersed again in his research at Kenwood Physical Observatory, with his telescope, spectrograph, and the improved spectroheliograph mechanism Brashear had built to his order. With it Hale discovered he could not only photograph prominences in the light of the ionized calcium lines in the ultraviolet spectral region, called H and K by astronomers, but could record related disturbed areas all over the disk of the sun.[38] This was an important new result. Following it up meant long hours at the telescope, day after day. Hale never had the time nor the inclination for this dogged type of observing that is the backbone of astronomical research. In 1892 he hired Ferdinand Ellerman as his first full-time, professional assistant. Hale's father paid his salary. Ellerman, then twenty-three years old, had grown up in Centralia, in southern Illinois, and after finishing high school moved to Chicago. Working at a variety of jobs in the rapidly expanding metropolis, he had become an expert in photography and in machine-tool work, skills that made him an ideal observer and handyman with the complicated technical apparatus of the Kenwood Observatory. Ellerman soon took over all the routine observing. He idolized Hale and was to remain with him in the role of an admiring assistant all the rest of his life.[39]

In the spring Hale went back to Beloit for a second week and gave another series of lectures on astrophysics. He found the students good, but he still did not like teaching.[40] Nevertheless, Hale was rapidly becoming involved with the new University of Chicago, which was springing into existence in Hyde Park, little over a mile from his home and observatory. The earlier institution of that name, of which the Dearborn Observatory was a part, had gone bankrupt a few years before. The new university, financed in large part by John D. Rockefeller, was being organized by William Rainey Harper, an amazing bundle of energy then thirty-five years old. He had finished high school at age ten, graduated from college (where his favorite subject was Hebrew) at fourteen, and earned his Ph.D. at Yale in Sanskrit, Greek, and Chaucer at eighteen. He became a Biblical scholar and teacher, started the Chautauqua movement, returned to Yale as a professor, and in 1891 agreed to head the university which Rockefeller wished to build in Chicago. Harper wanted to make it a great research university, and with Rockefeller's money he was able to do so. He could offer much higher salaries than other universities of the time were paying and was able to hire away from them such eminent scientists and scholars as Albert A. Michelson, Thomas C. Chamberlin, Albion W. Small, and William I. Knapp.

Even Rockefeller's fortune was not limitless, though his gifts to the university were huge. Harper was a great university president because he was a great fund-raiser. He was tireless, relentless, and hugely successful in getting contributions from the wealthy people of Chicago.[41] He soon heard of Hale, whom he considered a "young man of promise," no doubt for his skills in research as much as for his valuable private observatory and his father's fortune.[42] Harper arranged to meet Hale on a short trip to Chicago in the spring of 1891, even before he officially became president. He offered Hale a faculty position on condition that he move his Kenwood Observatory, at that time completed but not yet dedicated, to the campus and throw it open to the students and public. The observatory's research program would be widened to include classical astronomy as well as astrophysics. Hale would have charge of the whole department and of the spectroscopic research; another "young man" would be appointed to take over the purely astronomical work. The observatory was to be endowed with enough money to pay the salaries of two professors and their expenses by an unnamed benefactor, whom Harper clearly intended should be Hale's father. The president promised Hale that if he accepted appointment under these terms he would be promoted "as rapidly as your age and work might seem to you and to us to call for advancement."[43]

Hale indignantly rejected this offer, went ahead with the dedication, and left for Europe. He wrote Harper that he and his father could not agree to get a faculty position for him by giving or leasing the observatory to the university. He said that if he were not competent enough to be appointed on his own merits, he would prefer to continue his own studies and research. Harper smoothly replied that Hale had misunderstood him and that he had had no intention of making the gift of the observatory to the university a condition of appointment.[44] In fact, that was exactly what he had intended, and he had not given up.

In the spring of 1892, after Hale had returned from Europe, Harper, who had now moved to Chicago (the first students were to begin classes that fall), approached him again.[45] Asaph Hall, the famous American astronomer who had discovered the two satellites of Mars in 1877, advised Harper, "Again, it was said that you have in Chicago a young man, Mr. George E. Hale, who is devoted to his branch of astronomy, and who has already shown good ability. It was said that Mr. Hale's father is a man of great wealth, and is very generous in the support of his son's investigations. It may be well for you to know Mr. Hale."[46]

Within less than two weeks after receiving this letter, Harper met Hale again, and this time he was successful in signing up the young faculty member. The terms were fairly similar to the earlier ones but omitted the endowment. Hale agreed that the university would have the use of Kenwood Observatory, but only for students under his own supervision and not for the public at all. He would be appointed associate professor of astrophysics but would receive no salary "until such time as the University has the necessary funds or at the latest after three years from October 1st, 1892." Since Harper maintained to his dying day that the University of Chicago never had enough money, there is little doubt that Hale did not go on the payroll before October 1, 1895. Hale was to be in charge of all research at the observatory, but would have no required teaching duties. The university would pay up to $1,000 per year toward the operating expenses of Kenwood Observatory, but Hale's father agreed to give it, the telescope, and all the instruments to the university if his son still wanted to keep the job after one year, and if Harper succeeded in raising $250,000 or more for a larger telescope within two more years.[47] Thus the University of Chicago got an observatory and a young research associate professor, George E. Hale got his first job, and William E. Hale paid the bills.

Within just a few months of Hale's appointment, Harper, with his help, suddenly succeeded in raising the money for the big new telescope they both wanted. In August Hale, at a meeting of the American Association for the Advancement of Science in Rochester, had learned of the existence of two partly finished glass disks, forty inches of diameter, intended to be made into the lens of what would be the largest telescope in the world. They had been produced in France after tremendous effort, for making such large pieces of clear, perfect glass was then an art, not a science. The glass disks had been ordered by a group of men connected with the University of Southern California who intended to put up the telescope on Mount Wilson, to outdo the 36-inch Lick Observatory telescope in Northern California that Hale had visited. However, their financial plans were based on land speculation in the Los Angeles area; in 1892 this bubble burst and the trustees of the projected Spence Observatory could not meet their payments on the lens. They defaulted, and Alvan Clark & Sons, the optical firm that had made all the large American refracting telescopes and was working on this lens, was left holding the bag.[48]

Hale recognized this situation as a wonderful opportunity for the University of Chicago. He hurried back to Chicago and alerted Harper. They could get the glass disks and build the largest telescope in the world if they could only raise the necessary money. Harper decided that Charles T. Yerkes, the unscrupulous, flamboyant street-car magnate he had been unable to interest in any other major University of Chicago project, would surely be interested in the largest anything in the world. Harper had Hale write a letter explaining in simple terms this exceptional opportunity for the University of Chicago, how much more light the planned 40-inch would collect than the Lick 36-inch, and how the new instrument would attract visitors to Chicago.[49] The letter caught Yerkes in a euphoric mood; he had only a few months previously married a chorus girl whom Harper described as "the most gorgeously beautiful woman I have seen for years."[50] Yerkes invited Harper and Hale to call on him, and a few days later one of Harper's aides could telegraph

Frederick T. Gates, the Baptist minister who was Rockefeller's personal adviser on University of Chicago matters, "Yerkes builds observatory with largest telescope in the world." Harper himself wrote Gates, "The whole enterprise will cost Mr. Yerkes certainly half a million dollars. He is red hot and does not hesitate on any particular. It is a pleasure to do business with such a man."[51] Yerkes exulted in the knowledge that he was to be the "owner" of the largest telescope in the world. Hale, the twenty-three-year-old MIT graduate, was to be its director. He quickly resigned his honorary professorship at Beloit.[52]

The young astrophysicist found himself more and more immersed in the organization of science. Already he had to fit his research into times left over from his committees and editorial duties.[53] Hale was not forced into these activities; he went out and sought them himself. The first scientific organization he founded was the Chicago Section of the Astronomical Society of the Pacific. Hale called the first meeting of it at his father's home on November 20, 1890, soon after his return from his honeymoon trip to Lick Observatory, and had himself elected secretary, and his older friend Gayton A. Douglass the figurehead president.[54] The A.S.P. had never before had any separate sections or any meetings anywhere but in San Francisco or Mount Hamilton, but Hale had no problem in persuading Holden, the founder and guiding spirit of the society, to let him start one. In its first year the Chicago Section had eight meetings, three of them at Hale's Kenwood Physical Laboratory or Observatory, and he gave the main talk or scientific discussion at five of them.[55] The culmination was the formal dedication of his observatory in June 1891, after which Hale left on his European trip.[56] The July meeting was a distinct letdown, with no lecturer and no energetic secretary to organize a discussion. The Chicago Branch nearly died. It held no further meetings for two months, until Hale returned and got it started again with a meeting at Kenwood at which he described his trip through Europe and the various laboratories and observatories he had visited. Then after missing two more months, the Branch met once more, again at Kenwood, and heard another talk by Hale.[57] Then he let it expire.

Hale was also busy getting his new scientific journal into operation. He had planned to start a journal devoted completely to astrophysics, but either he or his father, who was financing him, found the market was too small. So Hale joined forces with William W. Payne, who was publishing the *Sidereal Messenger*, a semipopular astronomy magazine

George Ellery Hale, director of Kenwood Observatory, 1892. Courtesy of the American Institute of Physics/Niels Bohr Library.

intended chiefly for teachers and advanced amateur astronomers. Their new journal *Astronomy and Astro-Physics* started publication January 1, 1892. It was divided into two parts, the astronomical section, edited by Payne, still largely popular in nature, and the astrophysical section, edited by Hale, much more professional. He intended from the beginning to build *Astronomy and Astro-Physics* up for a few years and then get it away from Payne and make it over into a true research journal.[58]

Once he had *Astronomy and Astro-Physics* well into operation, Hale started a new organization to take the place of the Chicago Section of the Astronomical Society of the Pacific. He attended a meeting of the Chicago Academy of Sciences and proposed that it set up a Section of Mathematics and Astronomy. Hale's motion carried and the new section held its first meeting at his Kenwood Astrophysical Observatory on June 15, 1892. All the members of the Chicago Section of the Astronomical Society of the Pacific were taken into it, and though this time Hough was named president, Hale again was chosen secretary. No doubt, he had arranged both nominations.[59]

Very probably one of the reasons Hale had let the Chicago Section of the Astronomical Society of the Pacific die was that he had begun to learn that the professional astronomers whom he knew regarded Holden as more of an administrator and a poseur than a serious research scientist. More important, Hale wanted to organize the astronomical part of the World Congress Auxiliary of the Columbian Exposition, the educational part of the Chicago World's Fair of 1893. He had himself named secretary of its Special Committee on Astronomy as early as February 1891, and he clearly recognized that the Chicago Academy of Sciences was a better base for this committee than the Astronomical Society of the Pacific. Hale wrote literally hundreds of letters in organizing the World Congress of Mathematics, Astronomy and Astro-Physics, which was held in Chicago at the Fair's Palace of Art in August 1893. It was the first international astronomical meeting held in the United States.[60]

As part of the World Congress, all the participants were present at the opening of the display of the Yerkes Observatory 40-inch refractor tube and mounting, which had been completed and erected in the huge Manufactures and Liberal Arts Building. There, at the base of the telescope, Hale, Worcester Warner, of the Warner and Swasey firm that had built it, and Alvan G. Clark, who was making the 40-inch lens, gave short talks about the instrument to the assembled astronomers and other interested onlookers.[61]

Clark's talk was a paean of praise for refractors. His father, Alvan Clark, had been a portrait artist in Boston until, at the age of forty, he developed an interest in optics and astronomy. He had taught himself to make telescope lenses, and his firm, Alvan Clark & Sons, had become world famous for their products. They had made hundreds of smaller telescopes and the lenses for succeeding larger and larger refractors, from the 18½-inch of Dearborn Observatory (ordered originally for the University of Mississippi before the Civil War), through the 26-inch of the Naval Observatory and the 30-inch of Pulkovo Observatory in Russia, to the Lick 36-inch. The old man had died just after completing that lens in 1887, and his son Alvan G. Clark, now sixty-one, who had worked with him on all his lenses (some thought he had actually put the final touches on the 36-inch lens), was now figuring the 40-inch, Yerkes Observatory lens.[62]

A telescope lens is in reality a compound system made of two simple lenses, one convex (focusing), the other concave (defocusing). A single convex lens alone, such as Galileo had used in his very early telescope, brings light of different wavelengths, or colors, to different foci. This chromatic aberration, as astronomers call it, is a characteristic property of all glass lenses. Though it is not noticeable in eyeglasses, it would make a telescope built around a single lens practically useless. This is what Hale had discovered as a boy when he tried to assemble his first telescope. However, a pair of lenses, made of different types of glass, can be shaped so that their individual chromatic aberrations approximately cancel each other. This is the basis of "achromatic" or color-corrected telescopes. The Clarks, working by almost completely empirical methods, had made a series of successively larger compound lenses of this type, unmatched anywhere in the world. Each time they made a successful lens, they simply scaled up the same basic design a few inches for the next one. Contemporary optical theoreticians scoffed at them. For instance, physicist Charles S. Hastings of Yale University wrote, "It seems to me that going to Clark for a telescope in these days is like sending to Persia for a plow or rifle because of there being some famous mechanic there. The making of a telescope objective is no longer empirical, and with the coming of the scientific method Clarks' supremacy passes."[63] However, Hastings was decidedly in the minority, and Alvan G. Clark, when he gave his talk at the foot of the 40-inch telescope in Chicago, was the most highly respected living lens maker.

In his speech, entitled "The Great Telescopes of the Future," he emphasized that the most important discoveries in astronomy had almost all been made with the largest telescopes in use at the time. He traced the history of observational astronomy from the two 15-inch telescopes, made in 1846, which had then been regarded as monster telescopes, to the 36-inch Lick refractor, completed in 1888, and the Yerkes refractor, whose 40-inch lens he was figuring in his shop in Cambridgeport, Massachusetts. He favored the step-by-step increase in size, and although he did not directly say so, he evidently did not foresee any limit to the size of telescopes of this type that could be built.[64]

Hale and Ritchey undoubtedly both listened politely to Clark's speech that August day in Chicago, but with tremendous inner skepticism. Ritchey had first met Hale at the charter meeting of the Chicago Section of the Astronomical Society of the Pacific back in the fall of 1890, soon after the young MIT graduate's return from his trip to Lick Observatory.[65] Henry H. Belfield, the director of the Chicago Manual Training School, was another charter member of the Chicago Section, and he probably introduced to one another his shop teacher and his rich young former student, whom he was proud to consider his friend.[66] Hale and Ritchey immediately recognized their usefulness to one another, Ritchey for his skills in the shop, and Hale for his ownership of an observatory. They hit if off famously. Soon Ritchey was making slides for Hale's astronomical talks and was dropping by Kenwood Astrophysical Observatory to use his rich friend's power lathe.[67] Hale was recommending Ritchey's reflecting telescope to an interested buyer.[68]

In contrast to the elderly Clark, the thirty-year-old Ritchey and twenty-five-year-old Hale believed in reflecting telescopes, not refractors, as the great telescopes of the future. Reflectors, invented by Isaac Newton in the late seventeenth century, use curved mirrors rather than lenses to focus light. A parabolic mirror, or paraboloid, has the exact form needed to bring all the light of a distant star to a perfect focus and thus has no chromatic aberration whatsoever. Newton had made the mirrors for his first small reflecting telescope of polished metal, ground to parabolic shape. Larger and larger telescopes of this form had been built in England, up to one with a four-foot diameter mirror by William Herschel, and a six-foot by the Earl of Rosse. These telescopes were quite poorly constructed mechanically, and their mirrors were far from perfect paraboloids. They were made of speculum metal, an alloy of relatively high reflectivity, which was impossible to grind completely smooth and to the correct form. Probably, the best reflector of this type was the four-foot Melbourne, Australia, telescope, made by the firm of Thomas Grubb and Sons in Dublin, but it never performed satisfactorily. Years later Ritchey was to write of it: "I consider the failure of the great Melbourne reflector to have been one of the greatest calamities in the history of instrumental astronomy, for by destroying confidence in the usefulness of great reflecting telescopes it has hindered the development of this type of instrument, so wonderfully efficient in photographic and spectroscopic work, by nearly a third of a century."[69]

Even before Grubb made the telescope, glass had begun to replace speculum metal as the preferred material for telescope mirrors. Glass could be worked to a perfectly smooth surface, then coated with silver, which was a significantly better reflector than even the best alloy. Andrew A. Common, an English engineer and amateur astronomer, had made a three-foot diameter reflecting telescope of this type in 1879;

Ritchey had read his paper describing it while he was a student in the Cincinnati Observatory.[70] Common also had accomplished little with this telescope, partly because of its flimsy construction. Thus American professional astronomers did not take reflectors seriously.

Many amateurs, however, were making small, silver-on-glass reflecting telescopes by the 1880s. Only one glass surface must be ground for a reflector, rather than the four required for a refractor, two on each component of the compound lens. John A. Brashear was making mirrors and small reflecting telescopes in his shop in Allegheny, Pennsylvania. Ritchey had bought a mirror from him and built a telescope around it, then had learned to grind, figure, and polish mirrors himself. He had already begun to improve the time-tested methods he had learned for making mirrors, applying the same principles he had learned from his father and grandfather in woodworking: careful, intelligent consideration of the task and the materials, complete concentration on the work, advance preparation of tools designed for the job at hand, ruthless analysis of every failure, and application of the lessons thus learned in all future work.[71] He knew he could make better big mirrors than anyone else had before, and he knew that they would be better than Clark's lenses.

Hale's belief in reflectors as the great telescopes of the future was more cerebral.[72] He had traveled widely, particularly in England, and had discussed research telescopes with many experienced observers. He knew that Thomas Grubb's son and eventual successor as head of the family business, Howard Grubb, had predicted in 1877 that reflectors would take over from refractors as the "monster telescopes" of the future.[73] Even before that the young Scottish astronomer David Gill had made a careful study of the respective advantages and disadvantages of reflectors and refractors for the Lick Trustees and had concluded that in the long run reflectors would undoubtedly be best.[74] For the old visual astronomy, refractors had been convenient because they were easier to mount rigidly and to move frequently from one star to the next. But for the "new astronomy," astrophysics, large-aperture telescopes would be required to collect light for spectroscopy of faint stars. Reflectors could be made much larger, because their mirrors could be supported from behind and did not require clear, perfect glass blanks. They were automatically achromatic and allowed all the light of a star to be used. Hale had decided years ago that his lifework was to be astrophysics, and by

October 1892, just two months after Yerkes announced his gift of the 40-inch refractor, long before work on it had even begun, Hale had already decided that what he really needed at the new observatory was a 60-inch reflector.[75]

Ritchey and Hale were in tune with the advanced thinking astronomers of their time, such as Holden and Keeler, but not with traditionalists, particularly experienced visual observers such as Burnham and Edward E. Barnard. In a brilliant, sarcastic polemic published in 1885, Burnham had written that misguided astronomers who mistakenly believed that small telescopes were better than large ones were actually comparing small *refractors* with large *reflectors*, which he maintained were never as good as refractors.[76] In fact Burnham's mistake was that he was comparing well-built Clark refractors with amateur-made reflectors.

After the World Congress, Hale went abroad for the year 1893–94, while Clark continued working on the 40-inch lens in Cambridgeport. Hale's announced plan was to learn German and to study, but he did neither. Instead he worked on designs for the Yerkes Observatory building and on schemes for breaking away from Payne and *Astronomy and Astro-Physics* and founding a completely independent new *Astrophysical Journal*.[77] After spending most of the summer of 1894 in Italy and Sicily, he returned to Chicago, where he did start the *Astrophysical Journal* with publication of its first issue in January 1895. It is still to this day the most important research journal on astrophysics in the world.

Progress on Yerkes Observatory remained stalled, but at least the site had definitely been chosen. All the astronomers whom Harper had consulted had recommended locating the observatory outside of Chicago. Hale had favored Lake Forest, north of the city, but Harper decided on Williams Bay, Wisconsin, a little village on Lake Geneva just a few miles over the state line from Illinois. Many wealthy Chicago families had summer homes there, making it especially attractive to Harper, who hoped to raise money to support the observatory from them.[78] When Hale returned from Europe in September 1894, the newly chosen site had not yet been transferred to the University of Chicago, so he went back to work at Kenwood.[79] He was even more interested in a large reflector than he had been before and discussed the prospects for one often with Ritchey.

The budding optician had by now made one or more reflecting-telescope glass mirrors in his "laboratory," as he called his shop in his home. Probably, they were only ten or twelve inches in diameter, but to Ritchey and Hale these little mirrors were harbingers of the giant reflecting telescopes of the future. Ritchey treated every optical job he undertook as an experiment and subjected all his results to careful analysis, always seeking improved techniques for the future.[80]

It is hard to know when Hale began working on his father, but by the summer of 1895 he had persuaded him to provide the funds to buy a disk of glass large enough for a monster telescope. Ritchey was to make the mirror, and Hale hoped that when it was finished he would be able to raise the additional funds necessary to mount it in a telescope from his father, from Yerkes, or from some other rich man.[81] The first step was to get the necessary large piece of glass. It could only come from France, and Ritchey handled the negotiations with the St. Gobain Glass Company through its American agent in New York.[82] At Hale's behest, Ritchey asked for price quotations for disks big enough to make a 48-inch, a 60-inch, a 72-inch, and an 84-inch. The St. Gobain firm's general manager replied that they could supply the disks for a 48-inch or a 60-inch at a price of eight shillings sixpence per kilogram (in the English currency that was the basis of international commerce in those days), approximately one dollar per pound in the American money of the time. The other, even more "immense," disks would require special study. The firm was anxious to get the order for the prestige that supplying the glass for the large telescope would bring.[83] Ritchey estimated the weights of the necessary glass disks as about 700 pounds for the 48-inch and 1200 pounds for the 60-inch, but he had made a serious error in calculation; actually, the correct values are more like 1100 pounds and 2000 pounds respectively, and the quoted prices these same amounts in dollars.[84] Hale, on his father's advice, had Ritchey ask what guarantee the St. Gobain firm would offer against the contingency that the glass disk might prove defective.[85] Finally, in the summer of 1896, Hale ordered a 60-inch disk, on specifications provided by Ritchey and somewhat thicker than they had first planned, along with smaller disks for the auxiliary secondary mirrors for the telescope. Hale directed that all the bills were to be made out to him personally, not to the University of Chicago. His father paid $2,000 for the 60-inch disk and $150 for all the smaller disks together, and they belonged to him.[86]

By the time Hale bought the 60-inch disk, Ritchey was working full-time as his paid optician. He had begun as a part-time volunteer the previous summer, when Hale had tried a new method for detecting the solar corona outside of eclipse, using a small reflecting telescope at the Yerkes Observatory site. This was a favorite project of Hale's, which he had tried twice previously and to which he was to return several more times, always unsuccessfully, in the remainder of the decade. The corona is the faint, extended envelope of the sun, which ordinarily cannot be seen because the sky is so much brighter than it. At a total solar eclipse, however, the moon cuts off the light of the sun, and the corona becomes visible against the darkened sky. Several of the early pioneers of astrophysics, among them Samuel P. Langley and Hale's hero Huggins, had tried to devise methods to observe the corona on a routine, daily basis but had not succeeded. Hale dreamed of using the newest, most advanced astrophysical methods and of succeeding where these giants of the past had failed.

After he had gotten his spectroheliograph working successfully at Kenwood Observatory, he realized that by setting it to the wavelength of a strong solar absorption line, he could cut down greatly the brightness of the sun and thus of the sky, which is illuminated by scattered sunlight. This would enhance the contrast between the corona, which has a strong continuous spectrum without absorption lines, and the sky. Hale tried the method in Chicago in 1893, but the air there was laden with haze and smoke even under the best conditions, and the sky was too bright. The traditional method Langley and most of the other pioneers had followed was to observe from a high mountain peak, where the atmosphere was thinner and purer than at sea level. However, Hale had heard tales that at the bottom of the Grand Canyon in Arizona the daytime sky was very dark, "nearly black," so he planned to take his spectroheliograph to the banks of the Colorado River and detect the corona from there. When he investigated the stories, however, he quickly discovered that they were apocryphal, and he went to Pike's Peak instead.[87]

Hale planned this expedition at the last moment, in a great hurry. Clearly, his hope was to return from the wilds of Colorado and lay before the assembled savants at the World Congress in Chicago the evidence that he had succeeded in detecting the corona in the daytime sky, where Huggins, Langley, and the other legendary figures of the previous generation had failed. Hale tried to get a group of his astronomical friends to go with him, to share the work, but Burnham, Henry Crew, and Brashear all declined and in the end only Hale's wife Evalina and Keeler accompanied him. They were only at Pike's Peak for two weeks. Evalina was sick or suffering from headaches most of the time they were there, Hale tended to her, and Keeler, a skilled, experienced observer (he had been with Langley at Mount Whitney in 1881) did most of the work. They were hampered by clouds, smoke from forest fires, and even swarms of insects flying above the mountain, all scattering sunlight into the telescope. They saw no sign of the corona and returned to Chicago empty-handed on the Fourth of July.

The following summer, at the end of his year in Europe, Hale had gone to Sicily to try a second time, this time from Mount Etna. His party included Evelina, Antonio Ricco, who was director of the Catania Observatory, and Ricco's wife. This time they had to contend with smoke from the volcano, and again the results were completely negative. Nevertheless, Hale said he intended to try again. He wrote, "In conclusion, I may say that the investigation has been a fascinating one, in spite of its succession of failures."[88]

In 1895, the first time Hale attempted to detect the corona at Williams Bay, he tried a new method. Instead of the spectroheliograph, his plan was to use a bolometer, an energy-measuring device, with a small reflecting telescope. The bolometer is sensitive to radiation of all wavelengths, and by using filters Hale could restrict his measurements to the far infrared spectral region, where scattering by the air is very weak and the sky correspondingly dark. A difficulty was that bolometers are extremely sensitive and very difficult to construct and use. It took much longer to get the instrument working at Kenwood than Hale had hoped.[89] Ritchey helped with the optical work and silvering the small flat glass mirrors that were used in the galvanometer, which measured the current from the galvanometer. Ellerman assisted Hale with the observing, but again the result was failure. The corona was too weak to detect in the infrared.[90]

In 1895 Ritchey still had a full-time job at the Manual Training School and could only help Hale in his spare time. However, Hale wanted to hire him on the University of Chicago staff, and Ritchey wanted the job. He was anxious to visit the Yerkes Observatory site, and he hoped to work there someday.[91] In the meantime, he was making larger telescope mirrors in his shop in his home. He had bought several glass disks from France on his own, probably with borrowed money, before he had entered into the negotiations for a large disk on Hale's behalf.[92] After he went to work for Hale in the spring of 1896, Ritchey ground and figured one of the 24-inch disks into the mirror for the larger, long focal-length reflecting telescope that Hale used for his next trial to detect the corona, again at Williams Bay, that summer.[93] (All Hale's attempts were made in summer, when the sun is high in the sky and the effect of scattering in the atmosphere consequently minimized.) It was a fixed, horizontal telescope with a heliostat to bring the sun's image into it, much like the system Hale had used for his thesis at Harvard College Observatory. Again the detector was a bolometer, this time water-cooled and mounted in a specially designed room in the partly completed Yerkes Observatory building, much more solidly than in the experiments of the previous year. Nevertheless, the result was failure once again; Hale did not succeed in detecting the corona's heat radiation.[94]

By that time Ritchey was working full-time as Hale's paid employee. Early in 1896, while he was still on the staff of the Manual Training School, Ritchey had begun work in his home shop on his most ambitious project to date, the 24-inch parabolic mirror designed to be used in a reflecting telescope to photograph nebulae. For this purpose it had to be optically "fast"; its focal ratio was f/4, requiring steep curves on the mirror. It was, therefore, much more difficult to figure correctly than the slower f/31 mirror Ritchey had made for Hale's solar-corona bolometer attempt.* But Ritchey had mastered the technique of glass-mirror making and had gone well beyond previous workers in the field. He had perfected his method of zonal testing of the mirror, using a knife edge to cut the rays at the focus, which was far superior to the eyepiece method which Henry Draper, Common, and Roberts had used.[95] Ritchey had the mirror nearly finished by the fall of 1896 and was ready to start testing it on stars.[96]

By then he had already been working for Hale for six months. Hale's father had agreed to provide the money for Ritchey's salary. Very probably it was $1,200 a year; William E. Hale paid it directly to him by personal check each month.[97] Ritchey's main job was to work as Hale's optician, and his big project would be to make the 60-inch disk into a telescope mirror. But whatever optical or shop work Hale needed, Ritchey could do. During the summer of 1896, while Hale was at Wil-

*See the appendix and its Figure A.1 for brief descriptions of these optical terms.

liams Bay, Ritchey worked for him in Chicago silvering mirrors, packing them, shipping them up to the observatory site, and doing all the other little instrument jobs necessary to keep a research program going.[98]

Sometime just at the end of 1896, or the beginning of 1897, Ritchey moved to Williams Bay. The observatory was nearing completion, and he was setting up its optical shop.[99] By this time his family was complete; he and his wife had two children, a girl, Elfleda, then seven years old, and a boy, Willis, just turned six.[100] They had both been born in Chicago and now, with their father and mother, they moved to the little town in Wisconsin where Ritchey was to make his first, hugely successful, professional reflecting telescope, and to begin work on his first big mirror.

Yerkes Observatory

1897 - 1904

THE YERKES OBSERVATORY to which George Willis Ritchey went in 1897 was an imposing building located on a level field overlooking Lake Geneva and the tiny hamlet of Williams Bay. A beautiful, green site in the summer, or a colorful one in the crisp autumn as the trees turn yellow, orange, and red, it was cold and forbidding in the icy Midwestern winter. The building was an impressive brick structure, designed by University of Chicago architect Henry Cobb following the general layout sketched by George Ellery Hale. He based it on his memories of Lick Observatory but modified them to include his ideas for the newer astrophysics which he intended to develop. The general plan was a two-story structure with offices on either side of a long hall on the first floor, and the large, high, impressive dome of the 40-inch refractor at one end of it. At the other end a shorter hall crossed, with a smaller dome at either end of the T. In 1897 one of these domes was not completed with a top, because there was as yet no telescope to put into it. The basement was given over to laboratories and shops, and the second floor of the long building was an attic, while the second floor of the shorter hall was designed for a fixed solar horizontal telescope and laboratory, with a roll-back roof.[1]

Getting a new observatory into operation is an exciting business, particularly if the observatory is built around the largest telescope in the world. Ritchey's first task in 1897 was to lay out the optical shop. It was located in the basement and was well insulated to protect it from sudden temperature changes. Ritchey installed his own grinding and polishing machine, on which he had made the two 24-inch mirrors, and started building the much larger machine on which he would shape the 60-inch mirror to perfect parabolic form.[2] The shop was situated so that long-focal-length mirrors, as they were being figured, could be tested frequently, tipped up on one edge, with a light source and knife-edge apparatus at the other end of the long basement hall.

Ritchey wrote his first scientific paper, "A Support System for Large Specula," describing the "flotation" system he had designed years before for supporting large mirrors in telescopes. In his paper he described how all previous reflecting telescopes had suffered from shifts in the position of the mirror. These lead to blurred images and loss of light. The problem is that the mirror cannot be rigidly attached to the mounting, because of the stresses that would result, and would cause distortion of the mirror and thus poor optical images. Ritchey's basic idea was to support the mirror from below on only three points, defining a plane, but with most of its weight balanced, or "floated," by a system of independent weighted levers. Similarly, the edges of the mirror were to be supported by steel arcs to prevent it from moving sidewise, but again most of its weight would be balanced out. He emphasized the importance of making the cell that supports the mirror very massive to insure rigidity but designing it so that the mirror was well ventilated on both its front and back surfaces. This would keep its temperature uniform so that it would not warp as the air temperature in the dome changed during the night. For the same reason, he declared, the mirror should be silvered not only on its front surface but on the back as well. As Ritchey stated in his paper, preventing thermal and stress distortions of the mirror was a problem that must be solved in order to achieve large-aperture reflectors.[3] He had not made a working example when he published this paper, but he used these principles when he built the 60-inch a few years later, and they have been used in all subsequent large telescopes.

Soon after the Yerkes Observatory building had been completed, Hale and his wife, with their four-month-old daughter, Margaret, had moved to Williams Bay in November 1896, just a few weeks before

Ritchey had come. In December Hale had his 12-inch refractor brought out from Kenwood and erected in one of the smaller domes, as his father had earlier agreed with President William Rainey Harper. With it Hale was soon observing the sun.[4] One of his first observations was another attempt to detect the solar corona without an eclipse.

In February 1897 he learned in a letter from William Huggins in England of a rumor that someone had photographed the corona in X-rays, which had been discovered only a little more than a year previously by Wilhelm Röntgen in Germany. Hale was not at all sure he believed the report, but he could not afford to ignore it. He quickly assembled a pinhole camera of slabs of lead, with a piece of cardboard glued over the pinhole to block optical light. Then he mounted this X-ray camera alongside the 12-inch refractor, which he used to track the sun for a 70-minute exposure. When he developed the plate he found, as he had more than half expected, the same old result, no sign of the corona.[5] Hale and other astronomers did not then realize that solar X-rays are all absorbed in the earth's atmosphere, but this futile attempt is one of the first X-ray astronomy experiments on record.

By May the 40-inch mounting had been erected in the large dome and Alvan G. Clark personally delivered the lens, which had been completed and tested two years before. No doubt Ritchey helped Clark install it in the telescope. Although there were some problems with the system for mounting and pointing the giant telescope, they were minor and the instrument was soon in operation.[6] On May 21 a group of University of Chicago trustees and officials, led by Harper, came out for the first night of observing with the big refractor. They were pleased with it and with the globular clusters and nebulae they saw through it.[7] The astronomers could begin testing it seriously and using it.

Then, little more than a week later, on the morning of May 29, the huge rising floor of the 40-inch dome broke loose from its cables and fell. As it came crashing down, it tore itself apart against the iron column that supported the telescope. Half of the floor ended up there, warped and canted at a steep angle. The other half fell all the way to the ground, forty-five feet below, smashing itself into a pile of shattered lumber. If it had happened a few hours earlier, Edward E. Barnard and Ferdinand Ellerman, the observers who had stopped work at daylight and gone home to sleep, would have been killed; if it had happened a week earlier, the high command of the University of Chicago would have been eliminated. Investigation revealed that the cables had not been securely fastened to the floor; once one of them broke loose the others, overloaded, almost immediately gave way, too, and the floor came down. Warner and Swasey, the makers, had a new floor built, with the cables very securely fastened this time, and the telescope was back in operation in little more than a month, but the astronomers must have stepped on it gingerly at first.[8]

Their number was very small; neither Harper nor Charles T. Yerkes, the donor of the observatory, was willing to put up the money to hire a significant research staff. The faculty astronomers were Hale, Sherburne W. Burnham, Barnard, and Frank L. O. Wadsworth, with Ellerman and Ritchey as assistants. Burnham, Hale's first mentor in Chicago, kept his job as a court clerk there but came to Williams Bay as a volunteer on the weekends to observe double stars. Barnard, a fabulous observer, had very little education but extremely good eyesight and fantastic dedication to astronomy. He had begun his career as an amateur, discovering so many comets that he earned a position on the Lick Observatory staff when it opened in 1888. There he pioneered in wide-field photography, especially of the star clouds of the Milky Way, with their associated bright nebulae and dark features, which we know today result from extinction by interstellar dust clouds. In 1892 the keen-eyed Barnard discovered Amalthea, the faint fifth satellite of Jupiter, with the 36-inch refractor, making himself briefly famous. But, like Burnham, he quarreled violently with Edward S. Holden, the Lick Observatory director, and finally, at the end of 1895, resigned and took the job at Yerkes Observatory that Harper and Hale had been urging on him. He arrived at Chicago long before the observatory was completed, as he had at Lick almost a decade before, and he observed at Kenwood until he could move to Williams Bay.[9]

Wadsworth was more of an engineer, an expert in optics and astronomical instruments, than an astronomer. He had been physicist A. A. Michelson's assistant and had come with him from Clark University to the University of Chicago. Hale was determined to have a mechanical shop in the observatory, where telescopes, spectrographs, and other instruments could be modified and improved, rather than having to send them back to their makers, as at other observatories. Hale was fascinated by some of Wadsworth's instrumental ideas and recognized his design skills. He hired Wadsworth to be in charge of the shop and to supervise the two instrument makers who worked in it.[10]

The glass disk for the 60-inch mirror finally arrived from France in July.[11] Ritchey had completed the large grinding and polishing machine for it and began rough grinding it immediately. Thus when the formal dedication ceremonies were held at the observatory in October, the visiting astronomers who came to Williams Bay from all over the United States to attend it saw Ritchey hard at work on the giant mirror. The astronomers gave papers on their latest scientific results, and the high point of this week-long meeting was the prophetic address by James E. Keeler, director of Allegheny Observatory, on "The Importance of Astrophysical Research, and the Relation of Astrophysics to Other Physical Sciences," delivered in the 40-inch dome with the floor safely down on blocks. This was the first large astronomy meeting in the United States and led directly to the founding of the American Astronomical Society (originally called the Astronomical and Astrophysical Society of America), in which Hale played a very major role.[12] He was already spending so much effort on organizational activities and fund-raising trips that he had little time left over for astronomical research.[13]

Hale wanted to add Keeler, the outstanding young American stellar spectroscopist of the time, to the Yerkes faculty. Hale had invited him to give the main address at the dedication partly to show him off to Harper. But the president was adamant that, no matter how good Keeler was, there was no University of Chicago money available to hire another astronomy professor. Hale went to New York on a fund-raising expedition and succeeded in getting a promised $15,000 to pay Keeler's salary for five years from Catherine W. Bruce, a wealthy recluse and supporter of astronomy. However, just as she agreed to the gift, Keeler was offered, and ultimately accepted, the directorship of Lick Observatory, where Hale had seen the 36-inch refractor on his honeymoon eight years earlier.[14]

Instead of Keeler, Hale then hired Edwin B. Frost, a Dartmouth graduate who had studied astronomical spectroscopy with Charles A. Young at Princeton, and with the German masters H. C. Vogel and Julius Scheiner at Potsdam. Frost, two years older than Hale, was from another old New England family; his father had been a professor and ultimately the dean of the Dartmouth Medical School. After his two years in Europe, young

Frost returned to Dartmouth as an assistant professor of astronomy, translated Scheiner's book on astronomical spectroscopy into English, and settled down to a comfortable life of teaching. He first met Hale at the World Congress in Chicago in 1893 and also attended the Yerkes dedication in 1897. After Keeler turned down the Yerkes job in favor of the Lick directorship, Hale, ready for any eventuality, managed to convince Catherine Bruce that Frost was "better qualified than anyone else we could secure for the place." She agreed to allow her gift to be used to pay his salary. Hale had made certain in advance that Frost would accept the position, and he did, in spite of his feeling that Williams Bay was a frontier hamlet in the wilds of Wisconsin, far from the New England he knew and loved.[15]

One year later Walter S. Adams, whose career was to be even more closely intertwined with Ritchey's and Hale's than Frost's, appeared on the scene. Thirteen years younger than Ritchey and eight years younger than Hale, Adams was one of the first graduate students at Yerkes Observatory. Another New Englander, he had actually been born in Antioch, Syria, where his parents were missionaries in what was then part of Turkey. They returned to Derry, New Hampshire, when he was eight years old, and he was educated in academies (private academic high schools) in New Hampshire, Vermont, and Massachusetts, and then at Dartmouth College, where he studied under Frost and graduated with highest honors in 1898. Then he went to the University of Chicago as a graduate student, dividing his time between the campus and Yerkes Observatory.[16] The spare, ascetic Adams was highly intelligent, tactful, an extremely hard worker devoted to science, and a skilled, careful observer. Hale and Frost were greatly impressed with him. They encouraged him to go to Munich for a year of additional study and work with Hugo von Seeliger and Karl Schwarzschild and then brought him back to Yerkes as an assistant in 1901.[17] Adams never bothered to get an earned Ph.D. (he was at the end of the time when a research career was possible without one), but he was to work closely with Hale all the rest of the latter's life, and to extol his memory after his death.

Though Adams was outstanding, Wadsworth proved to be less so. He was quarrelsome, difficult to get along with, and had an extremely inflated opinion of his own importance. When he demanded a promotion to associate professor and a large salary raise, Hale passed his near-ultimatum on to Harper with only a weak recommendation. The president would always support outstanding research workers but was only too happy to get rid of the lesser lights in any department. He rejected Wadsworth's appeal. The latter angrily resigned and took a job at Allegheny Observatory, where he subsequently did no research of any importance.[18]

Meanwhile, Ritchey had been doing excellent work in the optical shop and was obviously expert in all kinds of mechanical work. Now

Yerkes Observatory group 1898. Sitting on the wall and second from the right is Ritchey; on the steps, in front, are Ellerman and Schlesinger; in the rear, left to right, are Ernest F. Nichols, Harry M. Goodwin, Barnard, Frost, and Hale. Courtesy of Yerkes Observatory.

Hale recommended that he be hired on the Yerkes staff, at the rank of assistant at a salary of $1,400 per year, with part of the salary money released by Wadsworth's departure. (Frank Schlesinger was also hired as an assistant with the other $1,000). Ritchey's duties were to include "assistant and computer, . . . investigator in astronomical photography, and . . . optician." Under the last title, Ritchey's main work was to be to complete the 60-inch mirror, which Hale optimistically described as "nearly finished. This great mirror is one of the largest and most important pieces of optical work ever undertaken." Hale gave Ritchey credit for doing the optical work on it, designing and constructing the grinding machine, devising the new optical test methods used in figuring the mirror, and "completing designs for the entirely new type of mounting required for the 60-inch [telescope]."[19] Harper approved the appointment, Ritchey went off Hale's father's payroll and onto the University of Chicago's, and Hale immediately put him in charge of the Yerkes mechanical shop, as superintendent of instrument construction in place of Wadsworth.

At that time Ritchey had already begun a program of astronomical photography with Hale's 12-inch refractor, first at Kenwood and then, after it was moved, at Yerkes. This telescope had an especially long focal length for its aperture (in technical terms, a large or slow focal ratio) and hence was particularly well suited for photographing the moon. According to Hale, Ritchey had obtained better pictures of the moon with it than anyone else (meaning Hale himself and Barnard) had.[20] Soon Ritchey was using the giant 40-inch refractor for this same program. With it he took many excellent pictures of the moon, unmatched for years, that were widely admired by astronomers,[21] and later magazine and book readers everywhere.[22]

As astronomical objects go, the moon is relatively large and bright. From the earliest days of photography it had been a favorite subject for astronomers. In principle the longest-focal-length instruments, which produce the largest images, should therefore provide the best pictures. Edward S. Holden had begun a lunar program at Lick Observatory with the 36-inch refractor, when it was the largest in the world, but the results were not particularly good, and this failure was one of the counts against him when he was finally forced to resign in 1900. Holden was not half the perfectionist that Ritchey was and was far below him in ability to understand and analyze a technical problem.[23] All large astronomical refractors are corrected for *visual* light, the band of wavelengths in the yellow and green spectral region to which the human eye is most sensitive. This means that their lenses are designed and made so that they work best at these wavelengths. Their focal lengths are essentially constant for this visual spectral region but as a result necessarily vary much more rapidly in the blue and violet regions. Ordinary photographic plates, however, are most sensitive to blue and violet light, have very low sensitivity to green, and are practically insensitive to yellow. Holden, his assistants, and previous observers had all realized this; Lick Observatory had a third "photographic corrector" lens, 33 inches in diameter, to mount in front of the 36-inch lens to convert it to a system corrected for photographic light. It was difficult to align and center the corrector lens perfectly, and this was probably one of the main problems with the Lick moon photographs.

Ritchey, on the other hand, conceived the idea of using orthochromatic plates, which are sensitive to green and to some extent yellow light. They were just coming into use at the end of the last century. With them he needed no photographic corrector. He put a thin yellow glass filter directly in front of the plate, removing entirely the out-of-focus blue and violet light. Instead of trying to focus the telescope with an eyepiece—the traditional method—Ritchey focused on a star using a knife edge, a more difficult but much more critical and exact method. Furthermore, since he focused and took the picture with light of exactly the *same* color, there was no offset between the visual and "photographic" (blue-violet) position of the best focus. Finally, Ritchey concentrated his efforts entirely on photography. He had the telescope many nights and took many plates of the moon. Among them there were a few good nights, in which the "seeing," or atmospheric steadiness, was unusually good. On those few nights there were a few short instances of exceptionally good seeing. The plates he took during those brief moments had the best images and showed the finest detail. He could identify them by painstaking inspection and comparisons of all the plates taken of the same region on the moon, generally near the terminator, or division between the illuminated and dark parts of the moon, where the contrast is highest and the surface features can best be seen. Everyone who saw Ritchey's pictures marveled at the fine detail they revealed, the myriads of craters, the mountain ranges, the rills and rays. Yet no one had much of an idea as to what direct scientific information could be determined from these photographs (except the heights of mountains and depths of craters, from the measured lengths of the shadows), and Ritchey himself had only very primitive physical ideas about the moon.[24] This summary of his first lunar photography program with the 40-inch refractor was to apply almost equally well to all of Ritchey's future observational research programs: superb technical skill and instrumentation, striking photographs, and very few hard scientific results.

In addition to the moon pictures, at the 40-inch Ritchey was also photographing globular and galactic clusters, rich aggregates of stars. He used exactly the same yellow filter and orthochromatic plates, but to record the images of the faint stars, long exposures, several hours in duration, were necessary. It is crucial that the star images not be blurred in the slightest; spreading them out makes them fainter and therefore loses all the faintest stars. Ritchey, like all other skilled observers, rated the driving clock of the 40-inch telescope carefully so that the huge refractor would track the cluster accurately across the sky for hours. But he also developed a very precise double-slide plateholder, with accurately machined ways and two evenly moving screws, so that he could eliminate all the little periodic errors and irregularities in the telescope motion, as well as all but the most rapid of the tiny apparent motions caused by "seeing," irregular refraction of light in currents of air above the telescope. This double-slide plateholder had a high-power eyepiece with an illuminated cross hair. He would set the cross hair on a star image just outside the cluster and then, all during the exposure, with one hand on each of the two knobs, by turning the screws he would "guide" out all the motions by keeping the image of the guide star centered. It required intense concentration and dedication. There was also a flap shutter, which an assistant could close or open at Ritchey's order, to interrupt the exposure if the seeing suddenly went bad, or whenever the observing floor was raised or lowered, to be certain that no disturbance was communicated to the telescope and blurred the image. Sometimes Ritchey took exposures many hours long, spread over two nights, leaving the plate sealed in the plateholder during the day.[25] Again the results were excellent, far better than any previous photographs of clusters.[26]

Ritchey's ideas on photography with long-focus telescopes were further developed at the solar eclipse of May 28, 1900. The eclipse's path of totality swept over the southeastern United States, and every major observatory in the country, and many minor ones, sent parties to make special observations during the eclipse. Hale set up a national committee, with himself as the secretary, to try to organize the work of the various observatories, but it had little actual effect.[27] He only succeeded in raising the necessary money for the special Yerkes expedition to the eclipse at the last moment.[28] The group that went included himself, Frost, Barnard, Ritchey, and Ellerman. They located their eclipse camp at Wadesboro, North Carolina. Hale's program was to measure the heat radiation of the corona with a bolometer, this time at an eclipse. In 1898 Hale had tried it again without an eclipse, using the new 40-inch refractor to collect the radiation but once more without success.[29] At the Wadesboro eclipse, assisted by Ellerman, he was to use a fixed horizontal reflecting telescope. Its mirror, made by Ritchey, was 20 inches in diameter and its focal length was 27 feet; it was fed by a large two-mirror coelostat. However, Hale's delicate bolometer was somehow damaged on the way down from Williams Bay to Wadesboro and was inoperable. He managed to get another one made at the last moment by a friendly instrumentalist who was with another eclipse party. However, in the excitement at totality, someone kicked or knocked one of the small mirrors that fed the light to the bolometer out of alignment, and it was impossible to get any measurements. Hale's attempt had turned into a disaster.[30] It was Charles Greeley Abbot of the Smithsonian Astrophysical Observatory who succeeded in measuring the corona with a bolometer at this eclipse.

The other Yerkes observers got excellent results. Frost obtained a large number of spectrograms of the chromosphere and corona. Barnard, using another horizontal telescope, built around a 6-inch lens, corrected for photographic light, and fed by another coelostat Ritchey had made, obtained a number of excellent direct photographs of the corona. The long focal length of the lens, 61 feet, almost exactly the same as that of the 40-inch refractor, provided an excellent large image of the sun, and the long, shaded horizontal telescope tube that Barnard and Ritchey assembled at the site minimized thermal air currents and consequent poor seeing. Ritchey had built a large plate carrier, 15 feet long, mounted on

Yerkes Observatory solar eclipse group, Wadesboro, North Carolina, 1900. Ritchey is third from left; to the right are Ellerman, then Hale (standing). Frost and Barnard are standing on the far right. Courtesy of Yerkes Observatory.

ball bearings, so that seven large plates could be exposed one after another during the brief eclipse without changing the plates or shaking the apparatus.[31]

Barnard noted in his report on the eclipse that the excellent photographs obtained with the 61-foot–focal-length lens showed that much longer horizontal telescopes could be used to secure even better direct photographs of the corona at future eclipses. Ritchey generalized the idea and put forward the plan of building a long-focus fixed horizontal telescope for direct photography of the moon, planets, double stars, and clusters. His specific proposal was for a 24-inch aperture mirror with a focal length of 200 feet, providing an f/100 focal ratio and an image three times as large as the 40-inch refractor or the Wadesboro horizontal telescope images.[32] Hale also wanted such a permanent horizontal telescope for continuing his attempts to measure the heat radiation from stars, planned in collaboration with physicist Ernest F. Nichols, who had developed a sensitive radiometer. Such delicate detectors as bolometers and radiometers could not be mounted on a moving telescope but were ideal for use at the focus of a fixed instrument. Hale put Ritchey to work on a 24-inch long-focus mirror and applied to the American Academy of Arts and Sciences for a grant of $500 to help complete the instrument, whose cost he estimated as $2,000 in addition to the work of the Yerkes staff that would go into it. It would be much less expensive than a 24-inch long-focus refractor would have been. Charles R. Cross, the chairman of the Academy's Rumford Committee, which made the grants, was one of Hale's former teachers at MIT; he got the money toward the end of 1900. The telescope as built had a focal length of about 165 feet, making it an f/82 rather than an f/100. By now Hale had found that precision high-dispersion spectroscopy, the astrophysical research on stars which he wanted to do, was impossible with an instrument mounted on the 40-inch refractor. As the telescope moved, the necessarily long-focus, large, heavy spectrograph bent and warped slightly, differently at different positions, destroying the definition of the spectrograms. The fixed focus of the horizontal telescope would be ideal for a large spectrograph, just as in his undergraduate thesis work at the Harvard College Observatory. Thus high-dispersion spectroscopy soon became the main aim of the project. Ritchey and his mechanics built the horizontal telescope and erected it in a specially constructed 170-foot-long wooden house, mounted high above the ground to minimize heating and cooling and consequent distortion and poor seeing. The instrument was completed in the late fall of 1902, and Adams and Ellerman began preliminary spectroscopic tests with it. But in December a fire, caused by an electrical short circuit, completely destroyed the building and left the telescope and spectrograph a twisted wreck. It had been Ritchey's main project for more than two years.[33] Ultimately, Hale raised the money to build a new horizontal telescope, but it only came into real operation at Mount Wilson. Ritchey never got to use the fixed horizontal telescope for direct photography, but the ideas for the giant fixed vertical telescopes that he planned in France a quarter of a century later no doubt originated here.

The most important instrument built by Ritchey at Yerkes Observatory was the 24-inch reflector. He had made the primary mirror—a short focal ratio f/4 paraboloid—in his shop in Chicago. In June 1898 he sold it to Yerkes Observatory for $200. Apparently, Hale had paid for the glass disk earlier, and the $200 was for Ritchey's work in turning it into a mirror, but this is not certain from the one surviving requisition.[34] Ritchey designed a short, fork-type mounting for the telescope, but Wadsworth, who was still in charge of the Yerkes mechanical shop then, vetoed his plan and substituted a more traditional design of his own. Wadsworth had purchased most of the parts for the mounting and had put the shop to work on it before he left Yerkes in early 1899, so Ritchey, who succeeded him, had to complete the telescope according to a basic design he did not really believe in. However, he modified many of the details of Wadsworth's plans to fit better with his own ideas.[35]

Ritchey's assistant, who began helping him on the 24-inch reflector in June 1901, was Francis G. Pease. He had just graduated from the Armour Institute of Technology (now the Illinois Institute of Technology) in Chicago, where one of his teachers had been Ritchey's father, James. Three years after Ritchey had moved to Chicago in 1888, his father moved back to Evansville and again became the superintendent of the Armstrong furniture factory. Then in 1895 James Ritchey got a new job as an instructor in woodworking at Armour Tech, where the first classes had begun only two years previously. His son Ned (S. Edward Ritchey) had previously come to Chicago and started his long career as a woodwork and shop teacher at Lane Technical High School. The father and Ned and their families formed a tight little support group for Ritchey in Chicago. James Ritchey was a good craftsman and a good teacher; he wrote a textbook on pattern making that was in use up to 1940.[36] He was one of the teachers in the Shop Studies course that all the Armour Tech students, including Pease, took. It included carpentry, turning, pattern making, forging, founding, and machine-tool work.[37] There is a persistent legend among opticians that Pease, when he first came to the little village of Williams Bay as a twenty-year-old bachelor, lived with Ritchey and his family in their home, almost as a son.[38] This story certainly rings true, for James Ritchey had undoubtedly recommended Pease for the job, but there is no documentary evidence to support it. If true, it is especially poignant because in later years Pease was to become one of the chief instruments of Ritchey's downfall.

Pease worked only on the Cassegrain secondary mirror for the 24-inch reflector; the primary mirror had been finished and the basic telescope was near completion by the summer of 1899.* Ritchey was using the 24-inch as a finished instrument two years later, set up for photography as a Newtonian reflector.[39] In this mode a small, flat mirror brings

*See the appendix and its Figure A.1 for brief explanations of the optical terms.

the light from the primary mirror out to a focus at a double-slide plateholder, similar to the one on the 40-inch, at the side of the tube. Later the telescope could also be used for spectroscopy, at the Cassegrain focus, or for photography at the prime focus. Ritchey's first published papers reporting results with the 24-inch were on his photographs of the "new" star Nova Persei, which suddenly flared up in brightness and appeared in the sky in February 1901. His first plate of it, taken on September 20, showed several faint wisps of nebulosity close around the nova. Another photograph, which he took less than two months later, showed a considerably larger area covered by the wisps. He appeared to have observed the birth and rapid expansion of a tiny nebula. Ritchey and Charles D. Perrine, who detected the same features on his photographs taken with the Crossley 36-inch reflector at Lick Observatory, both interpreted it this way. Few astronomical objects are ever seen to change, because of their great distance from us; this rapid "expansion" was unique and hence of great interest.[40]

Dutch astronomer Jacob Kapteyn, however, soon put forward the interpretation that actual motion of material was not being observed in the Nova Persei 1901 nebulosity. Instead, he argued, the light flash from the suddenly brightening star, moving out with the velocity of light and illuminating and reflecting from wisps of nebulosity already present about the star, gave the impression of an expanding nebula. In his last paper on the subject, Ritchey published a series of drawings, based on all his and Pease's photographs with the 24-inch, from September 1901 through March 1902, showing the changes and development of the appearance of the nebula over the six-month time interval. As Ritchey stated, these drawings strongly supported Kapteyn's interpretation, for the individual wisps did not appear to move, but the area in which they were illuminated and hence visible was steadily growing larger. Ritchey promised to measure all the plates and to analyze and discuss them thoroughly, but he never did so.[41] He was soon too busy building new instruments and planning and bringing into existence a new observatory. But his published drawings provided the basis of many subsequent theoretical discussions of this phenomenon, which we now know as a "light echo."[42]

Ritchey's papers on Nova Persei 1901 led to a sensational newspaper article about him, the first of several in this vein to be published over the years. This one appeared in the *San Francisco Chronicle* of November 20, 1901, under the headlines:

How Worlds Are Made
Yerkes Observer Proves Truth of
Herbert Spencer's Theory of Evolution
He Notes Changes in Nebula of Perseus

The text of the article went on to say that Ritchey had "made a discovery of startling significance and importance in the history of the evolution of the universe. . . . [T]he brilliant astronomer ha[d] just finished observations which prove the truth of the celebrated theory of creation announced by the famous Frenchman, La Place. . . . This is one of the greatest discoveries in the history of astronomy, and Professor Ritchey alone deserves all the credit. It is the first work of absolutely great magnitude done at the Yerkes Observatory, and it will insure Professor Ritchey a foremost place among the world-renowned men of science."[43]

This article no doubt appeared in the *Chronicle* because of its relevance to nearby Lick Observatory. Perrine had taken his first photograph of Nova Persei 1901 after Ritchey, but he had been the first to see the apparent expansion of the "nebula."[44] If anyone deserved credit, Perrine did, or the German astronomer Max Wolf, who had first glimpsed the nebulosity and urged the California observers to train their "big" reflectors on it. Besides, astronomers were not supposed to release sensational stories to the newspapers, glorifying themselves and their discoveries. W. W. Campbell, who had succeeded Keeler as director of Lick Observatory after the latter's untimely death in August 1900, assured Hale that he and Perrine did not consider either the Yerkes director or Ritchey "even morally responsible" for the newspaper story, and Hale denied that he, Ritchey, or anyone at Yerkes has "even the remotest idea" of where the "very absurd stories" had originated.[45] Yet they obviously had come from Ritchey. Most probably his father or brother had passed on a garbled account of what he had told or written them about his discovery to a reporter friend in Chicago, who added some touches of his own about Herbert Spencer and evolution.

Ritchey also began taking photographs of comets with the 24-inch reflector whenever a bright one appeared. His first was Comet Perrine in 1902. Comets move with respect to the background stars, and the telescope must track them accurately during an exposure to prevent blurring the image. Previous observers had guided on the nucleus of the comet with an auxiliary visual telescope. The problem with this method is that the nucleus is often faint and difficult to see, especially if the guide telescope is a small one. Ritchey developed the much better method of using his double-slide plateholder, with one of its axes oriented in the direction of motion of the comet. With it he could guide on a nearby bright star, offsetting it by a known amount at precalculated intervals to compensate exactly for the comet's motion. The rate of motion could be taken either from a preliminary orbit of the comet, if one were known, or could be measured from two exposures taken just before the guided exposure. This method immediately became standard and was used everywhere for many years. Characteristically, in his paper describing it Ritchey sketched several possible improvements based on his analysis of the results he had obtained up to the time he wrote it.[46]

With the 24-inch reflector Ritchey photographed many nebulae, extended "clouds" of emitting gas, and reflecting dust particles as we know today. Keeler, in the two short years before he died in 1900, had taken over the 36-inch Crossley reflector at Lick, made it into a serious scientific instrument, and obtained a pioneering series of photographs of nebulae. These pictures were the first opportunity astronomers had to see the true forms and complicated structure of objects like the Orion Nebula, the Trifid Nebula, the Ring Nebula in Orion, and all the other nebulae that are so familiar to us today. They also included the Andromeda "Nebula," the Whirlpool "Nebula," and other spiral "nebulae," which we know to be galaxies of stars, not clouds of gas or dust. Keeler's photographs, first exhibited at the first American Astronomical Society meeting in 1899 and published in his papers in early volumes of the *Publications of the Astronomical Society of the Pacific* and the *Astrophysical Journal*, were a revelation to most astronomers of his time. They convinced the researchers, as no theoretical arguments could, that reflectors, not refractors, were the monster telescopes on which future research would be based. After Keeler's work no serious astronomer suggested building larger refracting telescopes, only reflectors. The 40-inch Yerkes telescope was already in operation, but it was the last of its kind.[47]

Ritchey, starting in 1901, was able to take better direct photographs than Keeler had. The Crossley reflector was a rickety, poorly designed, and poorly built telescope that Keeler had fixed up as well as he could. The 24-inch reflector was much better designed, especially in its rigidity, essential for long photographic exposures. It was optically faster than the Crossley reflector. Everyone who saw Ritchey's photographs was even more impressed with them than with Keeler's. Hale, always skillful in putting forward the results of his observatory, pressed copies of Ritchey's pictures on important astronomers everywhere. The most important in the United States was Simon Newcomb, the "grim dean of American astronomers," an expert on planetary motions and the first president of the American Astronomical Society, who served six years in that office. Hale aroused Newcomb's interest in Ritchey's 40-inch and 24-inch photographs and arranged to show them at a New Year's Eve reception during the meeting of the American Astronomical Society held in Washington at the end of 1901. All the assembled astronomers studied and admired them.[48]

Ritchey himself gave popular lectures, built around slides of his moon and nebular pictures, at Chicago, Milwaukee, Cincinnati, and St. Louis in the winter of 1901–1902. The following autumn he went further afield, giving similar illustrated popular lectures in Washington, Philadelphia, New York, and Boston.[49] Two semipopular articles he wrote, published in the high-quality *Harper's Magazine,* and illustrated with his photographs of the moon and nebulae, excellently reproduced, spread his name and results to a wider audience.[50]

Ritchey became more self-assured. His reports to Hale, while the director was traveling, were now confident letters to an equal, rather than fawning messages to a superior, as in the earliest Chicago days.[51] The earliest surviving written record of Ritchey's dissatisfaction with his subordinate position, subject to Hale's orders, dates from 1901. He complained that the director, without consulting him, had offered to lend some of his photographs of the Pleiades cluster to John K. Rees at Columbia University for measurement. Ritchey wanted to measure his own plates himself, not send them off to someone he did not even know. As he had not yet taken a plate of the Pleiades, he dragged his feet, and Rees had to do without a 40-inch refractor plate of it.[52]

In 1901, on Hale's recommendation, Ritchey was promoted from assistant to instructor (in "Practical Astronomy"), still a relatively low rank for his responsibilities, especially as his salary remained unchanged at $1,400 a year. It represented the reality of his position, outside the regular academic ladder. He was a worker, not a researcher, and hence a second-class citizen in the eyes of Harper and Hale. The next year Ritchey did receive a raise, to $1,600 a year.[53] His talents as an optician and telescope builder were very highly regarded by other astronomers. Campbell, who was having a 37-inch reflector built by John A. Brashear and Company, consulted Ritchey frequently about telescope design. He liked very much what he saw of the Yerkes optician's work during a short visit in January 1902.[54]

That summer Ritchey received his first request to make a large reflector for another observatory. Aristarchos Belopolsky, the director of the Pulkovo Observatory in Russia, wrote Frost and asked if Ritchey could make a 40-inch reflector for a projected astrophysical observatory to be built in his country. Ritchey was eager to take the job, and Hale got permission from Harper for him to do it on his own time, as an independent contractor. Ritchey recommended a massive, rigid Cassegrain (two-mirror) reflector along the lines of his own published designs and estimated the cost as $40,000. This was too expensive for the Imperial Russian government to consider, and the plan fell through, even though Ritchey tried to salvage it by relaxing his design requirements and cutting the price to $30,000.[55] Throughout his life his standards of perfection were to be too costly for most astronomers. They usually settled for a "realistic" alternative design and frequently lived to regret it.

Ritchey met Samuel P. Langley, the pioneer astrophysicist who had become Secretary (in reality, director) of the Smithsonian Institution in Washington, through his photographs of the moon. Langley had measured the moon's infrared radiation himself, in his earlier days at Allegheny Observatory. He maintained a vigorous policy of leading scientific research by using Smithsonian funds to publish results in fields he considered important. He wished to get out a detailed map of the moon, based on the best available photographs, which geologists or selenographers (students of the moon) could study and use for their research instead of trying to observe its features themselves. Langley recognized Ritchey's pictures as the best available (he had published two of them in the Smithsonian Annual Report for 1901) and asked Hale for permission to use them for the map. He was an old-fashioned director who would not have thought of communicating directly with a researcher at another institution without writing to his director first. Hale was strongly for the idea. A more modern director, he recognized clearly that distributing Ritchey's spectacular results widely was the best way to spread the fame of Yerkes Observatory and to build up support for his future fund-raising efforts. Ritchey was strongly for the idea, too.[56] Getting reproductions of the moon pictures made to the uniform scale Langley wanted, and up to the high standards Ritchey set, was difficult, and though in March he expected to have them made "very soon," he was not able to send the first six pictures until May. The other four, which he had promised would "certainly" be ready less than a week later, were not finished until July.[57] Nevertheless, they were excellent pictures and Langley published them in the appendix to a long *Smithsonian Contribution* comparing the terrestrial and lunar surfaces by the old-time "naturalist" and field geologist Nathaniel S. Shaler.[58]

In the course of arranging for the Smithsonian to publish the moon pictures, Hale told Langley that Ritchey was an expert in making astronomical mirrors and that he could write a good article on his new methods for testing them. Scientists' interest in reflecting telescopes was rapidly growing, but there was little information about them available in print. Forty years earlier Henry Draper had written a pioneering article on the methods he had used to make his small reflecting telescope. His article was a Smithsonian publication that had been out of print for years. Langley asked Ritchey to correct it and bring it up to date, perhaps adding a few thousand words of his own and any illustrations that might be needed. He emphasized that he wanted to get the publication back into print very soon. Ritchey replied that instead he would prefer to write a complete treatise on modern methods of making and testing telescope mirrors. Langley insisted that he wanted a quick correction of Draper's article plus a short supplement by Ritchey, and that he needed it in a month or six weeks at the most. The optician agreed but pleaded that he was very busy putting the finishing touches on the long-focal-length horizontal telescope and was about to leave for his lecture tour of the East but that he would surely get the manuscript to Langley in less than three months. The Smithsonian director accepted this arrangement but "trust[ed] it may prove possible for you to abridge the time mentioned."

As soon as Ritchey started working on the project, he realized that he had made a mistake. He could only do justice to his ideas in an article of his own. "Without egotism I can say that the article will be an important one in the scarce literature of the subject, and will be regarded and used as authoritative for years to come," he wrote. His article would be "incomparably more complete" than Draper's, and he did not want it published as a supplement or appendix but as a separate paper. Langley could see the justice of Ritchey's position and agreed to publish his and Draper's papers together as independent contributions. Nearly two years later, after innumerable delays, excuses by Ritchey, rejections of reproductions that did not meet his very firm standards, and hectoring letters from Langley and his underlings, Ritchey's "On the Modern Reflecting Telescope and the Making and Testing of Optical Mirrors" was finally published.[59] It was bound in the same volume with Draper's old paper but was independent of it. And it was worth every bit of the delay.

In his 51-page paper, with fifteen line drawings and thirteen photographs, many of them of the Yerkes optical shop, Ritchey described in clear detail all his methods.[60] The titles of some of the chapters, "Grinding and Polishing Machines," "Grinding Edge of Glass—Rounding of Corners," "Testing and Figuring Convex Hyperboloidal Mirrors," and "Silvering" give an idea of the contents. The paper begins "No greater mistake could be made than to assume that cheap and poorly annealed disks of glass, or those with large striae or other pouring marks, are good enough for mirrors of reflecting telescopes. . . . If it were not necessary to consider the questions of cost, I should advise the use of the finest optical (crown) glass always." The whole paper reflects the themes of knowing the materials, using only the best, and getting the most out of them. Ritchey's article was much like his father's and brother's books on shop work, except that he had invented or adapted or developed many of the optical methods himself. He emphasized the great need for absolute cleanliness, control, respect for the material, and doing successive parts of the job "perfectly."

The methods are simple. The optician adjusts the machine so that it grinds the glass disk symmetrically and adjusts the tool so that it shapes the mirror into a section of a sphere. After reaching that figure, he goes on with the small additional figuring needed to transform the sphere into a paraboloid. The optician uses successively finer grades of carborundum as the cutting agent and finally rouge as a polishing agent. He tests the figure at each stage, readjusting the tool to bring it closer to a sphere or paraboloid as necessary.

The tests are all-important. Ritchey used the Foucault knife-edge test, as he had modified and developed it himself. The principles are very simple. If a point light source is put at the center of curvature of a perfect spherical mirror, all the light rays from it are reflected back on themselves to the source. Ritchey used a pinhole, with a small, bright lamp behind it as a source. Very close next to it, and in the same plane with it, he placed a knife edge, mounted on a fine screw so that he could move it smoothly and cut the rays returning from the mirror at their focus. Putting his eye behind the knife edge, he could see whether the entire mirror darkened simultaneously. If it did, it meant that all the rays had been cut at the same point. If so, the mirror was a sphere and the light source and the image of the knife edge were at its focus. If not, by the way in which the mirror darkened he would know which parts of the disk had to be cut down more, and he would adjust the grinding machine accordingly. He could repeat this process in finer and finer steps, until the mirror was a perfect sphere. The next step was to parabolize it. Here the principle is that a perfect paraboloid will bring parallel light to a focus, or will reflect all the light from a source at its focus as a parallel beam. For this test Ritchey needed a large flat mirror. He would put the pinhole with the light behind it at the focus of the mirror being tested so that light from it was reflected from the paraboloid to the flat. He would adjust the flat so that it sent the light straight back to the mirror, which would bring it back to the same focus. Again using the illuminated pinhole and the knife edge, he would make the same test there. The large flat mirror could be tested with the sphere, before it had been parabolized, by another variant of the knife-edge test. It all sounds simple on paper. Ritchey could make it all work.[61]

In his article he also emphasized the great importance of making the telescope mounting rigid. Flexure would ruin the images of the stars and nebulae. In the two-mirror Cassegrain systems that Ritchey and Hale advocated, the demand for rigidity is especially high. A Cassegrain telescope consists of a primary paraboloid combined with a smaller, convex, secondary hyperboloidal mirror that reflects the light back through a hole in the primary to a focus. The focal points of the paraboloid and the hyperboloid must coincide for the system to work, and any misalignment or miscentering is amplified strongly in blurring the images of stars. Ritchey's *Smithsonian Contribution* was an authoritative treatment of reflecting telescopes and how to make them that did become a classic.

He does not deserve all the blame for the delay in completion of the article. He had been in poor health for some time, and eventually, in

January 1904, his illness was diagnosed as appendicitis. He had to undergo an operation, at the Hahneman Hospital in Chicago. An appendectomy was a serious, dangerous operation in those days. Ritchey had a particularly close call but survived. His recuperation was difficult and painful. His hospital stay was long, and from his sickbed he had to fight off what he regarded as an attack on his territory. R. James Wallace, the photographer who had been newly hired to make prints and transparencies from Ritchey's original negatives of the moon, clusters, and nebulae, tried to take over the 24-inch reflector for some nebular photography of his own. Ritchey regarded the 24-inch as *his* telescope and photographing nebulae as *his* subject; he complained vigorously that Wallace should get to work on the prints that Langley was demanding instead of poaching on *his* nebulae.[62]

After Ritchey finally got out of the hospital in March, his wife and parents persuaded him to go south for a visit with his three aunts, his father's youngest sisters who lived in Denison, Texas. There he could exercise himself back to health, beginning with short walks in the balmy southwestern spring, gradually increasing the length of his walks as his strength returned. Probably, his family wanted to get him away from his work as much as to get him out of cold, snowy Wisconsin. Hale, knowing Ritchey was worried about the cost of the operation and hospital stay, gave him $100 of his own money, roughly the equivalent of $1500 today. In the hospital in Chicago, and at his aunts' home in Texas, Ritchey completed his manuscript corrections and read his proofs, using Pease as his legman to deal with the engraver of the illustrations. In mid-April he returned to Yerkes, ready to start work again even if not completely healed.[63]

During his illness Ritchey had also completed another paper, for the University of Chicago Decennial Publications. These were a series of ten bound volumes, celebrating (a little late) the tenth anniversary of the founding of the university, and containing research papers by faculty members in all its departments. One volume was devoted to astronomy and astrophysics. Ritchey's paper, "Photography with the Forty-Inch Refractor and Two-Foot Reflector of the Yerkes Observatory" was a condensed, simplified version of his two earlier papers from the *Astrophysical Journal,* with minor revisions and additions. Handsomely illustrated with his photographs, and widely distributed by the university, this paper probably brought understanding and appreciation of his work to a larger number of American and British astronomers than his previous, more specialized publications.[64]

Ritchey had now become very well known. In 1903, at the request of the Royal Astronomical Society, he had sent a collection of his pictures for exhibition, in the form of glass transparencies, at its meeting rooms in London. The English prided themselves on accepting and understanding reflecting telescopes long before the Americans had, but Ritchey's photographs were clearly much better and showed far more details than the photographs that their heroes, A. A. Common and Isaac Roberts, had taken.[65] They recognized his work and in 1904 elected him a foreign associate, or honorary member of their society.[66] That same year Ritchey was promoted, again on Hale's recommendation, to assistant professor. His salary went up to $1600 per year, in addition to a supplemental $200 which he had been getting for at least a year or two previously.[67]

Hale had been receiving even more recognition than Ritchey. Beginning with the Janssen medal of the French Academy of Sciences, which he was awarded in 1894, at the age of twenty-six, he had received the Rumford Medal in 1902, the Draper medal in 1903, and the Royal Astronomical Society medal in 1904. In 1901 he was offered, but declined, a professorship at MIT, his alma mater, and in 1902 he was elected to the National Academy of Sciences at thirty-four, one of the youngest members in its history.[68] Yet he found himself with less and less time for research as his attentions and efforts turned increasingly to organizing new institutions and programs. He claimed he was forced into these organizational activities, but in fact he sought them. Hale knew that if he did not undertake them, no one else would, or if someone did, that person would not do it as well as he himself could. Thus, for instance, Hale personally organized a Congress on Astronomy as part of a vast Congress of Arts and Sciences held at the St. Louis World's Fair in 1904. The National Academy of Sciences provided the institutional sponsorship, and Hale invited the speakers, emphasizing astrophysics and especially solar physics. The time and energy he put into writing all the necessary letters shaped the meeting along the lines he thought it should have but left him little of either for research. This conference led to others in later years, to the founding of the International Union for Cooperation in Solar Research and finally to its transformation into the International Astronomical Union. They were all Hale's children.[69]

Ritchey's biggest project at Yerkes Observatory, for which Hale had hired him from the beginning, was the 60-inch telescope.[70] As soon as the disk arrived in 1897, Ritchey had gone to work on it so that all the University of Chicago officials who came to Yerkes that first summer and fall, and all the astronomers who came to the dedication in October, would see the reality of the giant mirror. It was a visible symbol that there would be a 60-inch telescope someday. Hale was one of the first observatory directors to grasp the importance of buying a glass blank and starting to work it into a mirror, even if there was no money to build the rest of the telescope, as the best way to attract the remainder of the necessary funds. Ritchey's real responsibility at Yerkes was to design the telescope, its optics, mounting, and dome.

It was a long-term program. One of his first tasks was to provide cost estimates of the telescope and drawings of what the finished instrument would look like, which would be used in the fund-raising. He planned an especially rigid mounting but proposed a light, cheap corrugated iron structure for the lower part of the dome, which would cool off at night and reach thermal equilibrium much more quickly than the traditional, impressive brick building.[71] Hale's aim was to use the 60-inch mainly for high-dispersion spectroscopy, the astrophysical research that was his chief interest.[72] This would require a large, fixed spectrograph. The 60-inch was, therefore, designed so it could be used as a coudé telescope. This is a long-focal-length Cassegrain system in which the light from a star is reflected by the primary parabolic mirror to a secondary hyperboloid, which reflects it back toward the primary. Before the light gets there, however, it is reflected again by a small, flat mirror, up or down the polar axis of the telescope to the focus. No matter how the telescope turns and moves, the final path of the light along its polar axis remains fixed. The English astronomer Arthur C. Ranyard had invented (or popularized) this arrangement. In his design the light came *up* the polar axis to a focus above it. Ritchey originally planned to use this design for the 60-inch and showed a preliminary design for it in his paper on the 24-inch.

Wadsworth modified Ranyard's general scheme, however, to a better one in which the light rays were directed *down* the polar axis to a focus below the bottom end of it. This arrangement had the advantages that the spectrograph was below the telescope where it could be solidly mounted in the ground, and that the light path along the polar axis did not go over the telescope in the dome where the seeing could be very bad. Instead the light went below the telescope, along a path that could be built into a thermally insulated, protected tube.[73] By 1904 Hale had decided to use this Wadsworth form of coudé mounting for the 60-inch, and Ritchey included a schematic general design of it in his Smithsonian publication.[74]

When Hale's father bought the glass disk for the mirror, their plan had been to erect the completed telescope at Yerkes Observatory, in a new, separate dome.[75] Hale knew that it, or any site in the East or Midwest, was not nearly as good as a site in California. The number of clear nights and the seeing are much better there, as experience at Lick Observatory had demonstrated, but Yerkes Observatory had been built close to Chicago. Then on November 16, 1898, William E. Hale, old and worn out, died in Chicago. It was a terrible blow to Hale. He had been very close to his father and looked on him almost as an Old Testament patriarch. Hale's parents had brought their son up in a strict, old-fashioned Christianity, to believe in the literal truth of the words of the Bible. While he was a student he was not allowed to study or do research on Sunday. That day was reserved for church (two services, morning and evening), quiet family conversation, reading uplifting magazines or books, or letter writing. As a result, nearly all of his letters from home to his college chum Harry M. Goodwin were written on Sundays. At MIT Goodwin, a scientific rationalist who was "nearly an agnostic," according to his son, had chipped away at Hale's faith. But his father's death brought him back to the fold. He wrote Goodwin: "I do not know how I could bear Father's death if I had no hope of seeing him again at some future time. We may not be able to find scientific proof of immortality but I think we *must* believe in it."[76]

While his father was in his final illness, just before he died, Hale wrote a formal letter to President Harper, stating that the 60-inch mirror, on which Ritchey had been working for two years, was now approaching completion and no doubt would be successful in every way. His father had bought the glass disk and paid Ritchey's salary; he owned the mirror. He estimated its value as $10,000. He offered it to the University of Chicago on condition that it provide the money for a mounting, building, and endowment to operate the telescope.[77] In fact the mirror was far from completion, and Hale knew it, but he wanted to place the facts of ownership on record.[78]

His father, in life, had been the last tie holding Hale to the Yerkes

Observatory site. William E. Hale had been willing to consider Williams Bay, Wisconsin, as almost a suburb of Chicago, but even his adored son would not have been able to persuade him to let him move his observatory to the far West. Soon after he died, though, young George had decided that the 60-inch must be erected in California as a remote observing station of Yerkes Observatory. In 1899, when Keeler invited him to put the telescope on Mount Hamilton, Hale replied that he could imagine "nothing more satisfactory" than the prospect of an annual trip to California to observe and work with Keeler but that the weather was better in southern California and that was where he intended to erect the 60-inch.[79] He had already asked N. B. Ream, a wealthy Chicagoan and friend of his late father, for a contribution of $130,000, which he estimated would be enough to complete the telescope, put it up in southern California, and provide the money to operate it. Each of the astronomers at Yerkes would spend a few months there each year to obtain observational data, returning to their headquarters in Wisconsin to reduce, analyze, and interpret it. Ream did not come through with the money. A few months later Hale wrote a similar request for President Harper to take to Yerkes, but the financier and street-car magnate was no more willing than Ream to provide the necessary money for a southern California branch of his observatory.[80]

Hale undoubtedly tried other prospective donors without success. Not until 1902 did the situation begin to change. That year the immensely wealthy steel magnate Andrew Carnegie set up the Carnegie Institution of Washington with an initial gift of $10 million. It was to support research. Hale immediately recognized his opportunity. He touched up his five-page statement, "A Great Reflecting Telescope," a proposal for funds to complete the 60-inch telescope and erect it "at a high elevation in southern California or Arizona," and sent it to the new Carnegie Institution executive committee. This document, which Hale had no doubt used in some of his earlier unsuccessful attempts to get the money, was centered around the spectacular pictures taken by Ritchey with the 24-inch reflector. They included the Orion nebula, the Andromeda "nebula," and the Pleiades. Hale also included Ritchey's photographs of the first two of these nebulae taken with the 40-inch refractor. This comparison showed graphically the great superiority of the smaller but much faster reflector for capturing images of faint extended objects. Hale stated that in addition to direct photography, the proposed

60-inch telescope would be used for high-dispersion spectroscopy and infrared measurements of stars, but these subjects are much more difficult for a wealthy banker, corporation lawyer, or steel-mill owner to grasp than a vivid picture. Hale also included with the proposal pictures of the 40-inch refractor, the 24-inch reflector, and the "immense" 60-inch glass disk, ground and polished by Ritchey in the Yerkes optical shop. The last picture was Ritchey's design of the 60-inch telescope.[81] Hale supplemented his proposal with a thick sheaf of letters of support from twenty-four American and European astronomers and physicists. He had more than one string to his bow; nearly simultaneously, he submitted the same proposal through Harper to Isabella Blackstone, a widow who had recently inherited her husband's considerable fortune. She might very fittingly build the telescope and observatory in southern California as a memorial to her late spouse, Hale thought. However, she declined.[82]

The Carnegie Institution, however, set up an Advisory Committee on Astronomy, with Edward C. Pickering as chairman and Hale as one of its members. The others were elderly high priests of astronomy—Langley, Lewis Boss, and Simon Newcomb. The younger, more energetic Hale was at his best in a situation like this, playing off one against the other as they cut up the cake. He attracted the attention of Charles D. Walcott, the geologist whom Carnegie had named as secretary of his advisory committee. Walcott visited Yerkes Observatory, and Hale was only too glad to show him Ritchey's pictures, taken with the 24-inch reflector, and his design for the 60-inch reflector, which would do so much better. The influential secretary was impressed and soon reached the conclusion, according to Hale, that "it will be a great mistake if we do not recommend the establishment of a solar observatory and a large reflector at some particularly favorable site." Walcott "asked" Hale to have some large, spectacular transparencies of the nebula pictures sent directly to Carnegie, to help convince him to put some real money into astronomy.

Hale claimed that he had not pleaded especially for his own observatory but for astronomy in general, and that he had told Walcott "that the Lick Observatory would doubtless be able to do quite as good work with a large reflector as we could." Its director, W. W. Campbell, was hoping to get funds for a 48-inch or 54-inch reflector, and the diplomatic Hale did not want to make an enemy of him. Campbell had not realized that the Carnegie Institution would make grants for astronomical research and was busy getting a 37-inch reflector completed and shipped to Chile. He had the transparencies of the Lick nebular photographs made and sent to Carnegie also, but he was not nearly as persistent as Hale in pressing his case.[83]

In the end the Yerkes director so manipulated the situation that in theory he had not even asked for funds but had simply provided information at the request of Walcott as to what a 60-inch telescope could do and what it would cost to build one. Likewise he had written into the record, with President Harper's approval, that the mirror belonged to him personally, and that its ownership could be transferred to the Carnegie Institution sometime in the future (and thus free him from the University of Chicago). Hale wrote to Campbell: "I withdrew our application, and have made no effort of any kind to secure assistance from the Carnegie Institution in mounting our five-foot reflector." That he had withdrawn the application was literally true but that he had made no attempt to get the funds from the Carnegie Institution was blatantly false.[84] Campbell hoped for the best and thought that "there certainly ou[gh]t to be room for several large reflectors, especially since the reflectors are so much more useful than refractors, astrophysically, and cost so much less."[85]

The Carnegie Institution did make a grant, to a committee consisting of Boss, Hale, and Campbell, to investigate sites for a "southern and solar observatory." Hale, who wanted to set up the 60-inch reflector in California, always referred to it as part of a solar observatory, even though it would never be used to observe the sun. He emphasized that it would enable him and his staff to study the stars spectroscopically for *comparison* with the sun, which was true, but it also deflected the Lick Observatory astronomers' fears that he planned to take over their California territory. Campbell, still in Pittsburgh at Brashear's shop, trying to push his 37-inch mirror through to completion, learned of the result from Hale. He professed himself "delighted" to be on the committee with Hale and Boss and ready to go anywhere to work with them on the project.[86]

The Carnegie Institution also made separate grants to Boss, Campbell, and Hale, enabling them to hire additional astronomers at their institutions. At Yerkes Hale added Philip Fox, John A. Parkhurst, and Frank Schlesinger to the staff.[87] There, with the 40-inch refractor, Schlesinger began his career as an astrometrist, measuring the parallaxes (distances) and proper motions of stars that he brought to fruition later at Allegheny and Yale observatories.[88]

Hale, speculating on his future, felt that with John S. Billings, whom he called "a most excellent friend at court" as the new president of the executive committee of the Carnegie Institution, and Walcott, "a hard-headed business man, who takes a very practical view of things" on his side, he would go far. His friends in the East still hoped to get him to come back there, and he believed that if his "California scheme" went through, he would probably be able to get it "thoroughly organized in five years." But, at the end of that time, he would not be willing to turn it over entirely to someone else to direct. He would be willing to accept a position "furthering research in general, perhaps through the administration of a fund," but he *"could not give up research of my own in astrophysics"* (his italics) and would also want to retain the directorship of the California observatory. In the nearer future, if his plans for the Carnegie-supported observatory went through, he was unsure as to whether he should retain the directorship of Yerkes also, leaving Frost in charge on the scene, or sever his connection entirely with the University of Chicago.[89]

Hale and Campbell assigned Lick astronomer William J. Hussey to make a quick site survey in southern California for the proposed solar observatory. Hussey examined several locations, including Palomar Mountain, but recommended Mount Wilson as the best.[90] It is in the San Gabriel mountains east of Pasadena, just north of Sierra Madre. Back in 1889 the Harvard College Observatory had briefly had a station there, intended to be the nucleus for the Spence Observatory, to house the 40-inch telescope that was never built but nevertheless provided the glass disks that eventually started Yerkes Observatory. Hale and Campbell went up the 5900-foot mountain themselves in June 1903, and Hale decided it was the place for him. The Carnegie Institution delayed making any grant but passed unofficial word to him that he would be funded the following year.

Hale was fed up with the ice and snow of the midwestern winter and especially concerned about his young daughter's health. She suffered almost constant colds. He was tired of isolated Williams Bay. And no doubt he was tired of President Harper, who continued to lecture him on the needs of the University of Chicago as against what Hale saw as the more important needs of astronomy. He decided to go west, start the solar observatory as a Yerkes station with his own money if he had to, and by his results attract more financial support. On Christmas Eve he was sitting on the porch of his rented cottage on Palmetto Street in Pasadena. He was watching Margaret and her little brother William play happily in the sun, with a backdrop of twittering birds, flowers so numerous he could not name them all, and orange trees loaded with fruit. Hale, who had long been troubled with gastrointestinal problems at times of stress, such as his period of waiting to hear the result of his proposal to the Carnegie Institution, had been in bed for almost a month in Chicago with colitis. But now he was positive that "[a]fter a little loafing in this glorious place all the trouble should disappear." He was "almost . . . certain that *the* place to continue [his] solar work [was] the summit of Mt. Wilson." He could see it from his front window. He had made up his mind to go there, provided that the conditions in the mountains were what he expected them to be. He had been looking at the sun with his little visual telescope in his yard, and wrote: "So far the seeing has been very poor, but on the mountain the conditions will of course be different."[91] Not surprisingly, so they turned out to be, and Hale was established at Mount Wilson.

The instrument he planned to take to Mount Wilson was the long-focus, horizontal telescope and high-dispersion spectrograph that Ritchey had built to replace the one destroyed in the fire at the end of 1902. The new instrument, called the Snow Telescope, had a 24-inch mirror with a 60-foot focal length. Ritchey had worked long and hard on it, and the system had been tested in operation at Yerkes in October 1903. Now Hale planned to bring it to California and set it up on the mountain.[92]

Helen Snow of Chicago, however, had donated the telescope as a memorial to her father with the stipulation that it be erected at Williams Bay, and she refused to give Hale permission to take it to California. So instead he had the horizontal telescope that Barnard and Ritchey had used at the 1900 eclipse, with its five-inch lens, sent out to California. Ellerman came, too, to set it up and do the observing. To complicate matters further, just at that time Yerkes cut off the $2,000 per year he had been contributing to pay two assistants' salaries at "his" observatory. One of them was Adams. Hale considered him "the most promising of all the younger astronomers in this country." He shifted funds desperately to keep him on the payroll but let the other assistant, Herbert

M. Reese, who had "done almost nothing of importance since he came," go. In spite of all the financial problems, Hale was determined to stay in California. He had met the exciting young Alicia Mosgrove, who was to become his mistress. His daughter Margaret was ill again, but he was sure that the coming summer would make her well.[93]

In March 1904 came the first, real financial support from the Carnegie Institution. They gave Hale $3,000 for a project to attempt to produce fused quartz telescope mirrors. If such mirrors could be made, they would be far superior to glass mirrors. Glass expands and contracts as it heats and cools, so any mirror made from it changes its shape and thus ruins the image it produces. This is an especially severe problem for solar reflecting telescopes, since their mirrors, exposed to direct sunlight, heat up greatly. Stellar reflecting telescopes suffer from it also as their mirrors cool off at night. Quartz expands and contracts much less than glass, so it should be a much better material for astronomical mirrors. The problem is that very high temperatures are required to fuse quartz, which makes it very difficult to melt it, contain it in a mold, and cast it into a disk.

Hale's idea was to put Ritchey to work with Elihu Thomson, one of the pioneers of commercial applications of alternating-circuit electricity, on this problem. Thomson was a very imaginative engineer, whose company, Thomson-Houston, founded in 1882, had merged with its competitor, Edison General Electric, to become the General Electric Company. Thomson became the director of its industrial research laboratory at Lynn, Massachusetts. In his lifetime he successfully filed more than seven hundred patents, many of them connected with generators, motors, arc-welding, and lighting. Thomson had been interested in astronomy from boyhood, and after his financial success he had personally made a small reflecting telescope and observatory of his own. He longed to solve the problem of producing large quartz disks for astronomical mirrors. His plan was to melt the quartz by a continuous process in a high-current arc, and he believed that it would be "perfectly practicable to produce [large mirrors] of any reasonable size."[94]

Thus, while Ritchey was in the hospital in Chicago and then recuperating in Texas from his operation, Hale was writing to him from Pasadena. He put Ritchey on leave from the University of Chicago and began paying him $200 a month from the Carnegie grant to work with

Thomson on the fused quartz project.[95] Ritchey, with his practical skills and detailed knowledge of mirrors, was the ideal person for this program. In addition, Hale could use the university funds saved from his salary to pay Adams.[96]

By May, Ritchey was well enough to travel again. Hale had gone east and convinced Helen Snow to let him move the telescope west, and persuaded the Carnegie Institution to give him $10,000.[97] He decided to use part of this money to set up an instrument shop in Pasadena, where he felt it would be needed more than in Williams Bay. He summoned Ritchey to look over the situation, then sent him back east to order tools and to ship them and whatever else he could from Yerkes to California. Adams was already there, helping Hale plan the new observatory. The young director had sent Ellerman back to Wisconsin to pick up his family and return to Pasadena, to be ready to start observing as soon as a telescope was ready. Hale reported that he was "tremendously busy, but enjoying myself at a great rate."[98]

In early June, Ritchey was on his way east to order tools and machines, and to Lynn to meet with Thomson.[99] The inventor knew the younger astronomer's work well, for he had studied his lunar photographs closely. He thought highly of them and of Ritchey's optical work.[100] Ritchey only remained at Lynn a few days, and nothing much came of this project at the time, but Ritchey and Thomson got on well together. Ritchey cemented their friendship with a gift of a transparency of one of his moon pictures and a reprint of his paper from the University of Chicago decennial publication.[101] Their mutual regard endured, and in later years when, as we shall see, Hale and Adams brought Ritchey's world crashing down and banished him from American astronomy, Thomson remained his one important supporter in the scientific community of his own country.[102]

That summer Ritchey was promoted to assistant professor, at a salary of $1,800 a year. Hale had recommended him for this promotion the previous year, writing that

> no single piece of work has done more to increase the general reputation of the Yerkes Observatory than Mr. Ritchey's investigations in astronomical photography. The astronomical photographs which he has obtained with the Yerkes telescope, and also with the two-foot reflector, are superior to any hitherto made.

Harper did not approve the promotion then, so the next year Hale tried again, this time praising his work even more emphatically:

> It has created so much interest among astronomers and the general public that many institutions are already planning to secure reflecting telescopes constructed in accordance with his design. Mr. Ritchey is worth to the Observatory more than twice his present salary; during the past two years he has saved us fully this amount in the optical work alone. It seems to me of great importance that this promotion be made.

This time Harper gave him the promotion, but at a salary of $1,800 rather than the $2,000 that Hale had twice recommended.[103]

Ritchey must have been pleased and relieved to know he had a secure job on the university faculty, for in May, Hale's financial house of cards had briefly collapsed. Harper had adamantly refused to increase the Yerkes Observatory budget, and when the promised money from the Carnegie Institution did not arrive on time, several of the younger assistants were not paid. Hale's personal check came from Pasadena a few days later and rectified matters, but everyone was worried.[104]

In California, Hale was ebullient. He was counting on the Carnegie grant, for he had already committed himself to pay the bills personally, which by then amounted to nearly $30,000 more than the $13,000 he had received. But he still felt it would come through. He loved California and wrote his college chum Harry M. Goodwin:

> You can't imagine *any* place better than this for a quiet rest—I do all my writing (dictating) now lying in a hammock hung over the edge of a canyon under beautiful spruce and live oak trees. It is *just the place* for you to rest in—So get here as soon as you can—d____ the other places on the way. I am *very sorry* to hear about your indigestion, for I know only too well what it means.[105]

He announced publicly his plans for building up the Yerkes Observatory "solar expedition" on Mount Wilson but avoided mentioning his hopes to erect the 60-inch there also.[106]

In October Hale went east to Washington to present his case personally to the Carnegie Institution executive committee. He decided to try to sell them on what he called his "big scheme" for $65,000 per year for five years, to build and put into operation the solar observatory and the 60-inch on Mount Wilson. He held them entranced. No one blamed him for running up a debt of $27,000 on the expectation of the grant. Walcott, the secretary of the committee, told him that they were strongly inclined to give him most, or even all, of what he needed.[107]

In California Ritchey had set up the new instrument shop in Pasadena, with five men working under him. They were making several spectrographs and a spectroheliograph for Mount Wilson, which he had designed. He had also put an electric furnace into operation at the Edison power house in Pasadena to begin the fused-quartz experiments. He had finished a paper, based on but going beyond his Smithsonian monograph, describing his methods of testing mirrors almost constantly as he worked them into their proper forms. Most of all, he was "still hoping to have an opportunity to use a great reflector in this wonderful climate."[108] His Smithsonian publication on the modern reflecting telescope had just come out, with its magnificent pictures, and Hale had made sure that all the Carnegie advisory committee members got advance copies of it. John S. Billings, the chairman, wrote Ritchey:

> I have spent the last two hours in reading your paper, and with great satisfaction. I sincerely hope that the Carnegie Institution will take such action as will enable you to carry out your ideas, and that substantial and important additions to human knowledge will be the result.

Ritchey was entranced by the "superb climate," and the "wonderfully clear and transparent . . . sky" of southern California and aching to have the chance to finish and use the 60-inch reflector on Mount Wilson for "a magnificent series of photographs" of the moon, nebulae, and clusters.[109]

Though the Carnegie Committee's advisory committee chairman "hoped" that it would grant Hale the funds to build the observatory, the young director grew worried as autumn changed to winter. Some of his investments, which his brother Will always managed for him, were threatening to turn sour as municipal reformers tried to regulate their street-car franchises. Hale would be in a bad financial hole if he had to pay off the observatory debts himself. He thought often about a woman in Chicago whom he and his wife knew well whose "nervous trouble" developed into melancholia, "and she did what so many victims of this disorder do—killed herself."[110] But, three weeks later, Hale received a telegram from Walcott, which said: "Allotment made. Thirty thousand will be sent immediately." He could pay the bills and the "big scheme" would go full speed ahead. Hale was "*mightily* pleased at the outcome, because he would be able to carry out all [his] plans in fine style."[111]

Ritchey was euphoric, too, now that he knew he could go ahead and finish the 60-inch reflector and use it on Mount Wilson. "We are about to build a fine instrument and optical laboratory here, and will have much better facilities for both kinds of work than we ever had at the Yerkes Observatory," he wrote Thomson.[112] Only Frost, back in Williams Bay, seeing Hale's household furniture being packed for shipping to Pasadena, felt as if he were watching a funeral.[113] Campbell, who had hoped for money from the Carnegie Institution for a Lick Observatory reflector, got nothing. He congratulated Hale.[114]

Mount Wilson Success

1905 - 1908

GEORGE WILLIS RITCHEY began working for the Carnegie Institution of Washington, under its grant to George Ellery Hale to build the Mount Wilson Solar Observatory, on January 1, 1905. His salary doubled from $2,000 to $4,000. Hale had recommended the raise to Robert S. Woodward, the geophysicist whom Andrew Carnegie had named the president of his institution, arguing that Ritchey was the only man competent to do all the tasks that would be required of the superintendent of construction. He described Ritchey as "the first authority at the present time on the construction of reflecting telescopes." Ritchey's salary was to be reduced from $4,000 to $3,000 a year when the construction period was finished and he would become a staff astronomer; this did occur four years later at the end of 1908 when the 60-inch reflector was completed and in operation. Hale's own salary as director was $5,000 a year, and the other two staff members, Ferdinand Ellerman and Walter S. Adams, received $1,700 and $1,500 respectively.[1] In the days before World War I, there was little inflation, and scientists' salaries often remained constant for many years.

Although there was general jubilation in Pasadena about the Carnegie grant, the feeling was quite the opposite among the astronomers left at

Yerkes Observatory, deserted by the first team. President William Rainey Harper of the University of Chicago had spent the last six months of 1904 living in Hale's house in Williams Bay. Only forty-eight years old, Harper was suffering from cancer. He had already undergone one operation, and he knew that his chances of survival were poor, but he was trying to build up his strength at the quiet little lakeshore settlement. In spite of his illness, Harper considered the first month of his stay in Wisconsin "the best time I can recall in thirty years, absolutely without exception. It has consisted in uninterrupted opportunity to read and write." He was a Biblical scholar, master of Sanskrit, Hebrew, Greek, and Latin, who had to give up most of his research activities when he became president of John D. Rockefeller's university. But at Williams Bay, Harper began his working day at 6:30 A.M. with an hour's dictation to his secretary, followed by nine hours of reading and study, broken up by lunch, dinner, and two hours for golf. His last study period ended at 10 P.M. He wrote one of his aides whom he had left on the campus that he was turning out more work than he had dreamed he could and that he hoped to keep up this schedule indefinitely.[2] Alas, it was not to be.

Hale had warned Edwin B. Frost, in his absence the acting director of Yerkes Observatory, to get a commitment from Harper to continue or increase its budget *before* the Carnegie grant was announced. Frost was far less effective than the persuasive, determined, self-assured Hale in handling university presidents or wealthy prospective donors. Frost thought he had his budget nailed down, but when the Carnegie Institution made the big grant to Hale for Mount Wilson, its trustees naturally cut off the $4,000 per year they had been giving him for Yerkes. A near explosion occurred. Harper professed to feel betrayed; he claimed that Hale, by resigning, would breach an agreement not to leave the University of Chicago faculty that he had supposedly made before going to California. This was not true, but Harper still hoped to keep the Carnegie-financed observatory as part of the university, which was undergoing one of its recurrent budget crises. He was reducing all the departments' appropriations; in addition he intended to keep the money saved from Hale's, Ritchey's, and Ellerman's salaries in the general university budget and refused to use it to pay Frank Schlesinger, John A. Parkhurst, Philip Fox, and Louise Ware, who were left high and dry by the withdrawal of the Carnegie funds. Harper insisted that they were not his responsibility, and at one point he threatened to sue Hale to recover the

Hale, in his office at Mount Wilson, ca. 1905. Courtesy of the Observatories of the Carnegie Institution of Washington.

University of Chicago money that had been spent on the Mount Wilson expedition. Hale replied firmly that he had acted honorably and had kept Harper informed at all times. Furthermore, he had discussed what he had done with Martin Ryerson, the millionaire chairman of the University of Chicago Board of Trustees, who agreed with him, not the president. Finally, Hale's attorney had assured him that the Carnegie Institution had no legal or moral obligation to restitution. Hale backed up his statement with a five-page opinion from his lawyer. He used some of the Carnegie funds intended for Mount Wilson to pay Schlesinger's and Ware's salaries at Yerkes for a few months, requiring them to do only a minimum amount of solar work for him, and also helped pay Parkhurst and Fox with his own money. Meanwhile, Frost continued to reason with the president, who eventually saw the light.[3]

After his second operation, in February 1905, Harper knew that he was near death. He accepted the loss of Hale and the Mount Wilson Solar Observatory. He finally acted upon Hale's resignation, which he had been holding back from the trustees in the hope that he would withdraw it. Instead, Harper wrote Hale that "nothing has given me more satisfaction in the work of the University than the development of the Department of Astronomy under your supervision. I regard it as one of the most successful features of our work here." It was his valedictory, though he survived for nine more months.[4]

One appropriation that Frost, who succeeded Hale as director of Yerkes, had been unable to get from Harper was for Edward E. Barnard, to take his new, wide-field Bruce photographic telescope to Mount Wilson. He was photographing the Milky Way, and the high, clear California site was much better for this program than Yerkes. Harper turned down Frost's request for $500 to finance this expedition, but Hale then had no trouble in getting $1,000 for it from John D. Hooker. Barnard stayed at Mount Wilson for several months, happily taking his classic photographs of star clouds and nebulae, but his trip remained an expedition, and eventually he returned to Yerkes, bringing the Bruce telescope back with him.[5]

Another outstanding astronomer was lost to the Yerkes staff, however. Schlesinger, who had planned a long-term program measuring the distances of stars (more technically their parallaxes) with the 40-inch refractor was offered a permanent job at Allegheny Observatory. Sure that he would be paid every month there, he snapped it up and began his pioneering work in Pittsburgh instead of at Yerkes.[6]

Meanwhile Ritchey, in Pasadena, was hard at work on the design of the drive-clock for the 60-inch reflector. It is the mechanism that turns the telescope mounting about its polar axis at exactly the right rate to track the stars as they revolve about the north pole of the sky. "To save us some time and money," he got Frost to borrow from Warner and Swasey the plans their company had used to make the drive-clock for the 40-inch refractor and send them on to him. Two years later Ritchey still had the plans, and Frost had to pry them out of him so he could lend them to Schlesinger, who also preferred to pirate a working design rather than pay for it or make a new one himself.[7] Lillie G. Ritchey, who had moved to Pasadena with her husband and children, suffered an attack of appendicitis and had to undergo an operation in January 1905, but she recovered more quickly than he had the previous year.[8]

By February, Ritchey and Hale had laid out the general plans for a permanent shop building on Santa Barbara Street in Pasadena to replace the rented quarters they had been using, "building spectroscopes and spectroheliographs galore." Hale sent Ritchey east to buy more machine tools and to ship them west along with all the lenses, gratings, mirrors, and other instruments at Yerkes that were the director's own personal property. Ritchey stopped in Cleveland, Niagara Falls, New York, Philadelphia, and Plainfield, New Jersey, consulting suppliers of special steel about the requirements for the 60-inch mounting; at Lynn to report to Elihu Thomson on the fused quartz experiments; and at Washington to visit Charles Greeley Abbot at the Smithsonian.[9]

The most important item that Ritchey was shipping west, now that the Carnegie-financed observatory was a certainty, was the 60-inch mirror. It belonged to Hale, for his father had paid for the glass disk and Ritchey's salary until he had gone on the university payroll in 1901. However, some of the expenses of the work on the disk had been paid from the regular Yerkes budget. Hale originally estimated them as $1,000, but later, on the basis of records he had kept, gave them as $337.20. It had turned out to be, he said, "much less than I supposed, but this seems to be a liberal allowance, based on reliable data." The figure included only direct expenditures for the few materials which had been bought and for the time of the machinist who helped Ritchey build the optical machine, plus an allowance for the electric power consumed. There was no payment for space in the building, heat, or overhead. Frost was "rather shocked" that the total was so low and asked Hale to draw up a careful accounting, for he knew that President Harper would scrutinize it carefully. Hale then found "some possible loop-holes in my previous account," and authorized Ritchey, en route to Williams Bay, to settle them, but he warned Frost not to hold up shipment of the mirror. He assured Frost that he would pay whatever final accounting they agreed on. Frost took Ritchey to Chicago to explain the accounts to the university auditor, and the trustees, on Harper's recommendation, transmitted from his sickbed through Acting President Harry P. Judson, officially turned all their claims to the mirror over to Hale. They set the maximum amount of the settlement as $400, but in the end Hale's $337 prevailed.[10]

Meanwhile, Hale was forming links for his new observatory in southern California. He emphasized that it would not compete with Lick Observatory but would apply new methods of research under "the remarkably favorable atmospheric conditions of Mount Wilson" to study the physical nature of the sun and the problem of stellar evolution. Mount Wilson Solar Observatory would resemble a physics laboratory more closely than an astronomical observatory. The Snow horizontal telescope and the 60-inch reflector would be used to feed large, fixed spectrographs. With the Snow it would be possible to obtain a large image of the sun and thus to study small features of its surface, such as sunspots. With the 60-inch it would be possible to study the spectra of bright stars in great detail and compare them with the sun.[11] Hale described the new observatory as the natural culmination of his own solar astrophysics work at Kenwood, of Ritchey's work with the 24-inch at Yerkes, of the horizontal telescope at the Wadesboro eclipse, and of the Snow telescope at Yerkes. Always he emphasized high-dispersion spectroscopy.[12]

By April, Adams was starting research with the Snow telescope on Mount Wilson. He had already proved himself very valuable to Hale at Yerkes as a young scientist eager to do research, who knew and understood spectrographs and was willing to work hard. Although Hale had taken Ellerman with him to California first, Adams had made it clear that he wanted to go, too, and that he saw his future in astrophysics with Hale at Mount Wilson, not in radial-velocity research with Frost at Yerkes. Hale needed Adams; Ellerman was a good observer but had neither the knowledge of optics and spectroscopy to install a new instrument nor the creative imagination to do research with it in the director's name. Adams had been at Mount Wilson since mid-1904. He spent most of his time on the mountain, setting up the instruments and laboratory, and measuring the spectrograms he had taken before leaving Yerkes Observatory. By April 1905 he had the Snow telescope aligned and adjusted and was obtaining new observational data with it.[13] Some of the first published research to come out of Mount Wilson were papers by Hale and Adams on the spectra of sunspots and of the bright star Arcturus, based on spectrograms taken chiefly by the young assistant. The large solar image that the Snow telescope provided enabled him to isolate the faint light of the dark spots and thus to observe them relatively free of contamination by the brighter areas around them. Hale and Adams showed conclusively that the spectra of the spots were more like the spectrum of Arcturus than the spectrum of the main body of the sun. Just a year later, with the help of spectroscopist Henry G. Gale, who came out from the University of Chicago to work in the Mount Wilson laboratory with Hale and Adams, they proved that the observed spectral differences meant that the spots were cooler than the rest of the visible surface of the sun and that Arcturus was a cooler star than the sun. Mount Wilson Solar Observatory was beginning to get significant new astrophysical results.[14]

Hale had made sure to get a promise from Frost, as soon as the grant from the Carnegie Institution came through, that he could keep the Snow telescope on loan at Mount Wilson for two or three years. It belonged to the University of Chicago, but Frost, now the acting director, found it impossible to negotiate at arms length with his long-time superior. This was especially the case after Hale gave Frost, not a rich man, all his books that he did not take with him to Pasadena. Frost insisted that he would not let this gift interfere with his duty to represent the University of Chicago forcefully, but clearly he was aware of the possibility that it might. The Snow telescope proved so successful that Hale decided to keep it on Mount Wilson. Frost was willing to sell it. He made a hard-nosed calculation of the actual time, effort, and money that had gone into it. Frost included in its value an estimate of what keeping it on Mount Wilson would be worth to the Carnegie Institution and arrived at an overall price of $7,500. This was too high for Hale, who countered with his own figure of $5,700. George Isham, the friend of Helen Snow who had arranged her gift of the telescope, thought that it was worth $10,000 to Hale and that the Carnegie Institution ought to be made to pay it. In the end Frost compromised at $6,240, and the Snow telescope remained a productive research instrument at Mount Wilson for many years.[15]

All this was in the future as Ritchey returned from Yerkes to Pasadena in April 1905 with the 60-inch mirror. By May he and Hale had moved into their offices in the completed building on Santa Barbara Street, and work had begun in the machine and optical shops. Ritchey ordered the glass disks that he would make into secondary Cassegrain mirrors for the 60-inch telescope. On his trip east he had visited several companies, including Warner and Swasey, the makers of the 40-inch refractor, who wanted to provide the 60-inch mounting also. On his return, he and Hale decided that none of these eastern companies was good enough to outweigh the advantage of a nearby source. They decided to have the large parts made by the Union Iron Works of San Francisco, the largest steel fabricating company on the West Coast.[16] Among its products were several battle cruisers and destroyers for the navy. Ritchey hired his father, James, who had just reached the age of seventy, to work as his draftsman on the 60-inch design. James Ritchey left his job at Armour Institute of Technology, where he was still teaching, and with his wife came to Pasadena. There they lived very close to Ritchey's house and to the observatory office. A year later Ritchey's uncle John B. Ritchey, a physician, also moved to Pasadena with his wife.[17] The supportive family group was reforming around its current leader, the highly successful Mount Wilson superintendent of construction. Francis G. Pease had come out from Yerkes in 1904 and stayed on the Mount Wilson staff as Ritchey's assistant.[18]

Ritchey was working hard on the detailed plans for the 60-inch. The Union Iron Works had started fabricating the large parts of the mounting in the late summer, and he was going to San Francisco for one or two days each week to check on its progress. One of the features of his design was that the moving parts of the mounting floated on mercury, providing friction-free support. After completion and testing in Pasadena, the entire telescope would have to be taken up to the summit of Mount Wilson over a mountain trail. Ritchey had designed the mounting so that no part of it would weigh more than five tons, and he was starting to worry about how to get the pieces up to the top. He was also thinking about the design of the dome for the 60-inch. He wanted to make it as open as possible, with canvas sides that could be rolled away, instead of metal in several large sections of the wall, to allow free air circulation. This would reduce the temperature differences between the telescope and the outside air, and would reduce the turbulence or "dome seeing," as it is called today. This concept did not remain in the plan, but it indicates how far ahead of his time Ritchey's thinking was. He was so busy working on the 60-inch and the other mirrors for the solar telescopes that he had no time to write up his projected paper on M8, a wonderfully complicated nebula he had photographed with the 24-inch reflector before leaving Yerkes. Frost was hounding him for the manuscript so that he could publish it in the *Astrophysical Journal*, but Ritchey had a strong tendency to immerse himself completely in his most pressing current problem and to ignore cries for aid from others.[19] It was to cost him dearly later.

On one of his trips to San Francisco in September, Ritchey stopped at nearby Lick Observatory. Its director, W. W. Campbell, was then gone on an expedition to Spain to observe a solar eclipse, taking with him among his instruments a coelostat borrowed from Mount Wilson.[20] He had left instructions that Ritchey, whom he regarded as the outstanding expert in the world on reflecting telescopes, be welcomed to Mount Hamilton whenever he could come. Campbell had first met Ritchey in 1902, while he was restlessly waiting in Allegheny, Pennsylvania, for John A. Brashear to complete the 37-inch mirror for a Lick Observatory reflecting telescope to be erected in Chile. Campbell could not help but compare the triumphant Ritchey, whose 24-inch was a great success and who was confidently explaining his plans for future large telescopes, with the bumbling Brashear, who had already failed completely with the 37-inch mirror once and was having great difficulties completing it on his second attempt.[21] Characteristically, though the 60-inch mirror had not even been completed, Ritchey was planning ahead toward making a 96-inch reflector. The conservative Campbell was worried by the problems of the thermal deformation of so large a mirror and of the effects of atmospheric turbulence or "seeing" on such a large telescope. He believed that jumping from the 24-inch to a 96-inch without actually completing and testing intermediate sizes, a 48-inch (which he hoped to build at Lick), a 60-inch (Hale's disk, which Ritchey had begun working), and perhaps a 72-inch also, would be far too large a step. On the other hand, Ritchey was certain that he had solved all the problems. Large, reflecting telescopes were so important to astronomy that they *would* be built, he believed. He did not regard even a 96-inch as the upper limit that could be built *then*, but rather as a safe, conservative step forward.[22]

Finally, Brashear completed the mirror, and William H. Wright took the new reflector to Santiago and erected it there. Campbell, always cautious and conservative, carefully studied the reports on the thermal properties of the mirror and the effects of seeing that he had ordered Wright to send back regularly. The Lick director discussed them thoroughly with Hale, asking for Ritchey's advice.[23] By now Campbell was planning a 54-inch or 60-inch reflector, which he hoped to build on Mount Hamilton. When Ritchey published his paper on modern telescopes, Campbell studied it closely. He asked the author many detailed questions and got a complete cost estimate of a 60-inch from him. Ritchey gave $66,700 as the cost of the telescope alone, complete but with no dome or auxiliary instruments.[24] Campbell never succeeded in raising the money for a large new reflector, so he had no chance to find out if this estimate was correct.

Meanwhile, Hale was reveling in research and scientific meetings. Between the Snow telescope, the new spectroheliograph that Ritchey and Pease designed for it, and the spectroscopic laboratory he and Adams were setting up, Hale was busy and happy.[25] In September 1905 he traveled to England for the second "Conference" of the International Union for Solar Research that he had founded, this one held at New College, Oxford, in England.[26] On his return he found his wife Evelina in a state of nervous collapse. She had worried about him all the while he was gone, and, in his words, she "broke down under her heavy load." Her doctor put her in a sanatorium.[27] Hale felt guilty, partly no doubt because he was also enjoying life with his mistress, Alicia Mosgrove.[28] This was to become another recurrent theme of his life. Whenever he was furiously active, he was well, but Evelina was ill. When new projects thinned out and old ones soured, Hale became ill and his wife thrived.

Just at the end of 1905 a figure who had once been very important to Hale passed from the scene. Financier Charles T. Yerkes died. He had provided the money to build "his" observatory, but had been niggardly in supporting it, in spite of all Harper's and Hale's blandishments. In his will he left $100,000 as an endowment fund for the care and maintenance of Yerkes Observatory. Hale was pleased that "the titan" had remembered it in the end. Still, he wrote Frost: "[I]t seems as though he might just as well have given a much larger sum, but the amount named is not to be despised." A few weeks later Harper, wasted by the ravages of cancer, followed Yerkes in death. In contrast to the ruthless financier, he left no fortune and almost no money, but his monument was the great university he had built. John D. Rockefeller provided a pension for Harper's widow.[29]

In Pasadena Ritchey and his assistants had been working steadily on the 60-inch since the fall. Before starting it he had outfitted the optical shop to the highest standard. The Carnegie money made this possible. The room in which the 60-inch mirror was ground and polished had sealed double windows to keep out all dust that might scratch the glass disk. All the air entering the room was filtered, and canvas was hung just above the grinding machine to further reduce dust. Thermostated electric heaters were installed to keep the temperature uniform day and night.

Ritchey was also supervising the widening of the "new trail" up Mount Wilson in preparation for getting the 60-inch reflector to the top. He thought that they might be able to complete and test the telescope in Pasadena in 1907 and take it to Mount Wilson the following year.[30] But on April 18, 1906, the great earthquake struck San Francisco. At the Union Iron Works the partly completed mounting escaped damage, though not far from it two ships that were under construction were dashed from their stocks. Another ship, in the dry dock for repairs, crashed through it and sunk in the shallow water. A great fire swept through the city after the earthquake but did not touch the Union Iron Works on the waterfront.[31] Hale, Ritchey, and the Mount Wilson Solar Observatory had been lucky. They still had their telescope mounting. But the aftermath of the earthquake delayed its completion. Ritchey kept pushing the Union Iron Works, and finally the mounting was ready for his inspection at the end of July. He went north and inspected it, and in August the company sent it south. It arrived in Pasadena in September.[32]

Soon after the earthquake, while he was waiting for the mounting to be finished and shipped to Pasadena, Ritchey had gone to Arizona on his second abortive attempt to land an outside big-telescope contract. In 1894 Percival Lowell, the scion of a wealthy Boston Brahmin family, had built his own Lowell Observatory in Flagstaff to prove the existence of intelligent life on Mars. It was a sudden whim that he stuck with to the end of his life. A self-financed, gentleman astronomer, he could afford it. His observatory was equipped with a 24-inch Clark refractor, but by 1905 he had decided to get a large reflecting telescope to supplement it. He thought Ritchey would be the best person to make it for him but believed that he would not undertake an outside job. However, when Lowell learned from his professional advisers that Ritchey might be willing to do it, he wrote the Mount Wilson superintendent of construction. Lowell had already decided he wanted an 84-inch reflector, which would be bigger than the 60-inch and hence the largest working telescope in the world. He had a quote from Brashear, whose foreman said their company could build an 84-inch for him for $65,000.[33]

Armed with this information, Lowell wrote Ritchey that he wanted an 84-inch and that he would pay $55,000 for the complete optics.[34] Ritchey replied that he was interested in building the telescope as well as the optics but that it would be more expensive than Lowell thought. He asked what Lowell would pay him for making it, and the Arizona director answered grandly, "Whatever you think proper for your compensation as superintendent I should be only too glad to pay, as I consider scientific work as very underpaid as it is." However, he did not believe the rest of the telescope should be very expensive. "As to mountings my experience has shown me that they can be made as effective and much cheaper than those commonly put up." Ritchey could not have disagreed more with this statement. His entire philosophy, expressed in his papers, was that the mountings commonly used in the past were responsible for the failures of the old reflectors and that it was foolish to use any but the best materials and the most solid construction. Lowell also wanted to have a minimal dome so as to keep the telescope out in the open air, to preserve the quality of the seeing.[35] Ritchey sent his own ideas on the proposed telescope to Lowell, who was in Boston, where he lived nearly half of each year. Though Ritchey also favored an open dome, he was not prepared to go as far as Lowell. Some of the optician's new ideas for mountings appealed to the Arizona director, and he invited Ritchey to meet him at Flagstaff in May 1906.[36]

Ritchey came at the end of the month, stayed overnight, and had a long discussion with Lowell.[37] Only after Ritchey had left did Lowell write him that it was "essential" that the primary 84-inch mirror be made f/2.5, a very short focal length in comparison with the diameter. Almost certainly, they had discussed this as a possibility and Ritchey had told Lowell that it could not be done. The shorter the focal ratio, the more difficult it is to shape the mirror to the necessary exact paraboloidal figure. The shortest focal-ratio mirror Ritchey had made was the f/4 24-inch for the Yerkes reflector, and the 60-inch had been an f/5. No doubt even he did not wish to try to make an 84-inch which would be the largest, as well as the fastest, mirror in the world. Perhaps Lowell was fixed on his idea because a short-focal-ratio telescope is automatically a small one, and hence the mounting and dome are less expensive than for a telescope of the same diameter and a longer focal ratio. Lowell stated this as a partial reason for his decision but said it was also based on "what I should want the instrument to do." He gave no specifics, and it is impossible to know what he had in mind.[38]

Ritchey told Lowell the difficulties of making a first-class f/2.5 84-inch mirror were "insuperable," but the gentleman director replied that he regarded them as only "of a practical and not a theoretic kind." He terminated the negotiations with Ritchey and tried next to get the Alvan Clark & Sons firm, headed now by Carl A. R. Lundin, to make the large f/2.5 mirror. In the end Lowell settled for a 42-inch f/5 primary mirror, half the diameter he had originally planned and twice the focal ratio. Brashear built it for him, and it went into operation in 1910.

Ritchey by now was considered America's greatest authority on reflecting telescopes. At Harvard Director Edward C. Pickering's request, he provided a written recipe for silvering large mirrors that was preserved and followed for years.[39]

After his visit to Flagstaff, Ritchey had gone to the Midwest to visit several manufacturers, to find the best one from which to order a truck to carry the parts of the 60-inch mounting up the steep trail to the summit. An especially large and powerful vehicle would be needed. In the end he ordered it from a company in Milwaukee for $5,000, with a three-month delivery date, which was an optimistic guess, indeed, as to when the telescope would be finished.[40] Hale, meanwhile, was feeling tired and run down, and had gone for a long rest by the sea, in Santa Barbara. His wife had been released from the sanatorium, and they were both "improving very fast."[41]

However, Hale and Ritchey had both been working better than they knew, for on September 14, 1906, John D. Hooker agreed to provide the money to buy a 100-inch glass disk and make it into a mirror for the next telescope after the 60-inch. Once again Hale had succeeded in getting the funds for the mirror, which he hoped would attract the rest of the money necessary to build the telescope. Hooker was the wealthy owner of a hardware supply company in Los Angeles, an investor in oil wells, and an amateur astronomer. As a young man, he had come west from New England at the time of the Civil War and made his fortune in California. He belonged to the Astronomical Society of the Pacific, the joint professional-amateur society that had grown from the Lick Observatory support group originally founded by Edward S. Holden in 1889. Campbell had met Hooker through the society, and when Hale moved to Pasadena in 1903, the Lick director gave him a letter of introduction to the hardware supply company owner, who he emphasized was *rich*. Hale soon met Hooker and his "most charming" wife, and began trying to line up financial support from them. He read Mrs. Hooker's "very delightful book" and flattered her by asking her to teach him Italian.[42]

A few months later Ritchey, who was ordering many tools and supplies for the observatory from the J. D. Hooker Company, got in contact with its owner. It turned out that Hooker loved astronomical pictures, and Ritchey started him out with prints of some of the photographs of the moon and nebulae he had taken at Yerkes Observatory. He followed them up with spectacular large glass transparencies.[43] Hooker apparently identified better with Ritchey, who was closer to his own age and range of experience, than with Hale. Thus after months of diplomatic "stroking" by both of them, Hooker asked Ritchey to describe for him the "field of usefulness" of a 96-inch (8-foot) reflector. Ritchey wrote a long reply, emphasizing especially the photography of nebulae, and at the Cassegrain focus with its large scale, the photography of globular clusters, the moon, and planets. No doubt at Hale's urging, he also mentioned high-dispersion stellar spectroscopy and infrared measurements of the heat radiation from stars. Clearly, Hale helped with this letter, but Ritchey wrote it, and the emphasis is all on the topics in which he and Hooker were interested.[44]

It was in response to this letter that Hooker wrote Hale less than two months later, announcing his gift of $45,000 to buy the glass and make the 100-inch mirror from it. He had rounded the size of the telescope up to an even 100 inches (and from that day, reflecting telescope diameters have been stated in inches, rather than feet as had been the previous custom). Hooker also emphasized that he was not giving the money for the mounting and dome but "trust[ed] to the future that these essential adjuncts [would] become available."[45] Ritchey was ecstatic. He had already been in correspondence with the St. Gobain Glass Company, and Hale could order the disk immediately. Soon Hooker was visiting the observatory office and shop in Pasadena, bringing his little visual telescope for Ritchey to clean and align, and hectoring him to hurry up with the job.[46]

When Hooker's gift was announced, Frost wrote Hale congratulating him on the new gift "for the biggest thing ever." But he suggested that it would have been better if it had come after the 60-inch telescope had been completed and tested. Hale replied

> As to the 100-inch reflector, there were also reasons why it should be undertaken at once. Mr. Hooker is getting to be an old man, his oil wells happened to be giving an extraordinary flow, and there were other grounds for believing that the psychological moment had arrived. Of course nothing will be done about the design of the mounting until the 5-foot has been thoroughly tested. There is no doubt, in my opinion, that the 100-inch will accomplish enough results to warrant its construction. Mr. Hooker was fully informed as to all of the difficulties in the way, but it was he who proposed that we should increase the size, first from 7 feet, the size I suggested to him, to 8 feet, and ultimately . . . to 100 inches.

These two letters illustrate perfectly the differences in point of view between a highly successful director and a mediocre one.[47] Hale very probably hoped to get the rest of the money that would be needed to mount and erect the 100-inch telescope from Hooker eventually, but it was not to be.[48] However, Hooker did begin making payments of $5,000 at a time, as needed, for a building and a large optical machine to be erected within it to grind, figure, and polish the 100-inch mirror. Andrew Carnegie wrote him a letter, welcoming him "to the number of those who see in wealth only a sacred trust to be administered in the service of man." Each hoped the other would provide the funds to complete the 100-inch, and Hale did his best to play the two off against each other.[49]

Hale had become a very important figure in American science by this time. When Samuel P. Langley died in 1906, Simon Newcomb asked Hale if he would accept the post of heading the Smithsonian Institution as his successor. Hale declined but outlined for "the greatest astronomer of his time" his ideas for building up Mount Wilson Observatory as an outstanding research institution and of converting the National Academy of Sciences from a sleepy honorary society to an activist

Carriage on Mount Wilson road, 1907. From left to right are the driver, Hale, Ritchey, and John D. Hooker. Courtesy of the Observatories of the Carnegie Institution of Washington.

organization that would lead U.S. science.[50] In his lifetime he was to make these dreams come true. Ritchey, meanwhile, was reporting on progress on the 60-inch to Hale and in December hired a draftsman to help him with preliminary design sketches for the 100-inch telescope.[51]

In 1907 Pease left the Mount Wilson shop staff. He had worked well with Ritchey and probably asked for a raise but left when he did not get it. Pease returned to Chicago where he worked on optical and mechanical jobs at the Scientific Shop, a small company that catered to schools and colleges. He spent a month at Mount Hamilton, making a Cassegrain secondary mirror for the 36-inch Crossley reflector.[52] Soon he was looking for a better job with Bausch and Lomb. But Adams had come to appreciate Pease, who had a New England background much like his own. He kept in touch with him, praised the solar "tower telescope," or fixed vertical telescope, that Pease had designed, directed small Mount Wilson Observatory jobs to him in Chicago, and began quietly criticizing Ritchey's work. In 1908 Hale rehired Pease on the Mount Wilson staff, and he came back to Pasadena where, a decade later, he was to supplant Ritchey.[53]

Meanwhile, by August 1907 Ritchey had finished the 60-inch mirror. His men were assembling and erecting the mounting for the 60-inch in Pasadena. A strike at the Union Iron Works prevented completion of the dome for the 60-inch that year, but some of the steel was taken up to the top after the strike ended, before the winter rains came. Ritchey, as superintendent of construction, had supervised widening the trail, and in September word came from France that the 100-inch disk had been successfully cast.[54]

Adams, meanwhile, was writing a very important paper on the differential rotation of the sun, and with Hale as coauthor, papers on the difference between the spectra of the sun at the center of the disk and at the limb (edge), and on more detailed spectra of sunspots. All were based on spectra Adams had taken with the Snow telescope at Mount Wilson; he was a master at using whatever spectroscopic instrument was available and getting important new research results with it. By now he was serving as Hale's scientific proxy, writing their joint papers and seeing them through the press, and even corresponding with Frost for him about gratings, mirrors, filters, and Mount Wilson Observatory publications.[55]

Hale had become more and more involved in high scientific policy. In 1906, besides the secretaryship of the Smithsonian Institution, he had been offered the presidency of the Massachusetts Institute of Technology, his alma mater. He had thought long and hard about it but declined; he wanted to stay in scientific research, he said. In addition he prized the independence of being a nearly autonomous observatory director, not at the beck and call of a board of trustees, as a university president would have to be. He feared the changes he saw in himself; he was suffering from "sleeplessness" and "my own nervousness, which has increased in the last few years, and might become serious with heavily weighing responsibilities." Nevertheless he took an active part in building up the little Throop Polytechnic Institute in Pasadena. Hale visualized its future as a first-class technical university, an MIT of southern California. Its trustees elected him to their board, and he immediately began thinking of how to get money for it from the Rockefeller-sponsored General Education Board and from local California millionaires. He asked Harry P. Judson, Harper's successor as president of the University of Chicago, to help him find a new president for Throop.[56] Again, Hale's dream became a reality because he pushed it hard all the rest of his life. Throop Polytechnic Institute became the California Institute of Technology, and Californians now are accustomed to thinking of MIT as an eastern Caltech. But Hale's many activities left him less and less time for doing research himself, no matter how much he liked to write about it.

He had become a true California booster from the moment he first moved to Pasadena and spent Christmas day of 1904 under an orange tree in his garden. In winter he could not resist gloating to his friends back in the frigid Midwest, as in a January 1908 letter to Frost, which Hale wrote just after he returned from visiting his brother: "The glorious weather here is in striking contrast to what I found in Chicago. You would find it difficult to discover a more enthusiastic Californian than myself, even among the most hardened dealers in real estate."[57]

Ritchey, too, had become a confirmed Golden Stater, writing Barnard and his wife, when they were both suffering severe colds during a Williams Bay February: "The climate there is enough to wear anyone down. I wish you could both live in Southern California. I believe that the climate here will continue to bring people here in such numbers that in a few decades this Southern California valley will be built up from ocean to mountains both North and East." Ritchey's prediction proved all too true, and the accompanying light pollution was to begin affecting Mount Wilson in the 1930s. He also told Barnard that Hooker had recently given him some of the first commercially available, successful plates for color photographs. Ritchey, a dedicated photographer off the job as well as on, had obtained "some very good results" with them, pictures of flowers and people. He and his wife had spent their previous summer vacation at the Grand Canyon in Arizona, where he had taken numerous excellent black-and-white scenic photographs, and he hoped to return in the coming summer and try the color plates there as well.[58]

By the spring of 1908 the 60-inch telescope was nearly completed. Ritchey and his assistants had set it up and put it through its paces in Pasadena. He inspected the road to the top, making sure it was strengthened with sandbags so that the truck that would carry up the precious pieces of the mounting would not be in danger of overturning. Within a few months the 60-inch would be in place.[59]

All was not well, however. When Hale departed on a long trip to the east in April, he left a detailed list of priorities to be followed by Ritchey and the shop personnel. Completion of the 60-inch must come first. The other projects that Hale wanted Ritchey to work on were stated in order. He had no freedom left at all. Hale emphasized that Adams would have "full charge of all research work in progress" and that Ritchey should consult him on any questions that might arise. It was a cold, hard letter to a subordinate, not in the least friendly.[60] A few months later Adams suggested hiring R. James Wallace, whom he knew had antagonized Ritchey at Yerkes Observatory, as a photographic assistant at Mount Wilson.[61] Hale was depressed, and complained to Frost that new ideas came to him only slowly and that he had "accomplished nothing whatever in solar research during the entire period of organization of the Yerkes Observatory," which had ended only when he went to California at the end of 1903.[62]

Yet just a few months later Hale made his big discovery. On the Hα spectroheliograms of the sun, which Ferdinand Ellerman, Charles E. St. John, and occasionally he himself had been taking, Hale saw structures that he named "vortices," looking for all the world like gigantic whirlpools, or like the pattern of iron filings around the pole of a magnet. He could see, near the limb of the sun, that some prominences ended in vortices and appeared to be pulled in to them. The vortices were located in the chromosphere above sunspots. They looked suspiciously like magnetic-field structures on the sun. Hale and Adams's high-dispersion spectrograms of sunspots showed that in some cases many of their spectral lines were split into close pairs, or "doublets." This could be the result of magnetic fields, which were known from laboratory spectroscopy to split spectral lines. This phenomenon is called the Zeeman effect. If there were indeed magnetic fields in sunspots, different lines in their spectra should be split by different amounts, and the different components of the doublets should be polarized in opposite senses.[63]

Just the previous year, Hale had put into operation on Mount Wilson the new 60-foot tower telescope. It was essentially a vertical form of the fixed horizontal telescopes of his thesis, of the Wadesboro eclipse, and of the Snow telescope. The coelostat was high above the ground, at the top of the tower. It directed sunlight down through a lens to a focal plane a few feet above the ground, where it entered the slit of a large grating spectrograph built into an underground concrete chamber. Its focus and the plateholder were back at ground level. This vertical tower telescope had several advantages over the horizontal form for solar work, resulting from the better seeing high above the ground and the diminished heating effects of the sun on the lens and on the spectrograph.[64]

Hale already had a little polarization analyzer, which he had had made three years before, when he had realized the possibility of detecting magnetic fields in sunspots by the Zeeman effect. This analyzer, with the excellent daytime seeing of a Mount Wilson summer day, the new tower telescope, and the very high-dispersion spectrograph in the pit below it, formed the ideal combination to check Hale's hypothesis that there are magnetic fields in sunspots. As he exultingly wrote his close friend Harry M. Goodwin, "I went up the mountain on June 24, and bagged them two days later." Hale immediately put his team of assistants to work full speed on the problem. He, St. John, and Adams obtained many more sunspot spectra with the tower telescope on the mountain. Newly hired physicist Arthur S. King took a special series of spectrograms of iron and chromium arcs in magnetic fields in the spectroscopic laboratory in Pasadena. Comparison of these solar and laboratory spectrograms fully confirmed the magnetic nature of sunspots and showed the field strengths in them to be near 3,000 gauss.[65] It was a very important discovery, and working out all its consequences and byproducts was to be a very active field of solar research at Mount Wilson for the next century.

Yet later that summer Hale was depressed and ill again, and had to take several weeks off to recuperate. Adams was on vacation, leaving Ritchey temporarily in charge. When questions arose about the insurance on the 60-inch reflector, Ritchey said he did not have the answers, and Hale's secretary had to bother the director with the request for information.[66] Unlike Adams, Ritchey was unwilling to put aside his immediate technical or scientific concerns, to fill in for Hale as an administrator when needed, and thus to forge a closer relationship with him.

Ritchey was thinking of starting a telescope business of his own. He visualized making large reflectors to order for observatories and universities, but also having standard 6-inch and 12-inch reflecting telescopes, designed especially for schools and colleges, always on hand and available for immediate purchase. Hooker, he thought, would help finance the business. One of Ritchey's assistants on the 60-inch project, J. A. Barnes, would leave the observatory staff to go into business with him. The other three assistants could handle all the work on the 100-inch; Ritchey would stay on the observatory staff to supervise them until it was completed and at the same time act as consulting expert for his business with Barnes.[67]

By September the dome was nearly completed. Ritchey was pleased with the 60-inch, which he had planned, designed, and brought into being. He calculated how much better it would be for photographing faint stars than the Yerkes 24-inch had been, because of the increased aperture, the somewhat slower focal ratio, the better atmospheric conditions (seeing and transparency) at Mount Wilson, and the longer exposures that the still dark California skies would permit. Ritchey hoped to be able to photograph stars down to the 22nd magnitude in 15- to 20-hour (two- or three-night) exposures, and he estimated that in a globular cluster like M15 one of his plates might reveal 30,000 to 50,000 stars. Since this would be five magnitudes fainter than visual limits of the Lick and Yerkes refractors, he further calculated that his plates would show stars, on the average, ten times further away than the stars seen with these great telescopes, "and hence [would] reveal a surrounding universe one thousand times greater . . . [in volume] than that revealed by the largest refractors visually." Like nearly all astronomers of his time, he referred to what we call "our Galaxy" as "the universe" and did not realize that other galaxies existed. Ritchey's numbers were essentially correct, and indeed the 60-inch did penetrate ten times further out into space, making a volume one thousand times larger photographically observable than visually with the largest refractors ever built. The 60-inch would, in time, reveal much about the structure of our Galaxy, and of the existence of a universe of galaxies beyond it.

Ritchey wished the new reflector could be used completely for nebular photography. He dreamt of taking photographs of "a thousand or more" nebulae, to be studied and preserved for comparison with future photographs "as a basis for the study of the development of the nebulae for centuries to come." He knew this was unrealistic; he would have the telescope assigned to him for this project "only" two nights each week, which he estimated would permit him to obtain at least fifty "first-class" photographs per year. The rest of the observing time with the 60-inch would be devoted to spectroscopy.

Ritchey hoped that the photographs he obtained with the 60-inch would attract so much attention that some other institution would hire him to build not one but *two* great observatories completely devoted to direct photography. One would be in the northern hemisphere, the other in the southern, so that between them they would cover the whole sky. At each site would be a large reflecting telescope, "similar to the 60-inch or the 100-inch" (work on the latter had not yet started), as well as "duplicate horizontal telescopes [one in the northern hemisphere, one in the southern], with coelostat, of great focal length, say of 4 feet aperture and 300 feet focal length, for the photography of the planets and the moon." Ritchey had worked out a complete design for these horizontal telescopes based on his experience with the Snow telescope at Yerkes and Mount Wilson. Each would be mounted on two high towers, one carrying the coelostat and the photographic plate at the focal plane of the primary mirror, and the other the mirror itself. The two towers would be 300 feet apart, and the light path between them would be 60 to 75 feet above the ground. No tube or house would extend between them.

Ritchey hoped that the series of papers he wanted to write describing the 60-inch, together with a book of pictures he would take with it, to be published at Hooker's expense, would attract the necessary support for this project.[68] It was a vast, inspiring dream, but it was not to come true, even in part. Just as Frost was too timid to be a good director, Ritchey was too extreme in his plans, not considering the realities of raising the correspondingly vast sum of money that would be necessary to build and operate these four telescopes at two distant sites. Furthermore, his scientific program to take pictures of nebulae, planets, and the moon was too narrow and specialized. Only Hale, with his realistic visions, his practical plans, and his sure knowledge of wealthy men and how to separate them from their money, was to be remembered as a great director and a great observatory builder.

In October the dome was completed, and within it the final touches were being put on the 60-inch mounting.[69] Less than three months later, in December 1908, the telescope finally went into operation. It had been more than thirteen years since Hale's father ordered the glass disk for

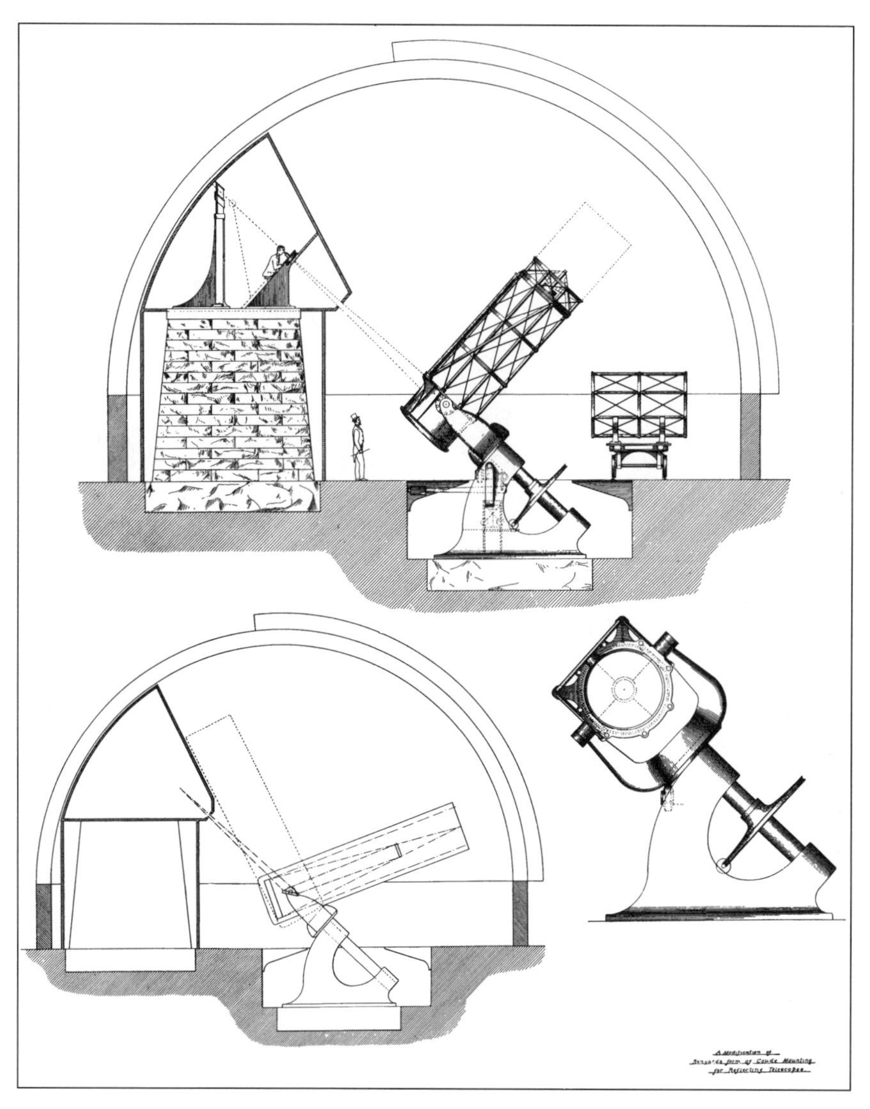

Ritchey's early (1901) design for the 60-inch reflector, with the coudé focus above the polar axis, rather than below it, as later built. Note the extra "cage," to be used at the prime focus, on a wheeled cart. From the *Astrophysical Journal.*

him, more than eleven years since Ritchey had begun rough grinding it in the basement of Yerkes Observatory. The first direct photographs, which Ritchey took just a few days before Christmas, showed that the 60-inch was indeed an excellent instrument. The star images it produced were "small and excellent" and the details of the Orion nebula were "beautifully defined." The 60-inch was a success.[70]

Ritchey added the last paragraphs to the paper he had been drafting, which described the telescope and its construction. All the principles that he had developed and explained in his previous papers went into this instrument. The paraboloid primary mirror was almost exactly f/5, with a focal length of 299 inches (nearly 25 feet). It was used in a Newtonian arrangement, with a flat mirror bringing the light to a focus at the double-slide plateholder at the edge of the tube. In addition there were three different hyperboloid secondary mirrors used to give Cassegrain and coudé focal lengths of 80, 100 and 150 feet. Each of these three secondary mirrors, and the Newtonian flat mirror as well, was permanently mounted in its own "cage," an assembly that could be bolted at the upper end of the tube, allowing all four systems to be rigidly aligned, yet rapidly interchangeable.* The dome for the 60-inch was built with a double-wall construction to enhance air circulation, and a small refrigerating plant was to be provided to keep the primary mirror and its cell close to the expected nighttime temperature through the long, hot, sunny Mount Wilson days.[71]

Ritchey completed the final revisions of his paper during a long train journey to New York and in stops along the way at Chicago and New Haven on essential observatory business. He was on his way to France. Serious problems had arisen with the glass disk for the 100-inch telescope, and he was headed for the source, to try to solve them.[72] He was not to succeed, and these problems were but forerunners of the far more serious problems he would face a decade later.

*See the appendix and its Figure A.1 for brief explanations of these terms.

Mount Wilson Problems

1909 - 1914

LESS THAN A MONTH after taking the first photographs with the newly completed 60-inch reflector, George Willis Ritchey left Pasadena, bound for Paris, to try to help solve a crisis. The crisis was the glass disk that had just been delivered, two years after it had been ordered, for the 100-inch mirror. The sole supplier for such large pieces of glass was the St. Gobain Glass Company of France. It had produced all the disks that Ritchey had made into mirrors in Chicago, Yerkes Observatory, and Mount Wilson Observatory, including the 60-inch disk, which had been perfect. As early as 1903 Ritchey had asked the company if it could make a 120-inch disk which, because of its correspondingly increased thickness, would be approximately eight times as massive. St. Gobain's agent in the United States, Lucien Jouvaud, had stated that it could try, but the results would be "somewhat uncertain." The problem was to melt such a huge mass of glass, to stir it well while molten, and to pour it into the mold without introducing any air bubbles or "stones," pieces chipped off the pot or mold. The glass had to be homogeneous throughout and kept that way as it annealed, cooling slowly for weeks in a regulated oven, so that no strains would be present in the disk.

Ritchey had been in contact with Jouvaud about the 84-inch disk for Percival Lowell in the spring and summer of 1906. He had had to cancel that order, but when John D. Hooker provided the funds for the 100-inch mirror, Ritchey ordered the disk with the assurance that St. Gobain could make it. He emphasized that it should be the same high quality as the 60-inch, with "perfect" homogeneity and annealing.[1] Nearly a year later, in June 1907, George Ellery Hale, in Europe on a trip to several scientific meetings, visited Lucien Delloye, the managing director of the St. Gobain Glass Company, and went with him to the works at St. Gobain outside Paris. By then tests and preparations to melt the huge mass of glass were well underway. The disk was finally cast two months after Hale's visit.[2] When the St. Gobain workmen opened the annealing oven and inspected the disk, however, they found that it was flawed, with many small cracks, indicating that it had cooled too rapidly. The only solution was to remelt the glass and start over again. Each attempt took months.[3]

Meanwhile, Hooker was providing the money to build a new Hooker Optical Building in Pasadena, in which Ritchey and his assistants would grind, polish, figure, and test the 100-inch mirror in a new "Hooker grinding machine." The wealthy donor had promised $45,000 for the mirror, and he paid it, usually $5,000 at a time.[4] By 1908 Hooker was becoming impatient. Ritchey, who was his main channel of communication with the observatory, had a mahogany display case made for the old man, to display the astronomical pictures he gave him, as well as his photographs of the Grand Canyon and color shots of Hooker's own garden. Ritchey, at Hooker's behest, bombarded Jouvaud with requests for information on the progress of the 100-inch disk but got little in the way of replies.[5]

In the summer of 1908, Ritchey gave Hooker a long statement about all the fine photographs that could be taken with *two* 100-inch reflecting telescopes, one in the northern hemisphere and the other in the southern. It was Ritchey's dream. His letter amounted to a proposal or request for funds, and Hale must have been angry if he learned that his staff member had plans of his own for the wealthy donor's money.[6] Meanwhile, there were still severe problems with the 100-inch disk in France. The St. Gobain managers realized their second attempt was far below the standard of the 60-inch disk and studied carefully the costs and delay that would be involved in starting wholly anew.[7] At the very same time they were making these studies, their agent in New York was assuring Ritchey that the mirror "appears . . . as perfect as possible," that it was almost ready to ship, then that it *was* ready to go but that there was no crane in the port of New Orleans that would be able to lift it off a ship, and even that it was in transit. Finally, in late November 1908, it left Antwerp bound, not for New York or New Orleans, but for Galveston, and it finally reached Pasadena just as the 60-inch went into operation.[8]

As soon as the workmen opened the gigantic wooden box, however, Ritchey and Hale could see the imperfections in the 100-inch disk. There were sheets of air bubbles within the glass. The disk was so large that no single pot could hold all the molten glass to make it; the St. Gobain glass workers had had to melt three huge pots simultaneously and pour them, one after another, into the mold. Myriads of bubbles had formed as the three potfuls successively flowed down troughs over one another into the mold. The bubbles destroyed the homogeneity of the glass. The mirror would not expand and contract uniformly as the temperature rose and fell and could not be counted on to form perfect images. Ritchey and Hale decided at once to reject the disk.[9]

Both of them had been starved for information during the more than two years the St. Gobain company had been working on the disk. Communicating through the New York agent was decidedly unsatisfactory. Ritchey believed that the French workmen had not taken the most elementary precautions in annealing the monster disk and that Jouvaud had failed to pass on his warnings to St. Gobain. Hooker was old and impatient to press the project forward.[10] He trusted Ritchey. They decided that Ritchey should go to France at once, see the St. Gobain operation with his own eyes, and help the glass experts there produce a "perfect" disk. By mid-January he was on his way. He called on Jouvaud in New York on January 18 and sailed for France the next day. His wife accompanied him, and Hooker's fund paid their expenses.[11]

Ritchey reached Paris before the end of January and soon was in conference with Delloye, the managing director of the entire French glass company. They had a very satisfactory discussion (through an interpreter). Delloye blamed himself for not giving more of his own time personally to supervising the work on the disk, but he said he was responsible for the company's fourteen factories in seven countries and had been too busy to do more. He was confident that now that his experts understood the problem they would profit by their experience and would make a new, greatly improved disk. The problem was pouring and stirring the molten glass; they would perfect these operations. Delloye accompanied Ritchey to St. Gobain and showed him all the processes. In spite of the fact that neither of them could understand the other's language, a strong bond of professional respect and friendship developed between them, and Delloye remained one of Ritchey's few supporters to the end of his life.

Soon Ritchey was immersed in experiments and plans for improving the methods of pouring the glass. His basic idea was to build a large auxiliary furnace so that all the glass for the 100-inch disk could be melted and poured in one continuous process. No doubt the idea had occurred to the French long before Ritchey appeared on the scene. It would be expensive, but as the price of the "perfect" disk, when produced, had been fixed at $11,000, he was all for pushing ahead as fast as possible, with Mount Wilson making some undefined additional financial contribution. The St. Gobain managers had to worry about costs, which might total far more than this amount. Hence they proceeded instead with time-consuming small-scale experiments.

Nevertheless, Ritchey was confident of success. He and his wife were living in the Paris residence of John Johnson, the highly successful Lake Geneva, Wisconsin, real estate speculator who had given the land for the Yerkes Observatory site to the University of Chicago.[12]

As winter turned to spring, Ritchey became impatient with the experiments and plans, and pressed for action, as did Hooker and Hale back in the United States. But they all learned that though Ritchey was welcome as an adviser, Delloye still gave the orders, and for all his friendliness his main responsibility was to the St. Gobain company. Ritchey could do little but wait.[13]

He took advantage of the enforced halt to go to London and describe the new 60-inch telescope at a meeting of the Royal Astronomical Society. As a Foreign Associate of the Society, he was especially welcome. Ritchey's talk, copiously illustrated with slides of the work on the mirror, the design of the telescope, the truck and the mule teams taking it up Mount Wilson, and the completed instrument and dome, gave the English astronomers their first, highly positive image of the 60-inch.[14]

On his return to Paris, Ritchey briefly met with Hale, who had just arrived in Europe for more scientific meetings and what was supposed to be a restful vacation. Ritchey hoped to see another pouring of a 100-inch disk at St. Gobain but had to be satisfied with a trial run of the new auxiliary furnace. He left for the United States in May, bearing a request from Delloye to Hooker for a payment of $2,000 to $4,000 for the work on the furnace, and only the hope of a pouring soon. By early June Ritchey was back in Pasadena.[15]

Hale had come to Europe in the hope of improving his health. In the summer of 1908 he had suffered badly from "nervousness," and his doctor had advised him to give up work altogether for a long period of time, a favorite "cure" often recommended for tired businessmen in that day. Instead Hale had decided to try to take it easy, get away from the observatory, and travel around Europe, but to continue discussing science.[16] Hale's letters to his wife from this period show a highly disturbed inner man, whose thoughts combined intense but repressed sexuality (also evident in those of his earlier letters to his close friend Harry M. Goodwin that he knew his own parents would not read), a great need to be loved, and severe guilt for past unfaithfulness, whether in the flesh or in the mind it is impossible to tell. He had great difficulty sleeping, and in concentrating while he was awake.[17] Probably, he constantly thought that he was about to die, a frequent phobia of successful, middle-aged eldest sons whose strong, successful fathers have died, as his had ten years before.

In France Hale met Delloye, saw the new furnace himself, and watched some of the preparations for the next pouring.[18] However, Hale could not stay in France, and by late June Ritchey was again sending telegrams to Jouvaud and letters to Delloye, begging for information on the progress of the work. Hooker was "very anxious" to be kept informed. All he learned was that the new disk had not been cast, but the St. Gobain officials "expected" that a new trial, late in August, would be successful.[19]

Ritchey found himself in a changed situation back in Pasadena. As the 60-inch went into operation, Hale had decided that some reorganization with clearer lines of authority would be desirable. Though almost all of his previous research had been on the sun, he told Walter S. Adams that his goal had always been to investigate stellar evolution and that he had studied the sun only as a typical star. Now with the 60-inch Hale intended to do high-dispersion spectroscopic work on stars and spiral "nebulae" to try to measure their rotation and to detect magnetic fields in them. Adams could work on his own on other topics. Hale promoted his former assistant to "First Astronomer, with duties as Chief of Staff," and said that during his own absences from Pasadena Adams would serve as acting director.

Very soon after Ritchey had left for New York in January, bound for France, Hale himself had taken a number of direct photographs of the Orion Nebula, some with red filters on red-sensitive plates, to investigate the strength of the Hα emission line of hydrogen in it. He knew that direct photography was Ritchey's own field and that he was jealous of anyone who tried it with "his" telescope. Hale's action did not augur well for the future. Hale quickly outlined a whole program of spectroscopic observations of the nebula and of the stars within it but never followed up on it or published the results of his direct plates.[20]

Ritchey knew that Adams was unsympathetic to him. He wrote from Paris asking Hale to assign him the 60-inch for eight nights each month, near the new moon (when the sky is dark), for direct photography. Ritchey hoped Hale would settle this matter before he left for Europe and Adams took over. Hale did give Ritchey eight nights for each month from July through December, but he assigned them in continuous blocks, rather than splitting them up into two four-night sequences with a few days for rest between them. He also vetoed Ritchey's request for an assistant to work with him on the long-exposure photographs that were his specialty. It was a regimen guaranteed to exhaust the forty-five-year-old Ritchey. Hale further notified Ritchey that in the future, if he wanted to buy any needed supplies or equipment for the observatory, he would have to ask either Adams or himself to sign a requisition. He indicated that he or Adams would assign the shop crew's work, rather than letting Ritchey do so himself as in the past. With the 60-inch in operation and the 100-inch on hold, Ritchey's usefulness to Hale was greatly reduced. The construction phase was halted, and the big-telescope research phase had started. These reassignments and limitations made sense to Hale and Adams but must have been galling to Ritchey.[21]

As Hale departed, he left written instructions which Adams was to show to Ritchey, assigning the jobs with highest priority to be done in the shop, naming specific men to work on them, and taking Francis G. Pease out of Ritchey's control. Hale was clearly still under severe strain when he returned in August, for in a complete about-face from his letter of seven months earlier, he now vehemently insisted to Adams that he intended to devote most of his own research activities to the sun and that his "First Astronomer" should stick to stellar observations.[22] This is one of the only two recorded real disagreements that Hale had with Adams in forty years of close association.

Ritchey observed each month with the 60-inch, taking long-exposure direct photographs. Because of the large scale of the telescope, its optical excellence, and Ritchey's own persistence and skill as an observer, they showed much more detail than earlier pictures. In September he was eager for another attempt at the 100-inch disk, and he wrote Delloye: "I am expecting to hear from you soon that an entirely successful casting has been made." To Edward E. Barnard, Ritchey described his work on the design for the 100-inch telescope. He planned to mount it high above the ground, in the open, with no dome but with a shelter on wheels to be rolled over it for protection in inclement weather and daytime, and pulled away at night. A windscreen would shield the telescope. The whole design was meant to keep it from heating up during the day and thus causing image deterioration.[23]

Hale was very positive about the 60-inch. He reported the excellent images obtained with it and had no doubt that the 100-inch would be even better. In his draft of the observatory's annual report, Hale gave comparative figures on image sizes, proving to his own satisfaction that the 60-inch was far more efficient than Lick Observatory's 36-inch Crossley reflector. However, Lick Director W. W. Campbell, to whom he sent the draft, complained that the comparison was not fair; he said that the Crossley had recently been improved and that Hale had compared the best recent 60-inch results with old, subpar Crossley-reflector photographs. Hale agreed to remove the offending passage from the report but proposed a controlled test, taking photographs of the same objects with the two telescopes under otherwise identical conditions. Campbell accepted the plan in principle, but the two rival directors could never agree on the details, and the test was postponed time after time and eventually forgotten. Each knew that the continued support of his observatory was dependent on its being "best" and was unwilling to enter it in a competition unless he was sure it would "win."[24]

By October Ritchey wrote Delloye again, reminding him that Hooker was impatient for news. He wanted a "full report" on the result of the experiments carried out during the summer and a "close estimate" of the probable date of delivery of "a perfect 100-inch disk." Delloye could only reply that St. Gobain had cast a disk that the workers had hoped would be homogeneous, but so many stones had broken off the mold and fallen into the molten glass that they had had to abandon it. They would have to start over once again. The repeated failures were expensive, and the company clearly was becoming reluctant to continue them without a change in the contract.[25]

Hale himself took some photographs of Mars with the 60-inch in its long-focal-length Cassegrain arrangement. The red planet was very much in the news because of Percival Lowell's purported discoveries of "canals" on its surface. Hale reported that his special red-sensitive plates and filter showed no canals, nor could he see any visually. He took a few moon photographs also and wrote his friend and successor as director at Yerkes Observatory, Edwin B. Frost, that they showed finer detail than any previous photographs. Though he was a serious solar astronomer, with the 60-inch Hale had become little more than a dilettante director. Nevertheless, he could report the 60-inch in use as "practically perfect" and described in glowing terms the "remarkably fine photographs of nebulae" that Ritchey was obtaining with it.[26]

As the delays continued in France, the situation in Pasadena appeared to be approaching a crisis. Hooker demanded to know when the mirror would be ready. Andrew Carnegie, the financial angel of the Carnegie Institution, the chief source of support of Mount Wilson Observatory, was coming on a visit. How could Hale explain the glorious future of his research center to him and the urgent need for more of his money to build the 100-inch telescope if he could not tell him when or even if the glass disk that was to be its centerpiece would be available? Ritchey wrote Jouvaud, the St. Gobain company's agent in New York, that it was "imperative that we have the latest and fullest information about the status of the 100-inch disk, the probability of an entirely successful casting, and the probable date of delivery." According to Jouvaud's replies, the workmen cast the new disk in February, and it remained in the annealing oven (and thus was impossible to inspect) all during Carnegie's stay in California, but when they opened the oven, the disk was broken. In May they were ready to try again, and "all possible w[ould] be done to make it a success this time." Ritchey said that he had "complete confidence that the present trial will be an entire success, but we are having considerable difficulty in keeping Mr. Hooker hopeful in the matter, as he was profoundly discouraged over the recent breaking of the disk in annealing."[27]

The 100-inch disk, intended to be a symbol that the telescope would be built, as the 60-inch mirror had been, instead had become a millstone that prevented the project from gaining financial support. And Ritchey, who had been so successful with the 60-inch, was now cast as the expert who had gone to France to solve the problem, had returned with no results to show, and was the continued bearer of bad tidings.

Ritchey chose just this time to make a bad tactical mistake. All his observational work depended on photographic plates. He had long believed that improving their speed and sensitivity was as important as increasing the apertures of telescopes. Now his friend, R. Benecke of the Cramer Dry Plate Company in St. Louis, the manufacturer of the photographic plates used by observatories throughout the United States, told Ritchey of a former employee who had an idea for producing more sensitive emulsions, which would be very useful in astronomy. Ritchey, without consulting Hale, impetuously asked Hooker to provide the money to hire this technician and set up a special photographic laboratory for astronomy. Hooker asked Hale for his opinion on the proposal, and the director exploded. Ritchey was tampering with the donor who he hoped would provide the funds to build the whole 100-inch telescope. Hale cast Ritchey's request into the form of a proposal to set up a commercial laboratory and indignantly rejected it as a ruse to divert Hooker's money away from "the benefit of the Observatory as a whole" for "the advancement of science." Hooker did not provide the money, and Ritchey had made Hale his enemy.[28]

Hale wrote a letter from Chicago severely chastising Ritchey. (His few harsh admonishments were generally delivered from afar, by mail, rather than face to face.) In Washington he met with Robert S. Woodward, president of the Carnegie Institution, and Carnegie's executive committee. Woodward and the committee expressed their "full belief" in the 100-inch and told Hale it must be built. Hale described Ritchey's disloyalty, as he saw it, to Woodward. The president supported the Mount Wilson director fully and advised him to tell Ritchey that "he is not indispensable to the Observatory," and "that if he is to continue his work with us, he must assume a different attitude."[29]

Unwisely, perhaps because of this episode, Ritchey chose this time to start his own outside telescope business. With the 60-inch telescope finished, and no 100-inch disk in hand, there was no big observatory project demanding his undivided attention. He had completed the preliminary design for the 100-inch, and, until St. Gobain produced an acceptable glass disk, there was no reason to go ahead with an expensive detailed design. He proposed to go on two-thirds time and salary and devote the other third of his time to making telescopes in his shop at home. Hale no longer trusted Ritchey. He accepted the proposal but drove a hard bargain. Ritchey's prestige had slipped, and Hale made the most of it. He had a three-page, highly legal contract drawn up. Under it, Ritchey agreed that his position as a regular member of the Mount Wilson Observatory staff be terminated and that instead he be employed in "a new relationship." He would work two-thirds time on the design and construction of the 100-inch, on observing with the 60-inch, and on maintaining its optical and mechanical quality. Hale would be "free at all times to specify or modify these services in any way that he [might] see fit." All Ritchey's designs, notes, manuscripts, and letters were to be the property of the observatory. The agreement was to terminate on the date the 100-inch telescope was completed and went into operation, when a new agreement was to be drawn up giving Ritchey "certain opportunities" to observe with it and "possibly providing other work for him to be done by him for the Observatory." However, if the mirror were delayed, or funds were not obtained, and it appeared to Hale that the telescope might not be completed by the end of 1915, he was to be free "to make such changes in salary and in the nature of the services required" of Ritchey.[30] Ritchey probably did not even consult a lawyer before he signed this contract, but it was to provide the perfect means of firing him nine years later. When it went into effect his salary, which had been reduced from $4,000 to $3,000 per year on January 1, 1909, was cut to $2,000. In addition, however, he continued to receive the full $1,000 a year he was getting from the fund Hooker had provided for the 100-inch mirror.

Hale carefully scrutinized the advertisement Ritchey had drawn up for his "G. W. Ritchey & Son, Pasadena, California, Reflecting Telescopes and Specula" business. Though Hale permitted Ritchey to say that he had designed the 60-inch reflector, the director insisted that it state that the optical and mechanical work on the telescope had been done under Ritchey's supervision at the observatory instrument shop, rather than by Ritchey himself.[31]

In March 1910 the thirty-one-year-old Henri Chrétien arrived in Pasadena. Born in Paris, he had been trained in mathematics, astronomy, physics, and engineering, and had become a rising young staff member at the Nice Observatory. In 1908 and 1909 he had visited and worked as a volunteer at the Cambridge Observatory in England, Pulkovo Observatory in Russia, and Potsdam Observatory in Germany. Now he was to complete his astronomical research training at Mount Wilson.[32] Hale assigned

Henri Chrétien at the slit of the solar spectrograph, Mount Wilson, 1910. Courtesy of Cercle Henri Chrétien.

Chrétien to work with Charles St. John on the solar observing program and with Ritchey on direct photography with the 60-inch. Chrétien was impressed by Ritchey's precise methods, fantastic accuracy in guiding, and wonderful results with long-exposure photographs.[33] In his first session at the telescope, Chrétien watched Ritchey guide a long exposure and listened to his schoolteacherish lecture on the care and precision required. Then Ritchey gave Chrétien a plateholder and invited him to try an exposure himself. When he had finished it, Ritchey took the plateholder and opened it. It was empty. The trial had been a dry run. Then Ritchey gave Chrétien another lecture, after which he let the young Frenchman take a real exposure. As related by Adams to Helen Wright forty years later, Chrétien had guided "several hours," and the story became an illustration of Ritchey's mean-spirited secrecy and vindictiveness. Adams had told the incident to Hale in this way as an argument for firing Ritchey. Yet Chrétien himself told the same story (giving three quarters of an hour as the exposure time) to David B. Pickering with a twinkle in his eye, as an amusing example of Ritchey's single-minded dedication.[34] Chrétien remained Ritchey's friend and supporter all his life.

In April and May, Comet Halley swung close by the sun, becoming a conspicuous object in the sky. Ritchey and Chrétien worked hard, obtaining an outstanding series of photographs of it as it changed in form and brightness, threw off segments of tails, and eventually faded as it receded back into the darkness of space.[35] Selections of their photographs were widely published in textbooks and were used by astronomers and plasma physicists for years in their research on comets.

Studying the long-exposure photographs of nebulae he had taken, Ritchey was impressed with the amount of detail seen for the first time on these large-scale images. He noted especially the thousands of "nebulous stars . . . apparently condensing . . . in great bunches" in the spiral arms of the great spiral nebulae M31, M33, and M81. He wrote to his friend Barnard at Yerkes Observatory, and to Sir David Gill, the president of the Royal Astronomical Society in London, of the many similarities between these objects and our Milky Way, emphasizing the "dark rifts" (interstellar dust clouds, we now know) so similar to those Barnard had photographed in our Galaxy. At times he thought M31 was a good model of our Galaxy, anticipating Heber D. Curtis, Edwin Hubble, and Walter Baade, who were to see this analogy increasingly clearly in the 1920s, 1930s, and 1940s, but at other times Ritchey thought there were too many "nebulous stars" in the spiral and that it was not another example of a Milky Way.[36] Ritchey sent a large set of glass transparencies of his photographs of nebulae and star clusters to London for exhibit at the Royal Astronomical Society. Again they created a sensation, and his descriptions of them, culled from his letters to Gill, were published as a paper.[37] Ritchey described his technical methods of taking the photographs in detail, emphasizing the use of fine-grained photographic plates to preserve every detail the telescope revealed. Such plates are by their nature relatively low in sensitivity, so the exposures were long.[38] He also completed the article on the making of the 60-inch telescope he had promised *Harper's Monthly* years before. It was very well written, and illustrated with pictures showing all phases of the telescope's construction.[39]

During this period Ritchey and Chrétien developed the idea of the system we now know as the Ritchey-Chrétien telescope. The main limitation to photography with a reflecting telescope is coma, the aberration that stretches the images of stars off the center of the field out into little v's or arrow-headed figures, or, technically, comatic images. The further from the axis (or center of the field) the star is, the more elongated its comatic image is. This is an inherent optical property, or technically, an aberration, of a parabolic mirror. The result is that the center of a picture is perfect, but further and further off-axis it becomes progressively poorer. This aberration thus limits the size of the good picture that can be taken in a single exposure. Ritchey was well aware of it from his experience with the Yerkes 24-inch reflector. Because of the larger scale of the 60-inch, its comatic images were larger, and the field of good definition correspondingly smaller. He knew the problem would be even worse with the still larger 100-inch telescope.

Ritchey noticed, however, that at the Cassegrain focus of the 60-inch, with one of the hyperboloid secondary mirrors in place, the coma was smaller (at the same angular distance from the center of the field) than at the primary or Newtonian focus. We understand this very well today; the coma of any paraboloid or paraboloid-hyperboloid Cassegrain system depends only on the focal ratio of the telescope, and as the Cassegrain combinations have longer focal ratios their coma is correspondingly smaller. Ritchey, along with most astronomers and working opticians of his day, did not realize this. He thought that the hyperboloid secondary somehow reduced the coma of the paraboloid primary (as in a way it does). He therefore suggested that Chrétien investigate mathematically whether some other shape of secondary mirror would reduce the coma even further, and if possible, do away with it altogether. Chrétien, with his Sorbonne training, was the ideal person to answer this question. He plunged into the complicated equations of ray-tracing for a compound reflecting telescope and soon found that though no secondary could do the trick alone, a new form of Cassegrain telescope in which both the primary and secondary mirrors are specific hyperboloids, the former convex and the latter concave (both fairly similar to the classical paraboloid and hyperboloid combination) can reduce coma to zero. Such a telescope would be able to take a photograph of a much larger field in a single exposure with "perfect" images all over it. The size of the field would be limited only by astigmatism, a different aberration that becomes important only much further from the axis than coma. Another advantage of this system would be that, as it would be used *only* at the Cassegrain focus, the primary mirror could have a relatively fast focal ratio, and the overall tube length would be quite short. This would have the very practical consequence that the telescope would be relatively small for its aperture, the dome or house would be small, and the cost would be small. Ritchey and Chrétien called this their "new-curves" design; in the 1920s it became known as the Ritchey-Chrétien system. It is also sometimes referred to, especially in France, as an aplanatic reflector, meaning that it is free of coma.[40] Ritchey, who could see all the advantages of the "new-curves" reflector especially clearly from his twin roles as direct photographer of nebulae and designer of the 100-inch, began almost immediately to try to convince Hale to switch plans and allow him to design what was to be the largest telescope in the world as, at the same time, the first Ritchey-Chrétien to exist. Meanwhile, the problem of getting a 100-inch disk remained. Hooker was becoming very restive. Hale feared that unless work on the telescope began quickly, the observatory might never get the funds necessary to complete it. (Hooker was seventy-two years old and probably Hale could see that his health was declining.) For the first time, the Mount Wilson director brought up the idea of trying to use the imperfect first disk that had been delivered in 1909, which was still resting in its box in the optical shop in Pasadena. The disk had bubbles throughout its interior and therefore would not expand and contract uniformly, but it could still be ground and figured into a mirror. Maybe it would perform well enough. Hale proposed that the observatory pay "a nominal amount" for this disk and try making it into a mirror. He emphasized this would not be a permanent solution; it would be a stopgap mirror, to be used only until the final, "perfect" disk was sent. The optical grinding machine was ready for work and standing idle, and the optical shop crew had time on their hands. Why not start grinding so that Hooker or any other prospective donor could see the reality of a massive 100-inch piece of glass being shaped into a telescope mirror?

The St. Gobain company, about to make a new pouring and once again "confident that this time they will reach success," responded that a nominal sum was not enough. However, it would sell the old disk for $5,000 and would refund $1,500 of this against the full agreed price of the replacement, "perfect" disk that it hoped to be able to ship soon. Hale learned "with great regret" that St. Gobain had tried once again the old method of pouring from three pots. He was sure it would fail and ordered that it not be used again. He was right; the resulting disk had fewer bubbles than the first one but still far too many. Finally, later in the summer, Adams, on Hale's behalf, proposed that the observatory be allowed to do the optical work on the old disk at no charge, but that if it proved successful it would pay the $5,000 price set by St. Gobain. The French accepted this offer in October, clearing the way to start the work.[41]

As these negotiations dragged on through the summer, the International Union for Cooperation in Solar Research met on Mount Wilson. Hale had brought this organization into existence at the St. Louis World's Fair in 1904. Its third meeting was intended to show the Mount Wilson Solar Observatory to the astronomers and physicists of the world. About one hundred scientists attended, the majority of them astronomers who actually worked primarily on stars and nebulae rather than on the sun. It was a great event.[42] Ritchey invited his friend Elihu Thomson, the General Electric director of research, to come, promising to show him the 60-inch in action and telling him of his plans for making "better photographic plates for astronomical photography, i.e., finer grained and faster plates." He hinted that he only needed $10,000 to carry out this development program, and he had "no doubt that greater advances in astronomical photography [could] be made in the next few years by this means than [could] possibly be made by means of larger telescopes in our generation." This plea for improved detection technology has been repeated after the completion of each large telescope, down through the years since the 60-inch, always followed within a few years by a plea for a larger telescope to utilize the new detectors more effectively. Thomson did not come to the meeting, and the General Electric Company did not send Ritchey $10,000. Nevertheless, the 60-inch was the great attraction of the meeting. Ritchey showed the assembled astronomers interesting objects with it each of the three nights they were on Mount Wilson and by day explained his transparencies of photographs taken with it.[43]

For Hale it was a different experience. His nerves had broken down almost completely. Just at the moment when he should have been preparing the mountain and meeting the visiting dignitaries, he could not face work or people. He departed precipitously with his wife for a Lake Tahoe fishing vacation, which he hoped would enable him to pull his nerves together. Adams and recently hired staff member Frederick H. Seares were left to do the best they could with the meeting. Jacob C. Kapteyn, the leading Dutch theoretical astronomer whom Hale had invited to come for the summer and advise him on stellar research problems for the 60-inch, sympathized with Hale by letter from Mount Wilson but made sure not to mention astronomy, which he believed would excite the patient.[44] Actually, the continued failures to get a satisfactory 100-inch disk, worry over what Hooker might do, and the struggle with Ritchey were probably the main immediate sources of Hale's inner tension. From Tahoe he moved restlessly on to Oregon for more fishing. His wife wanted him to enter a sanatorium, but he refused, saying he could not live without her. He "recognized" he could not "be completely well for some time." Only two things, reading and fishing, could keep his mind off the problems at the observatory. Travel abroad might be almost as good. Therefore, he proposed to go to Europe with his wife, "travel about as [they] might choose, to new places and old, avoiding all scientific men and institutions, and renewing [their] youth in a second wedding journey."[45] This plan, and close variants of it, did, in fact, become the pattern of much of the rest of Hale's life, interrupted by furious bursts of activity in crisis situations, which provided an outlet for his tremendous nervous energy.

All this was in the future, though, at the end of August 1910, when he pulled himself together for a brief one-day appearance at the meeting on Mount Wilson. Everyone sympathized with him; then he dashed off for Chicago with Evalina, en route for New York and Europe.[46] Meanwhile, Ritchey was bubbling with ideas. He had taken an excellent long-exposure photograph of the globular cluster M13, which he estimated showed 50,000 to 60,000 stars. He marveled at the number of "nebulous stars" he saw on his long-exposure photographs of the spiral M33, which he thought were stars in formation. Actually, Ritchey was misled by the bright background of nebulosity in M33's spiral arms; today we know these objects are mostly high-luminosity stars or clusters of stars, with a few nebulae among them. He was drawing up preliminary plans to make the 100-inch telescope as a Ritchey-Chrétien. He hoped to patent the design for this "short type of reflector," which he was sure would be ideal for photography, and was already trying to interest directors of other observatories in ordering such telescopes from him.

Finally, instead of using the imperfect 100-inch disk, Ritchey wanted to make a built-up replacement for it, consisting of three *thin* 100-inch diameter disks, each two-and-a-half or three inches thick, separated by spacers or ribs of the same type of glass cemented between them. The spacers would be arranged so that forced air, driven by a fan, could circulate between the three disks, keeping them at uniform temperature. The resulting built-up disk would be essentially as rigid as a single thick disk but much lighter and thus not nearly so subject to flexure. The three thin disks could each be cast from a single pot of molten glass and would be bubble-free. The main problem in producing such a built-up disk, Ritchey thought, would be to develop a cement which would hold the glass pieces together *permanently*, in spite of moisture or abrupt temperature changes. He had begun experiments making such disks at his shop at home with his son, Willis, and his father, James, as his assistants. According to one published report, he had succeeded in making a 30-inch built-up disk, but this news item was based on a misunderstanding. In his letters in September Ritchey reported that he was still searching for a good cement, and not until December did he produce a 20-inch built-up disk.[47]

Ritchey described all his ideas in glowing terms to Hooker. To the successful businessman they seemed the way out of his problems. He could bypass the imperfect disk and cut short the interminable delay in getting a new one. The "new-curves" reflector would be better for direct photography. Because it would be shorter it would be cheaper (Ritchey had estimated by 25 to 35 percent). Building no dome but using only a roll-off house to protect the telescope from the weather, as Ritchey still planned, would cut costs even more. Maybe Hooker could provide *all* the money necessary to build the whole telescope. He was certainly considering it.

Adams, however, was adamantly opposed to making the 100-inch as the world's first Ritchey-Chrétien telescope with the world's first built-up disk. Hale was willing to consider making it a "new-curves" reflector but rejected entirely the built-up disk concept. Hooker agreed to allow his 100-inch mirror fund to be used to enable Ritchey and his shop crew to make an experimental 40-inch built-up disk and grind and figure it into a mirror, but Hale and Adams opposed this as a diversion that would be too expensive in time and especially money. (Hooker's promised $45,000 was nearly all spent.) They wanted to push ahead on a conventional telescope with the disk they had. Hale, however, let Ritchey and his assistants start work on the optics for a test of the Ritchey-Chrétien system, using a 27-inch glass disk that was on hand. It would thus be essentially a quarter-scale model of Ritchey's plan for the 100-inch. Ritchey asked Chrétien, who had returned to Paris, to calculate the data necessary to figure and test the 27-inch primary mirror, which is not as simple optically in the "new-curves" design as a conventional paraboloid.[48]

One reason Hooker was favorable to the built-up mirror idea was that both Campbell and Barnard, whom he had consulted, had advised him that it would be a mistake to spend his money for work on the flawed 100-inch disk. Adams reported this problem to Hale, still in Chicago. Hale wrote a long, persuasive letter to Campbell (and no doubt a similar one to Barnard) explaining the technical and scientific problems involved and delicately hinting at the answer he wanted. They both saw the light and changed their opinions. It *was* a good idea to go ahead with the old 100-inch disk on an experimental basis after all! Hale also wrote Hooker arguing strongly against testing the built-up disk idea. Adams, meanwhile, told Hooker the problems that he and Hale foresaw with the Ritchey-Chrétien design, the opinions of other astronomers who had been at the Mount Wilson meeting "that the defects in the present 100-inch casting would by no means necessarily prevent it from working successfully as a telescope mirror," and the opinion of F. L. Drew, a newly hired engineer, that the savings with the short-form reflector would be only 12 or 15 percent, much less than Ritchey had estimated. Hooker pulled back from supporting Ritchey's plan. The hardware magnate said he would wait to hear what Hale learned on his planned visit to St. Gobain (as part of his European rest tour) but ominously added: "I cannot bind myself to further [financial] responsibility at this time." Coming on top of his previous statement, "It strikes me that a very large expenditure has been made too hastily since the very cornerstone—the disk—is not in evidence," it was obvious there was little hope for future support of Mount Wilson Observatory from him.[49]

However, Hooker did provide support for Ritchey's experiments on built-up disks in his shop at his home. There the optician, working with his son and father, succeeded in making a sound, permanent 20-inch disk and polished it to a plane mirror which was flat to very high precision. According to him, it remained stable against temperature changes, flexure, and aging.

At the observatory, under Hale's orders transmitted from Europe through Adams, Ritchey began work in October on the 100-inch disk. This in itself was a near commitment to make the 100-inch as a conventional reflector, for the first step was to grind the mirror to f/5 section of a sphere, which would later be parabolized, not to an f/2.5 sphere, which would have been the starting point of the Ritchey-Chrétien system. Ritchey was bitter against Adams, whom he considered "an unprincipled man who has made himself the favorite of the director." Hale, Ritchey darkly hinted, was peculiarly and exceptionally susceptible to the acting director's manipulation, "*particularly now, of late*, when he is probably not entirely sound mentally."[50] That same month Ritchey tried photographing some of the planets and the moon at the coudé focus of the 60-inch, which provided the largest scale of all. The experiment was not a success, and Ritchey never published any planetary images. The seeing, even at such an excellent site as Mount Wilson in the fall, was simply not good enough. The human eye remained superior to photography for recognizing and picking out fine detail on these small, bright objects down to the dawn of the space age. Ritchey planned a lecture tour of the East, showing his collection of slides of pictures taken with the 60-inch. He also spread the word of the new Ritchey-Chrétien type of reflector. Nevertheless, he was desperately unhappy. He felt that he was not getting enough observing time with the telescope, that Adams was thwarting his plans, and that he had to get out of Mount Wilson Solar Observatory.[51]

Incredibly, he proposed to Henry Norris Russell that Princeton

University build an observing station near Pasadena, with a large Ritchey-Chrétien reflector, devoted to astronomical photography and Ritchey himself as the resident astronomer. He included in his letter testimonials from Campbell, Gill, Barnard, Thomson, and others as to the excellence of his work. But Russell, who had visited Mount Wilson at Hale's invitation and who was immersed in drawing astrophysical conclusions from Adams's new spectroscopic results, had no intention of biting the hands that fed him. He did not answer Ritchey's unrealistic plea from the heart.[52]

Hale arrived in England in October. He felt his health was not up to a trip to Paris. The journey "had been very hard on [his] head." Instead of resting he collected expert opinions from Kapteyn, Gill, and others. They confirmed Adams's strong opinion, and Hale's own feeling, that it would be a mistake to build the 100-inch as a Ritchey-Chrétien telescope. Whatever doubts they had were resolved in their ill friend's favor. He sent their statements to Hooker in an effort to get him back into a giving mood, but it was no use. In November Adams requested nearly $6,000 from the old man to meet the expenses of the work on the 100-inch mirror still outstanding for 1910. He had already provided $35,000, and not one cent had been paid for the glass disk, which was being ground on approval. Hooker, in a final episode, asked Adams to show two of his friends, who were visiting California, around Mount Wilson. The acting director had no choice but to do so. Two days later he received Hooker's ultimatum, a check for $10,000 with a formal letter, addressed to Hale, stating that it completed his promised gift of $45,000 and demanding a receipt "in full satisfaction of the balance remaining unpaid on his gift to the Carnegie Institution of Washington."[53]

Just three days before Hooker had signed his check in Los Angeles, Hale had overcome his nerves, gotten to Paris, and met with Delloye. The St. Gobain director was full of hope. His men had cast a new disk just the previous week, and Delloye was "very confident of ultimate success." In an eighteen-page letter, Hale dredged up all the reasons Hooker should go on paying for work on the mirror and the design of the telescope. This appeal was querulous and quite unconvincing, totally unlike Hale's usual confident, inspiring salesmanship. He seemed to have known that he would not convince Hooker but to have written the letter to justify his own position for posterity. It was a traumatic crisis for Hale. He had been hoping to get more money from Hooker to go on

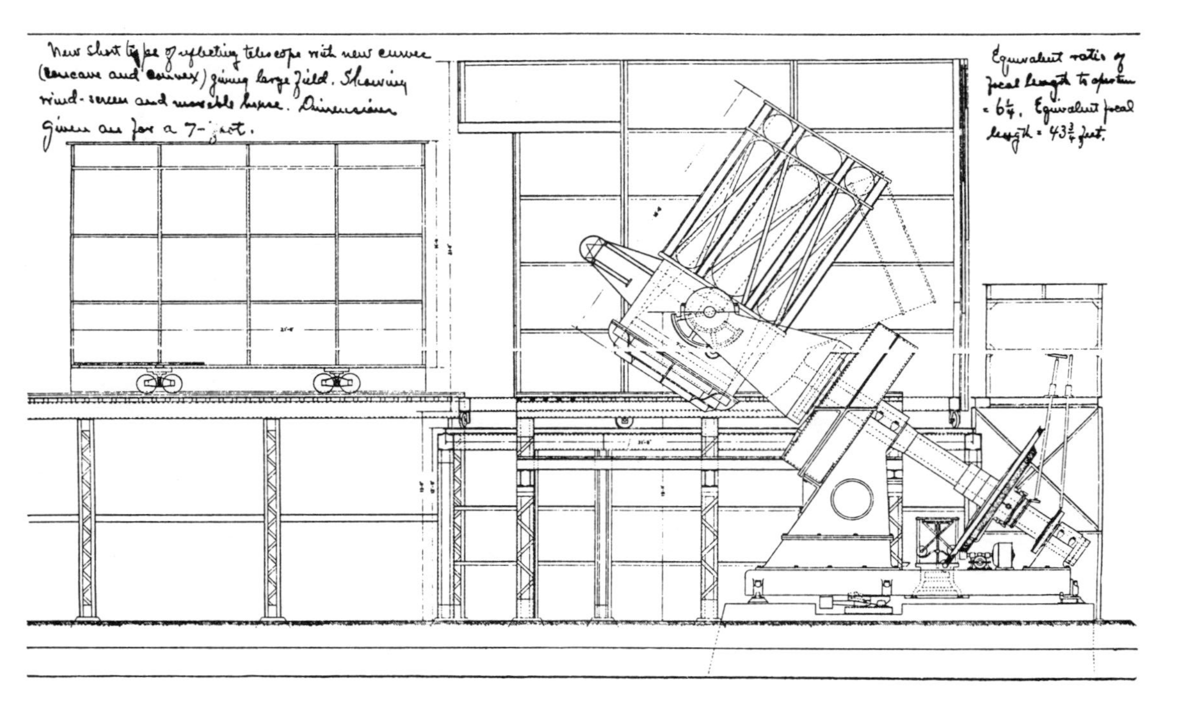

Ritchey's design for an 84-inch Ritchey-Chrétien reflector and its roll-off shelter, 1910. Courtesy of Princeton University Library.

with the 100-inch. His fund would soon be exhausted. Several men's salaries depended on it, and no payment had yet been made for the 100-inch disk. All Hale could see to do was to stop work on the 100-inch disk, let Drew go, and reduce Ritchey's salary. Adams agreed that he would hate to see Drew leave the observatory staff, but there was no alternative. The acting director had baited Ritchey, who was tired and depressed, into saying he did not want any observing time during the rest of the winter (the rainy season, when clear nights are few). Now Adams told Hale that he strongly favored cutting Ritchey's salary because he was doing so little observatory work. Finally, Adams advised slowing down drastically the optical work on the 27-inch Ritchey-Chrétien. In his opinion it would not be necessary now that Hale had decided to go ahead with the 100-inch as a conventional reflector. Adams poisoned Hale's mind against Ritchey in letter after letter at this time, depicting him as a selfish opportunist with little interest in science and no loyalty to the observatory.[54]

Hale agreed but appealed to the Carnegie Institution to provide extra funds to keep Drew on the payroll. He had been hired basically to make the detailed design of the 100-inch telescope mounting. Hale directed Adams, if they could keep Drew, to leave the design plans open and, in particular, to include a provision for using an additional 100-inch mirror "of the shortest possible focal length" (that is, a Ritchey-Chrétien primary) in case they ended up with two usable disks. He was vacillating and contradicting himself.[55]

Hale was in a pitiable state at this time, feeling "rather blue" and not knowing where to turn. Several doctors examined him in London, including the famous Sir William Osler. None of them could find anything physically wrong with him; they all advised him to rest, travel, fish, and not bother his head about the observatory or the 100-inch disk. In December Gill found him "looking exceedingly ill and suffering from severe nervous pains and noises in his head, symptoms which demanded rest from all excitement." Hale had gotten to Paris and was "in some anxiety in regard to the great mirror." He was badly depressed as he and his wife went on to the south of France and to Italy, on their way to Egypt.[56]

On the train between Mentone and Genoa, Hale learned great good news. He opened a newspaper and read that Andrew Carnegie had given $10 million more to the Carnegie Institution. Surely some of it would go to Mount Wilson Solar Observatory to build the largest telescope in the world. Soon Hale received confirmation from his friend Goodwin and then from Carnegie himself. The "good old boy" had heard of Hale's illness, and invited him and his wife to visit him at his castle at Skibo, in Scotland, during the coming summer. Hale felt much better. He and Evelina went on to Cairo, but in spite of the release of financial tension, he could not shake his depression and inability to concentrate. He searched desperately for something to do that would interest him.[57]

What Hale was really interested in was his observatory, in spite of his doctor's advice and his wife's orders to give up astronomy. As soon as he learned the good news of Carnegie's gift, he dashed off a thirteen-page letter to Adams, full of "schemes" for the future of Mount Wilson. He wanted to keep Drew on the payroll. He cautioned Adams not to give the Carnegie administrators a definite figure for the cost to complete the 100-inch but to strive for "a very liberal interpretation of the matter," under which auxiliary instruments, experiments on making optical glass and fused quartz, and developing new gratings would all be acceptable expenses. Hale sent word that the opticians should go on grinding the 100-inch disk and could also work on the 27-inch Ritchey-Chrétien optics "*provided* it can be done in such a way as not to interfere with more important work." Ritchey should not be left in charge of the 100-inch design; he could not be trusted. He could do the optical work on the 100-inch mirror, but an engineer would be hired to take over the general design. Ritchey should be paid only on a two-thirds time basis, less if he worked fewer hours. Hale was opposed to letting Ritchey go back to regular observing in the future. "It is perfectly clear that his sole purpose is to get a few fine plates of striking objects, rather than to use the telescope for a systematic study of stars or nebulae," wrote Hale, who had done just this himself, with his striking photographs of the Orion nebula, the moon, and Mars two years previously. Hale wanted Adams to have Pease start taking direct photographs with the 60-inch "and learn how to obtain the finest results. We can then prepare a systematic programme for him to follow."[58]

Adams was keeping a very close eye on Ritchey. In January 1911 he discovered that Ritchey had ordered letterhead stationery, giving his name as "Professor G. W. Ritchey" and with the regular observatory address. Adams objected to this. He thought Ritchey was using the stationery, and thus the observatory name, for his "private optical business." Ritchey denied the charge. He said the stationery was for his correspondence on observatory business. Adams thought this was wrong. "It seems to me that it would create a false impression in regard to the relationship of Mr. Ritchey to the Observatory [that he was the director], and if it were once started other members of the staff could equally desire the same privilege," he wrote Woodward.

The Carnegie Institution president supported the acting director, and Adams conveyed the news to Ritchey in brutal fashion that he must stop using his personalized stationery. The optician was about to leave on a lecture tour of the East. Adams ordered him in writing to return a number of reprints of his paper on the 60-inch telescope, which were observatory property, not his own. Secondly, Adams said, Ritchey was showing a poor attitude and not working hard enough on his two-thirds time for the observatory. Thirdly, Hale had decided that he probably would not give Ritchey any observing time on the 60-inch in the summer (the clear, good season at Mount Wilson). Instead, he intended to put someone else (Pease) to work on nebular photographs, who would "treat them more as scientific results and less as mere pictures." Lastly, Adams stated that Woodward had ruled against Ritchey's personalized stationery. For good measure, Adams added that he had received word that the latest St. Gobain attempt to make a new 100-inch disk had once again ended in failure. Ritchey scribbled a hot-tempered, emotional reply and departed for the East.[59]

He lectured in February and March at the Franklin Institute in Philadelphia and at several stops in New England, including the Massachusetts Institute of Technology. When Ritchey returned to Pasadena, he was ill and depressed. In the spring he made another trip, this one through the Midwest with a stop in Williams Bay to visit Barnard and lectures in Cincinnati, Chicago, and several other university towns. He could do little work, and suffered badly from insomnia.[60] Still Adams harassed him. In May the acting director sent Ritchey, at his home, a written order requiring him to return all the photographic plates he had taken of the moon, nebulae and clusters to the observatory plate-storage room. Ten days later Adams gave the sick man written notice that Hale, when he returned to Pasadena in September, would very probably modify his salary and work conditions, as the contract permitted him to do. In July Ritchey's wife Lillie appealed to Robert S. Woodward, president of the Carnegie Institution, to let her husband stay on the payroll and take his accumulated vacations, rather than being discharged. Independently,

Ritchey himself wrote a desperate letter to Woodward, begging not to be fired. He described his symptoms as "frequent fainting spells, continued insomnia, and extreme mental depression." Woodward forwarded Ritchey's appeal to Adams, recommending the optician be put on leave for three months at full pay, or for six months at half pay. Adams chose the latter and put Ritchey on half his salary, or $1,500 a year, on September 1, reminding him at the same time that Hale would modify his contract as soon as he returned. All of Adams's statements to Ritchey can be defended as actions that safeguarded the observatory's interests, but they were delivered in such a cold, heartless manner, and with such evident relish, that it is clear that the acting director wanted to drive the former superintendent of construction out of the organization. Ritchey had bought a plot of land in Azusa, in the San Gabriel Valley ten miles east of Pasadena, which he was converting into a lemon ranch, and there he recuperated.[61]

Meanwhile, Hale was traveling in Egypt, trying to keep his mind off astronomy. After the latest St. Gobain attempt at a new 100-inch disk failed, Adams cheered him up with word that as the work proceeded on the old disk, no problems arose. It seemed to Adams ever more likely that it would be usable for the mirror. Hale saw a paper of Barnard's, in which he published the best photographs he had taken of Mars with the Yerkes 40-inch refractor. The peripatetic director immediately wrote Adams to invite Barnard to come to Mount Wilson as a guest observer and use the 60-inch reflector for planetary photography, to supplant Ritchey in this field. Adams was only too glad to do so. Barnard eventually arrived in November. He obtained some images of Mars and Saturn, but they were not particularly good and he never published them. However, Hale did include the best of the pictures of Saturn in his annual report. Barnard also observed visually with the 60-inch, and its superior optical quality instantly converted him from a skeptic about reflecting telescopes to a true believer.[62]

In April, when Hale and Evelina returned to Europe, he wanted to discuss progress on a new disk with Delloye again, but he was so nervous that he could not bring himself to do it. Evelina, trying to protect him from astronomy, wrote Hale's brother William, the Chicago financier, to come to Paris at once. William came and met with the St. Gobain manager and with Gill, the president of the Royal Astronomical Society. William Hale reported to Adams that the French were doing all that could be expected but that the chances they would produce a better disk were slim unless someone was prepared to spend a lot more money. In May, Hooker died, but by that time everyone had given up on getting additional funds from him. Adams was by now very optimistic that the old disk would prove satisfactory. It had been ground deeply, nearly to spherical form, but no flaw had yet appeared.[63]

In June Hale, now ill, returned to the United States. Although he had been optimistic in the spring that he was getting better, by summer he was "all done up" again. William met Hale at the dock at New York and took him to Chicago, where he rested in a vain attempt to get over his "bad head." Then he agreed to enter a sanatorium in Bethel, Maine, for another attempt at a cure. Hale was apprehensive because he had heard that its proprietor, Dr. John G. Gehring, used hypnosis as part of the treatment.[64]

Once he got there, Hale found that the sanatorium was not so bad. The doctor declared that Hale's large intestine was displaced and enlarged, and that long use of one part of his brain had weakened it and allowed the poisons generated in his intestine to act upon it. Dr. Gehring called his problem "ptsosis." However ludicrous the diagnosis sounds, the therapy was beneficial: light massage of his intestinal area (this no doubt was designed to convince Hale that a physical effect was present and was being treated), light suggestion that he was getting well, and tremendous amounts of healthy outdoor exercise, sawing, gardening, and tennis. Soon Hale was a convert, writing Evelina, who had insisted he enter the sanatorium, that she would benefit from a course of treatment there, too. The "terrible haunted feeling" he had had "wholly disappeared, never to return I sincerely hope." Hale was released near the end of September and hurried back to Pasadena to prepare the observatory budget. By then he recognized part of his mental problems resulted from his "experience with Ritchey . . . last year, . . . a very hard and wearing one for me." Gehring's diagnosis, however, was that Hale's problems were largely the kinks in his intestine. The doctor sent his patient away with instructions on exercises and posture and with a battery of prescriptions for digestive aids and mild laxatives.[65]

In Chicago on the way back, Hale met with architect Daniel H. Burnham, who advised him that the planned windscreen and roll-off house for the 100-inch telescope would not provide adequate protection and that only a dome would be safe. Hale immediately stopped work on the tracks for the house, which had already begun on Mount Wilson, and ordered a design for a dome from Burnham instead.[66]

Ritchey, meanwhile, now trying to get over his depression at his home in Pasadena, started to correspond with Ralph A. Sampson, the director of the Royal Observatory in Edinburgh. He was interested in getting an old 24-inch mirror, which had never been satisfactory, reworked. Ritchey advised him to junk it and start over with a new, thick 24-inch disk, which would cost only $150. He would make it into an excellent paraboloid for $1,500. Sampson was an optical perfectionist, like Ritchey, and their correspondence on telescope mirrors was to blossom over the years. The director of the Scottish observatory became another life-long supporter of the American optician's radical ideas on reflecting telescopes.[67]

Soon after he returned to Pasadena, Hale gave Ritchey written notice of the modifications he was unilaterally making in his terms of employment. He had the right to do so under the contract, for Adams had previously notified Ritchey that he would. Beginning January 1, 1912, Ritchey was to be employed full-time on the *optical* work on the 100-inch mirror. His job was contingent upon the existing disk being usable. He would be an optician, not a regular member of the observatory staff, and he would be subject to the "direction" of Hale himself, Adams, "or any other person designated to represent me." Thus Ritchey would have no part in the subsequent design of the 100-inch telescope, nor would he be allowed to observe with the 60-inch he had built. Furthermore, Hale stated that whether Ritchey's job was to continue beyond the end of 1912 would depend on the work he accomplished during that year. Finally, he reminded Ritchey that all his astronomical photographs, optical and mechanical designs, and notes were the property of the observatory and must be turned in, except when he was actually using them for the book of photographs on which he was working. Ritchey was ill and should recuperate before coming back, Hale said, adding ominously that they could get along without him very well.[68]

At the end of the year Hale was still under his doctor's orders not to go to the office for more than three hours a day and not to try any serious writing for at least a year. The physician, J. H. McBride, claimed Hale was suffering from "brain exhaustion," and he believed it. Scientific discussion made his head ache.[69] When Ritchey came back from his lemon ranch to work full-time beginning January 1, it was obvious to everyone on the staff that he was in Hale's bad graces and did not get along at all well with Adams.[70] He was put to work on the 100-inch mirror but given no time to spend on the experimental 27-inch Ritchey-Chrétien reflector. Hale and Adams had decided it was not worth trying. Ritchey continued to suffer from insomnia and went back and forth between work and his bed. He was desperate for a chance to make a Ritchey-Chrétien telescope, writing Chrétien: "It will be necessary to demonstrate the entire proposition, with mirrors, curved plates and actual photographs, before we can expect astronomers to be convinced of the enormous gain with the new curves." That was to be the program that he strove to carry out for the rest of his working life, against great handicaps. Ritchey still hoped to substitute a built-up 100-inch disk for the solid "very bad" one on which he was working. He constantly emphasized its "bad internal structure." He desperately wanted to get out of Mount Wilson and into an observatory where he could follow up his own ideas, but there was no place else to go.[71] Ritchey told Sampson of the "new-curves" reflector system, and the Edinburgh professor quickly grasped the idea and worked out the mathematical details of it himself. He discovered one of its problems, namely that the focal surface is not a plane, but curved, and the photographic plates must be correspondingly curved rather than flat. Sampson thought this would be an insurmountable problem and designed a system of several lenses to flatten the field. No example of it was made at the time, though its principles are those used in field flattening lenses today.[72] Chrétien analyzed this problem correctly, stating that there was no difficulty in principle in bending the photographic plates and that they would be just as measurable as ones that had been exposed flat.[73]

By the summer of 1912 Ritchey's health had returned. He was working hard in the observatory optical shop and had several mirrors to make in his shop at home. The largest was a 30-inch for the Helwan Observatory in Egypt. His friend Thomson sent him $250 to support his experimental work and helped him apply to the American Academy of Arts and Sciences for a grant of additional funds. He asked for $2,500, but this was far more than anyone got. In December Ritchey received word from Charles R. Cross, the chairman of its Rumford Committee, that he had been awarded $500. He was ecstatic and decided to put the assistant he had hired to work on a 20-inch Ritchey-Chrétien telescope. Ritchey hoped to have it ready to try on the sky by the following summer.[74]

Hale, meanwhile, was not getting on so well. Soon after coming back to work for but a few hours a day, he had tackled the problem of trying to detect an overall, "general" magnetic field of the sun. This research soon became a vast project. If present, the field was certainly very small, and huge quantities of data would be necessary to prove its existence. Ferdinand Ellerman and other assistants took numerous high-dispersion spectrograms with the new solar tower telescope on Mount Wilson. Jeanette Lasby and, after he arrived, Kapteyn's protégé Adriaan van Maanen and still other assistants measured them, Seares supervised a battery of computers who reduced and averaged the measurements, and Adams himself made additional check measurements. The results ended up on Hale's desk for discussion, but they were not clear-cut. Sometimes he thought they proved the existence of a general magnetic field, sometimes he doubted it. He agonized over whether or not to publish. He did not want to lose priority for discovering the field if there was one; neither did he want to announce he had found it if there was none. With the pressure his headaches and haunted feeling returned. Eventually, Hale published a long paper stating that the measurements were indefinite but "probably represent[ed] the Zeeman effect due to the sun's general magnetic field." Today we know there is not a field of the pattern Hale and his assistants were looking for, but rather a more complicated arrangement that usually consists of several large bipolar areas. Nevertheless, his was a noble attempt. The tense uncertainty hurt his self-confidence and his mental equilibrium.[75]

Furthermore, problems arose with the 100-inch mirror. When Ritchey and his assistants put it up on its edge for testing, it showed astigmatism, meaning the mirror was not perfectly symmetric and the star images would be poor. Ritchey thought the astigmatism resulted from what he perceived as a disk weakened by the bubbles within it. Hale, and especially Adams, could not accept this interpretation. They devised tests with additional weights to increase the distorting force, which seemed to disprove this explanation. Usually, the tests were made quickly. On one occasion when the mirror was left up on edge for several hours, the astigmatism gradually disappeared. Adams interpreted this as a thermal effect and directed Ritchey to use more fans to stir the air in the shop and thus bring the mirror to equilibrium more quickly. Adams was convinced that he had thus solved all the problems. Hale was not so sure. He directed Seares, who was going to a meeting in Paris, to visit St. Gobain and investigate the prospects for finally obtaining a new disk. They were poor. The French had lost so much money already, $30,000 Delloye claimed, that they were unwilling to make a pot and furnace large enough to pour the 100-inch disk in one operation unless they were paid $10,000 for it, in advance, with no guarantee of success. Hale considered the alternative of trying to fuse two thin disks together, but it was not in the least attractive.[76]

Then, sometime in November or December someone cracked the problem of the astigmatism of the disk. Hale gave himself credit for the idea, but Ritchey disputed this and claimed it as his own. They discovered that in the vertical position the mirror distorted under its own weight because it was not properly supported but that when the edge support was applied precisely in the plane containing the center of weight of the mirror, the distortion disappeared. Thus it would be possible to use the disk, as long as it could be figured and tested in the shop and also mounted in the telescope with the proper support system. Hale declared the problem solved and paid the St. Gobain company $5,000 for the disk, thus implying it might someday be replaced by a "perfect" disk. This remote hope was never considered seriously, however, either in Pasadena or in Paris.[77]

In November 1912 Hale, no doubt at Adams's urging, put Pease, Ritchey's former assistant, in direct charge of the design of the 100-inch telescope.[78] The director was now very negative about Ritchey, whom he blamed for the faulty support system, his "hostile" attitude toward suggestions or criticisms while he was still responsible for the 100-inch design, and his now discredited idea of using a roll-off house and windscreen to protect the telescope instead of a dome. When Herbert H. Turner proposed nominating Ritchey for a prestigious Royal Astronomical Society medal, Hale counseled delay until after the optician had published a book and then recommended Adams or Charles Greeley Abbot, who he thought deserved the medal much more than his own staff member. Hale confided to Turner that Ritchey suffered from epilepsy, which the director evidently considered a form of insanity. He raked up all his charges against "Ritchey's conduct," going back "to his attempts to turn Mr. Hooker against us," and said that Adams had wanted to discharge him summarily but that he himself, as director, had decided to keep him to work on the 100-inch mirror, though he now believed that Pease would have done better.[79] Ritchey did not get the medal.

These struggles, scientific and personal, took their toll. Less than a month after Hale had denounced Ritchey to Turner, Dr. McBride sent the distraught director back to the Bethel sanatorium. After a few months of treatment there, accompanied by his doctor, he headed for Russia and a meeting of the International Council of Academics in St. Petersburg. Hale was tired before he started the trip, but he and McBride traveled through Russia, the Caucasus ("the ancient home of our race"), Constantinople and Anatolia, Greece and Italy. Hale had been worrying about an operation he thought he needed for hemorrhoids, but the excitement and exercise of travel made a new man of him and drove them away. He had arranged to give a talk on his work on the general magnetic field of the sun to the Royal Astronomical Society in London in June, but McBride "put his foot down . . . and said [he] must cancel it." In July Hale returned to the United States for a vacation with his wife and children at the old family home in Madison, Connecticut, before returning to Pasadena, where he vowed to remain in seclusion.[80]

Adams, who had been appointed assistant director on January 1, 1913, did a good job of running the observatory and shielding Hale from its problems. He had established good personal relations with Walter M. Gilbert, Woodward's assistant at the Carnegie Institution offices in Washington. To him Adams reported that the foundations for the 100-inch dome had been poured on Mount Wilson and that Ritchey was making progress on the mirror, though slowly.[81] For his own part, the optician was well aware of Hale's "mental trouble." He thought that the director was "insane, on certain subjects at least, such as his desire and determination to take for himself the credit for the work of other men." Ritchey claimed that he, not Hale, had solved the problem of the 100-inch disk by providing the proper edge support. (Very probably, the solution actually emerged slowly from a protracted series of arguments and discussions among Adams, Hale, and Ritchey, and all three of them deserve partial credit.) Ritchey reported that "two men, whom it will be almost impossible to replace, have been driven out by [Hale] during the past year—men with unique and special experience—an irreparable loss to the Observatory and to the Institution." (One of them was Drew, the other Edward A. Fath, who had proved observationally that spiral "nebulae" are composed of stars but who did not get along with Adams and left suddenly for a job at Beloit College just after van Maanen was hired.)

Nevertheless, Ritchey's own work was going well. He bubbled with ideas such as the book of enlargements of his best astronomical photographs, which he hoped would stimulate popular interest "in celestial photography and in the reflecting telescope as a photographic instrument." He had two assistants in his shop at home who were working on several telescope orders. He received additional contributions from Thomson and one of $250 from A. C. Dunham, a friend of Thomson's, to support his continued experimental work on the 20-inch Ritchey-Chrétien telescope. It was an f/6.25 telescope with an f/2.5 primary, faster than any mirror he had previously made. With this system he worked out in detail the methods of figuring and testing Ritchey-Chrétien optics that are still used today.[82]

Ritchey's largest order was the 30-inch conventional paraboloid mirror that he made for Helwan Observatory for $2,500. At Hale's suggestion Turner, who had arranged for the optician to take the job, asked Adams to test the completed mirror. He was a little worried that the personality clash between the two men would cause problems. But he need not have feared. Ritchey had made an excellent mirror. Adams's first test showed a slight deviation from the correct form near the center of the mirror. Ritchey examined the results carefully and accepted them; he then did some further figuring, and Adams's second test showed the mirror was "essentially perfect" with a "wonderfully good" figure as he reported. They *could* agree, when the question was a quantitative, technical one.[83]

Ritchey, ignored and shunned at Mount Wilson, appreciated the support, both moral and financial, he was receiving from Thomson and Dunham, as well as the great interest that Sampson and Samuel W. Stratton, the director of the National Bureau of Standards, expressed in his plans. Ritchey's astronomical photographs were creating awareness of his work in Washington.[84]

In his own observatory, however, his director was working very hard against him to keep word of his accomplishments from spreading. In the fall of 1913 Zoeth S. Eldredge, a San Francisco writer who was preparing a multivolume *History of California*, asked Campbell, the director of Lick Observatory, to contribute a chapter on the history of astronomy in the Golden State. He accepted and asked Hale to send him material he could use as a source for the part about Mount Wilson. Campbell and Hale by this time were the undisputed leaders of American astronomy, with the older Edward C. Pickering of Harvard. Campbell, then fifty-one, was six years older than Hale. He had started out as a very poor farm boy in Ohio. With an excellent mathematical mind and only an undergraduate education at the University of Michigan, he had fought his way up to become the outstanding observational spectroscopist of his generation. When Hale had founded Mount Wilson Solar Observatory in 1905, it ended Lick Observatory's monopoly of California astronomy, but Campbell had led his staff in welcoming their new neighbors to the state. With *his* staff Hale had reciprocated, pledging to cooperate in every way with the older institution to the north.[85] Campbell and Hale did work closely together for the rest of their lives, consulting very frequently by letter and occasionally in personal conferences. Physically, Campbell was more robust, taller, heavier, and balder than Hale. Mentally, he was more even in his disposition, more open in his actions, and more conservative in his ideas. Each respected the other, but Campbell considered the Mount Wilson director to be a little flighty, while Hale thought of him as a bit old-fashioned.

Hale replied to Campbell's request for information with a copy of his most recent annual report and a statement emphasizing that the Mount Wilson research program concentrated on the "study of the physical problems of stellar evolution, giving special attention to the sun, as a typical star, and to laboratory experiments for the interpretation of solar and stellar phenomena." Campbell used this material in writing the chapter. Not surprisingly, Lick Observatory turned out to be its major subject, but Mount Wilson also played a prominent role. Campbell sent a copy of the manuscript to Hale for his comments; the latter made a very few minor corrections and returned it, referring to it as "your admirable history of astronomy in California." Most of Hale's letter was devoted to a description of "the return of my old trouble." He felt he needed rest and intended to spend the coming summer, when Campbell would be leading a Lick Observatory eclipse expedition to Russia, on another family vacation in Madison, Connecticut. He described it as "a most tranquil old town, with many pleasant associations of my boyhood, as I spent some sixteen summers there at my grandmother's home." He planned on "some quiet writing," and looked forward "to building [himself] up by sea-bathing and golf."[86]

Hale evidently had not read the chapter attentively. Campbell was greatly impressed with Ritchey, as he had been since meeting him at John A. Brashear's shop in 1902. Campbell had carefully inspected Ritchey's 60-inch reflector at the international meeting on Mount Wilson in 1910 and had been one of the first to learn of the "new-curves" concept from him. The Lick director had been "much interested" in his experiments in making the 27-inch optics, and when Ritchey opened his own business, Campbell had extended his "best wishes in the undertaking. May you have many opportunities to supply astronomers with high-class mirrors."[87]

Thus Campbell had a very high opinion of Ritchey's work, and the chapter he wrote for Eldredge's history showed it. As the three "unique features" of Mount Wilson Observatory he listed the physical laboratory, the optical shop headed by Ritchey, and the arrangements under which temporary visitors such as Kapteyn could be hired to come and work in Pasadena and pass on their expertise. Campbell wrote

> The excellence of Mr. Ritchey's mirrors has contributed powerfully to the success of the Solar Observatory. There is no reason to doubt that the 100-inch reflecting telescope now under construction will equal in excellence the 60-inch reflector and give corresponding extension of power.

Then in three pages Campbell set down "a few [actually sixteen] of the most important results" of Mount Wilson Observatory. The first three all had to do with Ritchey, the excellence of the 60-inch reflector he had built, and the photographs of nebulae and clusters he had taken with it. The next three were concerned with astronomical spectroscopic measurements of the radial velocities of stars, the rotational velocity of the sun, and the spectrum of the Milky Way (by Fath). Only after these came ten of the astrophysical results that were dear to Hale's heart, the last being laboratory measurements. Except in the list of staff members, the only Mount Wilson astronomers mentioned by name were Hale and Ritchey.[88]

When the proofs of his chapter arrived, Campbell sent a copy of them to the Mount Wilson director for his corrections. This time Hale sat down and actually read the chapter carefully. He was horrified. It was not at all the picture of his observatory that he wanted to convey. He immediately sent a telegram, followed by a long, confidential letter, to the Lick director. In it Hale denounced Ritchey as "neither a scientific investigator, in the proper sense of the word, nor a man who can be depended upon to observe his due obligations toward the best interests of the Observatory as a whole. His point of view is wholly personal, and personal advantage is his prime consideration." In plain terms Hale considered Ritchey disloyal to him and hence to the institution. Hale wrote Campbell that the statements in his chapter were "deserved, but I think an outsider might gather a false impression from it." It did "a real injustice (though, of course, quite unintentionally) to Adams, Seares, King, St. John, Babcock, and others, who are true investigators and have contributed in the highest degree to our work of research." He asked Campbell to rewrite the history of Mount Wilson Observatory, playing Ritchey down and the others, especially Adams, up.[89]

Whatever he thought of Hale's request, Campbell was loyal to the director's code that whatever the head man says about his observatory goes. He counted on Hale's support in matters that affected Lick; in return, he expected to uphold the Mount Wilson director at his institution. Campbell "corrected" his history. He revised his chapter heavily, moving Ritchey's name down in the list of staff members and his results below the work of Adams and Hale in the list of accomplishments. He deleted the optician's name in the paragraphs describing the "excellence" of the optics and the "success" of Mount Wilson Solar Observatory.[90] Hale had succeeded in the beginning of his campaign to make Ritchey an "un-person."[91]

Mount Wilson Failure

1914 - 1919

IN THE SUMMER OF 1914 George Willis Ritchey was making good progress in his two main projects. At his home he had nearly finished the optics for the 20-inch Ritchey-Chrétien reflector on which he had been working. His assistant, C. A. Schrock, had completed the drawings for a temporary mounting for it, following Ritchey's design. To keep the costs down, he had specified the skeleton tube and some parts of the polar axis frame to be made of hardwood, rather than metal, and he planned to use a phonograph motor for the drive clock. With this instrument Ritchey hoped to take wide-field photographs of star fields with exposure times up to half an hour to demonstrate the system. He estimated the cost of the mounting as $600, far more than he could afford himself. He tried to raise it from Elihu Thomson's friend A. C. Dunham.

At the Mount Wilson Observatory optical shop Ritchey had brought the 100-inch mirror to essentially perfect spherical shape and was ready to begin the final stage, parabolizing it, which he estimated would take about a year.[1] He did not know that his director, George Ellery Hale, had been successful in keeping his name and accomplishments from receiving in the *History of Astronomy in California* the praise that its author, W. W. Campbell, had thought they deserved.

Hale himself felt that "my head has been so well that I worked at the office all day, with no bad effects" for two weeks. He completed a small, popular book, *Ten Years Work of a Mountain Observatory*, which he hoped would convert the Carnegie Institution trustees to a more open-handed policy of support for Mount Wilson. He hoped to get back to doing "practically nothing but research for some time," a wish he often expressed but something he had never actually accomplished. Hale even had a new plan for detecting the solar corona without an eclipse. His idea was to send up an automatic camera in a balloon to such a high altitude (he said 20 miles) that the atmosphere would be very thin and the sky essentially black, with no scattered sunlight.[2] Apparently, he never carried out his plan. In August Hale came down with appendicitis and had to undergo surgery. In the course of the operation his gall bladder was found to be infected and was removed along with his appendix. He recovered rapidly, and Walter S. Adams reported that with these two sources of infection removed "we all of us have great hopes that the operation may result in a permanent cure of his brain trouble."[3] Unfortunately, it was not to be that easy.

In 1915 Hale allowed Ritchey to go back to observing with the 60-inch reflector. He plunged into a program of taking long-exposure photographs of many large, spectacular, spiral "nebulae," including M31 (Andromeda) and M101 (Whirlpool). Francis G. Pease, Ritchey's former assistant who had been put in charge of the design of the 100-inch reflector, was also photographing spirals with the 60-inch. They carried out two essentially independent programs on the same objects with the same telescope. Ritchey remained in charge of the optical shop.[4]

Ritchey by now was considered a famous person, at least by the astronomy buffs who visited Mount Wilson. One of them proudly wrote to *Popular Astronomy*, the leading amateur magazine of the period, that he had glimpsed Ritchey "who is known throughout the world wherever astronomy is known." Another, visiting in Pasadena, "had the rare pleasure of enjoying a forty minutes talk with Professor G. W. Ritchey," who he thought was in charge of the observatory. "It seemed almost inexcusable to claim so much of his valuable attention, but one thing led to another" and before he knew it Ritchey, "perhaps the most expert of celestial photographers," was showing him his photographic plates of nebulae.[5]

Hale, however, remained negative on Ritchey. In 1915 the Mount

Wilson director asked Campbell to join him in nominating Ritchey, Adams, and Frederick H. Seares for membership in the prestigious American Philosophical Society. Election to it would be a feather in the cap of any scientist. In his letter to Campbell, Hale stated that all three were entitled to election, but that Adams and Seares should be preferred if only two of his observatory's staff members could get in. No doubt he was franker in writing other members of the Society who were not so committed to Ritchey as Campbell was. The Lick director joined in recommending all three. Adams and Seares both were ultimately elected to the Philosophical Society, with Hale's active support, and Ritchey was not.[6]

As he got back into actually doing research, Hale's head (which he frequently described to his close friend Harry M. Goodwin as if it were a separate organism) began troubling him again. When he worked on the mailing list for the new *Proceedings of the National Academy of Sciences*, which he was instrumental in starting, he felt fine, but when he went back to analyzing his assistant's experiments on vortex rings in water, intended as an analogy to solar phenomena, he felt tired and dispirited. Hale was beginning to suffer a financial pinch; he and his brother had a large part of their fortune invested in a street railway company in Toledo, which municipal reformers were attacking. The California scientist remained extremely interested in developments at the Massachusetts Institute of Technology, his alma mater. His friends Goodwin and Arthur A. Noyes were on its faculty, and he advised them as well as President Richard C. Maclaurin frequently, especially on matters connected with spectroscopy, physics, and astronomy. Throughout 1915 Hale remained morose and unable to do much science. In common with nearly all Eastern establishment scientists, by the end of that year he had developed strong anti-German, pro-British views on World War I, which had been in progress for a year and a half. He believed that the United States should have been in the war on the Allied side long before then. By February of the following year he included American pacifists, whom he scornfully referred to as "church members," among the enemy. He thought "those idiots," Henry Ford and William Jennings Bryan, had done the most harm (with their Peace Ship to Europe). "Both of them ought to be imprisoned as traitors or thoroughly chloroformed," he fumed.[7]

At the end of 1915, Robert G. Aitken notified Hale that he was to be awarded the Bruce Medal of the Astronomical Society of the Pacific and asked for some information on his career. Hale's reply was a modest but excellent scientific autobiography. He emphasized his lack of training beyond his MIT undergraduate education and said that he had accomplished what he had because of his "intense enthusiasm for research, especially spectroscopy" and his drive. He stated his indebtedness to his father but only implicitly mentioned how many large, expensive instruments he had bought for him. Hale was generous with his praise for Ferdinand Ellerman, "my devoted colleague since Kenwood in 1914," Adams, Seares, Harold D. Babcock, and even Adriaan van Maanen, his assistant for the solar magnetic field measurements, but did not even mention Ritchey, who had built two outstanding reflecting telescopes for him and had nearly completed the mirror of the third. He was working very hard on it and was so busy and so short-handed in his own shop that he could not take on any outside jobs, even for small mirrors.[8]

In mid-April 1916 Adams wrote Campbell that Ritchey had "essentially finished" the 100-inch mirror. By then the tests showed "almost a perfect figure," and only a little work remained to be done with hand tools. A set of optical shop notebooks on the 100-inch disk, full of cryptic notations on the tools, cuts, and tests used, has been preserved at the Mount Wilson offices in Pasadena. It documents that work on the disk had begun on November 10, 1910, and that the mirror was completed on April 26, 1916. President Robert S. Woodward of the Carnegie Institution sent his congratulations to Ritchey, who thanked him with a detailed technical description of how he and his assistants, W. L. Kinney and James Dalton, had done the work. Ritchey wrote Woodward that they both deserved "great credit . . . for their share in all of this work."[9]

In July Ritchey and his assistants silvered the 100-inch mirror for the first time. They got a good, highly reflective, strong coat onto the glass. They had begun working on the two secondary mirrors for the Cassegrain and coudé arrangements, and by October had them polished to spherical form, ready to be converted to the final hyperboloids. The steel pieces for the mounting had been made at the Fore River Works in Massachusetts. They were brought to Los Angeles by boat and were taken up the mountain and stored in the dome, now completed and painted, before the winter rains began. Ritchey had been very busy with the optical work on the 100-inch but, nevertheless, found time to carry out a full observing schedule on the 60-inch. He was concentrating on spiral "nebulae," "especially . . . some of the *smaller* spirals, *double* spiral nebulae, *triple* and *multiple* nebulae, [and] spiral nebulae seen *on edge*." He hoped to have the 100-inch mirror installed in the telescope and to be using it by the following summer.[10]

Andrew E. Douglass, at the University of Arizona, tried to persuade Ritchey to make a 36-inch reflecting telescope at a price of $60,000, but the optician was far too occupied at Mount Wilson to consider taking the job. Douglass then gave the order to Warner and Swasey but got copious advice from Ritchey, much of it very negative, about this company and other opticians.[11]

With the 100-inch reflector apparently approaching completion, Hale and Adams began looking for additional staff members to keep the two telescopes busy. In October Adams met Edwin Hubble, a graduate student at Yerkes Observatory working on his thesis on nebulae with the 24-inch reflector. Adams was impressed with him, and on November 1, 1916, Hale offered Hubble a job on the Mount Wilson staff, to be taken after he had completed his Ph.D.[12]

Hale himself had left Pasadena in June 1916. By then it was apparent to the leaders of the government, to the Eastern industrialists and financiers, and to President Woodrow Wilson that before long the United States would be in the war against Germany. Hale heartily approved. He went to Washington to take the lead, with Robert A. Millikan, in organizing America's scientists for the coming struggle. The National Academy of Sciences had been formed for just this purpose during the Civil War, but by 1916 it was too ossified for action. Hale was instrumental in setting up the National Research Council as the Academy's operational arm. He collected an activist, interventionist group of engineers and scientists, many of them from America's largest industrial laboratories, and forged them into an organization for winning the war. By August Hale was on his way to England to learn how British scientists had been mobilized. He and his friends improved greatly on the English experience. After his return from London, Hale spent most of his time in Washington until after the end of the war, with occasional forays back to California. He thrived on it. There are extant literally a hundred letters from Hale to his wife in the years 1916 and 1917, written when he was engaged almost fully in the NRC preparedness and then wartime activities. In not one of them does he mention his head, depression, or anxiety as he had in so many letters when he was trying to do research and run an observatory. Often he says he is "tired," but no more. Hale always maintained that research was what he really wanted to do, if only he could find time for it, and no doubt he believed it. But his symptoms belied his words. What he really thrived on was his "schemes," organizing new groups and setting up new programs. What his illnesses always prevented him from doing was the dogged, repetitive work and the daily discussion of results with other scientists on an equal basis that make up so much of actual research.[13]

On January 1, 1917, this was still all in the future, when Ritchey was still hoping to use the 100-inch in the coming summer. He was making progress on figuring the two secondary mirrors for the 100-inch and was planning a program of measuring his photographs of spiral nebulae to see if he could detect rotation over the time interval of six or seven years for which he had pairs of plates. He prepared a set of his best moon pictures for Thomas Jaggar, director of the Hawaiian Volcano Observatory, who, like Campbell and most scientists of the day, held to the volcanic theory of the origin of lunar craters. Ritchey was at peace with Hale, perhaps partly because his salary had been raised to $3,300 in a general 10 percent upward "adjustment" for all Carnegie Institution employees.[14] However, his good spirits were soon turned to sorrow, for both his mother and father developed bronchial pneumonia, and on April 19 James Ritchey died at the age of eighty-three. He had taught his son his skills as a craftsman and at the end of his life had served as his assistant. Ritchey mourned him.[15]

Earlier that month, America had entered the war on the side of the Allies. Ritchey wanted to put the Mount Wilson optical shop to work on the war effort by making lenses and prisms for binoculars, range finders, and periscopes. Hale had moved to Washington for the duration a month before the declaration of war, leaving Adams as acting director of the observatory. He was intensely pro-British and, at the age of forty, still eligible for a commission in a technical branch of the army, such as Chemical Warfare or the Signal Corps, in which other astronomers were becoming officers. However, Adams regarded it as his duty to free Hale by administering the observatory but no doubt was uncomfortable with this decision. He considered Benjamin I. Wheeler, president of the University of California, a "dangerous pacifist," and Armin O. Leuschner, Berkeley astronomy professor, who had been "an ardent pro-German up to the time this country declared war," a person not to be trusted.

Within the Mount Wilson instrument shop Adams sniffed out Ernest Keil, who was not only German but "becoming even more rabid these days." He recommended firing Keil, and Hale sadly concurred, knowing it would be difficult for him to get another job. "Except for his pro-German tendencies he could be useful in some shop, as he is certainly a very good instrument maker." Adams proposed to arm the Pasadena staff to guard the instrument and optical shops, presumably against German spies and saboteurs. However, when he learned that even Abercrombie and Fitch, the big New York sporting goods store, was completely out of rifles and could provide nothing more modern than 1894 Winchester carbines, he gave up his plan.[16]

By May, Ritchey was rapidly turning out small mirrors for optical range finders, which physicist A. A. Michelson had designed for the army. Hale personally had obtained this job. Ritchey had finished the two secondary mirrors for the 100-inch, so it was now possible to send the primary, which had been needed to test them, up the mountain. Adams himself thought the telescope would be ready to use by August. In July the 100-inch mirror was taken up to Mount Wilson and stored safely in the dome, where the mounting was being erected. Some of the necessary parts remained to be finished, but the prospects for using the telescope by fall were good.[17]

Ritchey was still observing with the 60-inch, and that summer he made an outstanding discovery. On a long-exposure photograph of the spiral "nebula" NGC 6946, which he took on July 19, 1917, he found a 14th magnitude star that had not been there on any of his previous photographs. He checked it carefully and was sure of the result. It could only be a nova or "new star" that had flared up in NGC 6946, as other, much brighter novae had flared up on our own Milky Way galaxy. To check, Ritchey, with Pease's help, obtained a low-dispersion spectrum of the new star. It showed a continuous spectrum with a few broad emission bands, just like the spectra of novae in the Milky Way and unlike the spectra of any other types of stars. This spectrum clearly proved that the new star in NGC 6946 *was* a nova, similar to those in the Milky Way. It meant that there are *stars* in NGC 6946, and that it is much more distant than the novae *within* our Galaxy. Ritchey had hard proof that this "nebula" is actually a remote star system, an "island universe," or, in our modern terms, a galaxy.[18]

He quickly examined his entire collection of photographs of other spirals. Finding a single new stellar image in these dense star fields is hard, demanding work, but Ritchey persevered. Soon he had located another nova in M81, two in M101, one in NGC 2403, and two in M31, the Andromeda nebula. This last, Ritchey, like many students of galaxies since him, correctly regarded as "probably typical of many or all spiral nebulae."[19] His discovery was a very important one.[20] Within a few months Ritchey had found another nova in M31 and before the end of that winter three more.[21]

Yet Ritchey himself did not immediately grasp the significance of his discovery. Heber D. Curtis of Lick Observatory did. He had been studying nebulae photographically for years with its Crossley 36-inch reflector, puzzling out their nature. He recognized immediately the implication of Ritchey's discovery, that it proved that the spirals are remote star systems, far outside our Milky Way, as Edward A. Fath had earlier deduced from their spectra. Curtis, though he was teaching in a wartime navigation school at San Diego, quickly checked his own photographic plates and found several more novae in the spirals NGC 4527 and M100. He speedily announced his conclusions in two excellent papers written very soon after he received the telegraphed announcement of Ritchey's first discovery and published even before it.[22] Harlow Shapley, who had recently joined the Mount Wilson staff, privately criticized Ritchey for not finding the novae earlier. He was a chronic complainer, and his own conclusion from the novae, that the spirals are objects within our Galaxy, was quite wrong.[23] Ritchey himself accepted Curtis's reasoning and agreed that "spiral nebulae are other stellar systems like our own."[24]

That fall Ritchey also returned to photographing Nova Persei, which he had studied with the 24-inch reflector at Yerkes Observatory in 1901. The faint nebulosity that his plates had then revealed had faded, but in December 1916, Edward E. Barnard had glimpsed another small faint nebula near the now quite faint central star. He asked the Mount Wilson observers to get a photograph of it if they could. Bad weather prevented an attempt in January or February 1917, but in September Pease obtained a photograph with the 60-inch that showed a small, fan-shaped nebulosity adjacent to the star. In October Ritchey got a much better picture, which showed a small, nearly circular ring centered on the nova. The fan was a bright condensation in this ring. The next month Ritchey obtained two even better photographs of the ring. He clearly enjoyed proving that he could outperform his former assistant at the telescope.[25] Hale confirmed that Ritchey's photograph showed the ring, or shell, about the nova most clearly. Barnard called it an "extraordinary photograph" and incorrectly thought that it disproved the light-flash theory of the earlier nebulosity. In fact, it did not; the ring was actually the shell thrown off by the nova when it exploded, expanding into space at high speed but not nearly as high as the velocity of light, and thus became visible only long after the light flash. Adams himself gave this interpretation at the time but did not publish it. Barnard was also "extremely interested" in Ritchey's discovery of so many novae in M31 and wrote that he was "beginning to believe that the spirals really are outside universes."[26]

Perhaps because the 100-inch mirror was going smoothly, perhaps because of Ritchey's burst of scientific activity, Hale became somewhat more sympathetic to him but never enough to permit him to receive outside recognition. In the years toward the end of World War I, Hale and Campbell were the dominant figures on the committee that awarded the Henry Draper Medal of the National Academy of Sciences. This was (and still is) the most prestigious prize in astrophysics awarded in the United States, intended to go to a scientist in any country who had made outstanding contributions in that field. Both Campbell and Hale had received it years before. In 1916, when they were considering candidates for the next award of the Draper medal, Hale wrote that several of the recent medalists had been Americans, that the most recent, Henri Deslandres, had been French, and that by tradition the next should preferably be a German or Austrian. The war should not affect their judgment, he said, but nevertheless he thought that instead it would be best to give the medal to an American. He suggested Ritchey, for "his optical work and his photographs of nebulae and other objects." As other possibilities he mentioned the French physicist Charles Fabry and Americans Theodore Lyman and V. M. Slipher. Campbell, who was even more anti-German and pro-Allies than Hale, as the latter knew well, named Fabry as his first choice and Michelson as second but stated that he also supported Ritchey. Campbell regretted, he said, that Alvan Clark had never gotten the Draper medal before he had died, because American refracting telescopes were the best in the world, and he hoped that Ritchey would get it someday, as American reflectors were the best in the world, too. However, he thought Ritchey and the others mentioned could wait.

Hale immediately seized on Michelson as a compromise choice and recommended him to the other committee members. They agreed, and the great University of Chicago physicist received the Draper medal in 1917.[27]

The following fall Campbell, then chairman of the Draper committee, polled the members again. Ritchey received two first choice votes (from Campbell and Hale), while Adams and Fabry each received one first and one second-choice vote. Campbell proposed that if the other members agreed, they give the medal to Ritchey in 1918, decide very soon afterward to award it to Fabry in 1919, and consider Adams for 1920. The Lick director was very strong for Ritchey for his accomplishments on the 60-inch and 100-inch mirrors, favored Fabry as a representative "of French men of science in the present trying times," and thought Adams had great promise for the future. Joseph S. Ames, one of the committee members, switched his first choice vote from Fabry to Ritchey to give him a clear-cut majority. Hale, however, belatedly realizing that Ritchey might actually get the medal, switched his vote to Adams, putting the two Mount Wilson staff members in a tie. He accompanied his change in vote with an impassioned letter to Campbell, giving a host of reasons why Adams and not Ritchey should get the medal first. Hale even stated that Adams might join the army and be killed in France, an argument calculated to appeal to Campbell, whose oldest son Douglas was a pilot in the nascent American air corps. Hale insisted that he had not changed his thinking. Ritchey deserved the medal, too, and he hoped he would get it in the near future—after Adams. Campbell was a firm believer in the principle of the director as boss. Hale wanted one of his staff members, Adams, to get the medal and not the other, Ritchey. Campbell followed Hale's lead and switched his vote from Ritchey to Adams. The Mount Wilson assistant director received the Draper medal for 1918.[28]

As part of this settlement, Campbell stated that he thought very highly of Ritchey's work but that he would be even stronger for Fabry to get the medal the next year. Recognition of France as one of the Allies and the idea that the award should not go to the same institution twice in a row no doubt both came into his thinking, as Hale had realized that they would. Thus in the fall of 1918 all five of the committee members voted for Fabry as their first choice, and the four of them who gave a second choice all named Ritchey. Amid great fanfare, Fabry personally received the Draper medal at the National Academy meeting in Washington in April 1919, soon after the armistice.[29]

On the face of it, Ritchey appeared to be an absolute certainty to be awarded the medal the following year. However, as we shall see, by October 1919, when the Draper committee began their deliberations, it was even more certain that Ritchey was only days away from being fired from the Mount Wilson Observatory staff. Campbell, true to the director's code to the end, withdrew from the "informal understanding" to give Ritchey the medal and advised Charles Greeley Abbot, Hale's close friend and supporter, to do the same. Not a single member of the committee voted for Ritchey. They all voted for English astrophysicist Alfred Fowler, and he received the Draper medal in 1920. By then Ritchey was in permanent disfavor in American astronomy, and he was never again considered for this medal, or for any other award in his native land.[30]

All of this was far in the future when Hale returned briefly to Pasadena in the late summer of 1917. In Washington he had put astronomy out of his mind and worked entirely on wartime "schemes." Even in California he was occupied largely with NRC administration, but was able to break away for the first test of the 100-inch telescope. The mirror was in the tube, and although the Cassegrain "cages" (secondary mirror support systems) were not finished and no instruments were ready, it was possible to observe visually at the Newtonian focus. Reports are fragmentary, but evidently the telescope worked perfectly and the mirror gave fine images once the telescope and dome had cooled down to the outside air temperature.[31]

Ritchey was growing increasingly concerned about the war. His son Willis was twenty-seven years old and either already had been drafted or was soon to be. Ritchey could not get him into a college Navy officers' training program that would keep him safe for a year or two, as Hale was to do with his son William at Princeton.[32] Young Englishmen, Frenchmen, and Germans were being relentlessly mowed down in the fields of Belgium and France where the American army was soon to be fully committed. Ritchey threw his talents into designing an immense "*land battleship*," a gigantic tank, which he hoped would "crush . . . Germany from the face of the earth for her atrocious crimes." It was to be sixty feet long, twenty feet wide, and eight feet high. He sent his plan to Hale in Washington, through Adams, and though it was pigeon-holed there, it undoubtedly drew these three middle-aged desk-bound warriors a little closer together.[33]

In March 1918 Hale, in Washington, called for Pease to come east for the duration, to work in the NRC offices designing military optics. It meant taking him off the 100-inch, but Hale and Adams now trusted Ritchey enough to put him in charge of the final alignment, testing, and preparation of the telescope. Pease was close to Adams and reported to him frequently from Washington. He noted that Hale was thriving under the wartime activity and seemed in perfect health.[34]

Ritchey, on his part, got into work for which his talents were far more suited than designing huge tanks. The army needed thousands of optical range finders. They were built around precisely ground prisms of which there was a great shortage. Hale got a contract for the Mount Wilson Observatory optical shop to make prisms for the Frankford Arsenal, where the range finders were assembled. Ritchey was soon tooling up for mass production and had his assistants at work. Adams welcomed the government job because it made the observatory staff members (including himself, no doubt) feel they were taking part in the war effort.[35]

In spring Ritchey broke away for a few days to go to San Francisco to deliver a lecture on April 19 on the 100-inch telescope, under the auspices of the Astronomical Society of the Pacific.[36] Adams was on the mountain in May and tested the 100-inch telescope. He reported the clock drive was much better, and only a slight misalignment of the mirrors remained to be corrected. Ritchey was working well, and reported: "[We] are making good progress and the end is in sight."[37]

In June 1918 Hale came back to California for another visit. He was brimming with plans for high-level scientific "statesmanship" and was beginning to organize an Inter-Allied Research Council. He recognized that the optical shop could handle much more war work and sent Ritchey east for a series of conferences with Ordnance Department procurement officers. Hale even had time to come up with another scheme to detect the solar corona outside of eclipse. This time he visualized using the 150-foot solar tower on Mount Wilson, equipped with two of the new, recently developed photoelectric cells, set up for differential measurements. He never followed through on this idea. In late August, before he left Pasadena, Hale reported to the Carnegie Institution that tests of the

100-inch telescope on the stars were "very promising" and that only a little more work remained to be done.[38]

In Washington, Ritchey made a strong impression on the Ordnance Department officers in the War Department. He clearly knew what he was doing and was backed by the immense prestige of Mount Wilson Observatory (the "Solar" had been dropped from its name on January 1, 1918), the Carnegie Institution, and Hale, the moving spirit of the NRC. Ritchey returned to Pasadena with a huge contract for lenses and prisms to be made in the observatory optical shop. As Hale had specified, the contract was a nonprofit one, and the Carnegie Institution contributed the Mount Wilson optical shop space and its employees' time free of charge. But the government agreed to pay for additional tools, machines, space, and salaries of additional opticians as needed. Ritchey immediately began ordering equipment, hiring inexperienced workers, and training them in mass-production precision optical techniques.[39]

Hale was still in Pasadena where he authorized the purchase of a large lot next to the observatory offices and had a two-story brick building for the Ordnance Department work erected within two weeks. (It was called the Optical Building at the time but was known for many years afterward as the "Government Building.") Elderly Robert S. Woodward, the conservative Carnegie Institution president, was horrified at the way the dynamic Hale was spending money without written authorization and sent him several long, tedious letters on the beauties of proper accounting and fiscal responsibility. Soon the director was on a train back to exciting wartime Washington, and Adams was left in the uncomfortable position of intermediary between the expansionist Ritchey and the dour Woodward. The optical shop was producing the prisms and lenses the army needed, and the Ordnance Department wanted to spend more money, hire more people, and build more buildings, but Woodward put on the brakes and slowed the proposed expansion.[40]

Willis Ritchey was drafted, but his father could legitimately call for him as a skilled optical worker in a critically needed area. Before long the Ordnance Department had Willis assigned to work in the Mount Wilson Observatory Ordnance Department optical shop rather than to an infantry division in France.[41]

Though he saved his son, Ritchey irrevocably lost his own future at Mount Wilson that last autumn of World War I. Abbot, Hale's close friend and the head of the Smithsonian Astrophysical Observatory, reported that Ritchey had "recently informed him that the 100-inch telescope is likely to prove a total failure, and that a great expenditure would be needed to put it in working order, at the best. It was possible that it might serve for spectroscopic work, but nothing better could be expected from it in view of the serious errors in design, etc, etc." Though this second-hand version (Hale's report to Adams of what Abbot had said that Ritchey had said) was undoubtedly overdrawn, Ritchey very probably *had* criticized the design. He was jealous of Pease, his former assistant who had become Adams's friend and displaced him, and, like all great craftsmen, Ritchey firmly believed that his own designs were the best and all others hence necessarily inferior.

Nevertheless, it was a most injudicious statement to make, particularly to Hale's trusted friend, who could be counted on to relay it to him. If Ritchey had told Abbot, he must have told others. If he was right and the 100-inch was a failure, the director who had raised the money to build it with his glowing promises of future discoveries, and who had overseen its construction, was wrong. This could not be. Hale immediately wrote Adams, notifying him of Ritchey's disloyalty. The absent director advised his lieutenant that Ritchey's statements, if believed, could do considerable harm and that to counter them the 100-inch should be gotten into "perfect working order" as soon as possible and put to use. Hale directed that John A. Anderson, a technical physicist who had joined the Mount Wilson staff in 1916, be placed in full charge of all instrument construction. Pease, Hale decided, had completed his NRC work and could return to Pasadena within a month. He would take over the 100-inch again. Adams and Pease should make all future tests of the alignment. Clement Jacomini, the chief instrument maker, should take charge of the final adjustments of the drive clock. Hale trusted all of them. He never trusted Ritchey again. All the wartime return of confidence had been lost by that one statement. The former superintendent of construction, superintendent of instrument design, and head of the optical shop had no responsibilities except the wartime Ordnance Department project. His days were numbered.[42]

By early October Hale had crossed the Atlantic and was enjoying "interesting discussions" with fellow scientists and dinners at the Athenaeum Club in London for four shillings (one dollar) "which you couldn't do at home." He was puttering around the Royal Astronomical Society library, reading old papers, and preparing a talk on sun-spots, "the most delightful experience I have had for months." This kind of astronomy, he thought, "is really *good* for my head," the first mention of his problem since 1916, and only in the context that it might trouble him sometime in the indefinite future.[43]

Adams, meanwhile, once he received Hale's letters denouncing Ritchey, made the most of his opportunity. He had never respected Ritchey, or in recent years trusted him. Now that he had been unleashed, the thin, ascetic New Englander knew how to get his own back. He had already written to Hale claiming that Ritchey was trying to subvert control of the two opticians who were not hired on the Ordnance Department funds. Hale had passed the letter on to Woodward, who had strongly backed the acting director.[44] Now the Ordnance Department decided to commission Ritchey as an officer to give him more authority. He was eager to become a major. When the army Chief of Ordnance notified Woodward of the intended commission, the Carnegie president routinely approved it. He very soon afterward received a polite but firm letter from Adams, laying out in detail what a "difficult situation" this would create. He emphasized that there could be only one master at Mount Wilson Observatory and that it must be himself as the representative of the Carnegie Institution, not Ritchey as the representative of the army. By this time in early November the war had nearly ended, and Woodward backed Adams completely. Ritchey was offered a commission as a major in the Ordnance Reserve Corps in February 1919, but when he requested the Carnegie president's permission to accept it, Woodward delayed any answer until the opportunity lapsed.[45]

By December 1918, with the war a month over, Adams was firmly in control. He recommended closing out the observatory's responsibility for the army optical work and letting the Ordnance Department continue it on its own, if it wished. Ritchey could be put on leave from his observatory job and paid by the army. Adams emphasized that the observatory's main job was to get the 100-inch into operation. To Hale, still in London, he made this point especially forcefully. The assistant director poured out his grievances against Ritchey. In particular, Adams claimed that during the war the optician had signed his letters as "Commanding Officer, Mount Wilson Observatory," a statement well calculated to arouse Hale's ire. (No record or copy of such a letter exists, and as Adams cited no proof, it is doubtful that there was any. Ritchey signed his name on official army correspondence as "Production Officer, Mount

Wilson Observatory," or "Officer in Charge, Ordnance Department, Mount Wilson Observatory," but no doubt others occasionally incorrectly addressed letters to him by the title Adams reported.) The assistant director hoped that the Ordnance Department would keep the optical plant alive "and that we shall be permanently rid of a man who never belonged in scientific work at all," but it turned out to be harder than that to turn him out. The plant was soon closed, and the observatory bought the government building and all its tools very inexpensively as "salvage."[46]

Ritchey was proud of the work that he had done, organizing the precision optical work so that barely trained employees were able to turn out 2500 prisms in ten weeks. At its peak, in October and November 1918, sixty people were working on the project. Ritchey would have liked to continue it and even to have gotten into making quality optical glass, but it was not to be.[47]

Hale returned to Washington in December 1918 after taking part in International Research Council meetings in London and Paris immediately after the armistice. He reported that his wartime "experience was the most interesting one [he had] ever been through." Now it was winding down, and he expected to be able to get back to California before summer. In February Adams made a definite request to Hale to discharge Ritchey. The construction phase was at its end (the 100-inch was nearly ready for use), "and the value of a man [was] going to depend upon his knowledge of scientific work and his willingness to handle things in a scientific way." Ritchey would never be a true research scientist, just a photographer. He would "prove dishonest in his relations to the Observatory by devoting a large part of his time to the care of his lemon ranch." Worst of all, Ritchey had been disloyal to the observatory by spreading word that the 100-inch was a failure. Adams completely wrote off the optician's war service, saying that he had not been doing observatory work for some months but had instead secured experience in practical optical techniques that would be useful to him outside the observatory. Ritchey could easily get a job with Bausch and Lomb, and the observatory could hire a real scientist with his salary. Adams realized that Hale hated getting involved in the messy business of firing an old employee who had once been his friend, and he offered to do it himself. Hale "recognized the force" of Adams's proposal and promised to take it up with Woodward. The old president respected Ritchey for his accomplishments, which he had heard so much about in earlier days. He also abhorred controversy. He waffled.[48]

Now a new element entered the complicated situation. Yale University received a large bequest, which it could use to build a new observatory. Its first choice to hire as director was Adams. Hale did not want to stand in his way, for it would be an important step up for the younger man, particularly in prestige, but he could not bear to think of Adams leaving. Now that the war was over the accountants were back in control at the War Department and the Carnegie Institution, and Woodward chose just this moment to send Adams a long, tedious lecture about how he *should* have handled the observatory's financial affairs when the Ordnance Department moved in. Nevertheless, after Adams had waited a decent interval to think over the Yale offer, he declined the new post. Mostly, he said, he did so because he wanted so much to remain with Hale at Mount Wilson. But, he reminded the director, he had found "that Ritchey [could not] be trusted at all to care for the interests of the Observatory. I must confess that I think it will be a distinct injury to the interests of the Observatory if Woodward decides we should keep Ritchey." This was the implicit bargain Adams struck. He would stay, but Ritchey would have to go.[49]

Hale was tremendously relieved. He had declined the presidency of the Carnegie Institution (Woodward was close to retirement) and a position as head of the National Research Council in Washington. He was anxious to get back to California and to research, he said, but he was now beginning not to feel so well and needed to get away for a rest at the beach. Adams grimly kept the pressure on to get rid of the Ordnance Department and drive Ritchey out.[50]

In May Hale threw himself into a campaign for raising money to buy a site in Washington for a building for the National Academy of Sciences, which until then had operated from borrowed quarters. The Carnegie Institution would provide $5 million for the building if the Academy would provide the land. Hale helped to choose the location on Constitution Avenue (then known as B Street) near the Lincoln Memorial, and with Millikan he quickly got the money for the land from several of America's top industrialists, including Charles F. Kettering, Henry Ford, and George R. Eastman.[51] Hale was to play a very active role in the design and building of this "temple of science," of which he is generally considered the father.

In June 1919 a delegation of leading astronomers of the United States, headed by Campbell, went to Brussels to take part in the transformation of the International Union for Co-operation in Solar Research into the International Astronomical Union, under the auspices of the International Research Council. Hale did not go, though he was one of the leading spirits behind the scenes, but Adams, Seares, and Charles St. John, all of the Mount Wilson staff, were in the group along with six other Americans, in addition to Campbell. They all sailed together from New York on the *Aquitania* at the end of June and spent most of July together on shipboard, in London, in Brussels, and on a tour of the battlefields of France.[52] Hale had sent Campbell a "confidential document" earlier in June to be burned after reading, which very probably was a denunciation of Ritchey, laying out all his faults from the Mount Wilson director's point of view.[53] Before departure, Adams stopped in Washington where he "discussed [the] Ritchey matter thoroughly with Woodward," and in Princeton to see Henry Norris Russell to talk over the latest Mount Wilson spectroscopic results. Almost certainly, however, he took the opportunity to pass the word that Ritchey's days at the observatory were numbered.[54] On the long voyage over to Europe and back, the astronomers had many hours to while away. They must have discussed many subjects, and one of them was surely Ritchey's future. By the time they returned to their observatories and universities at the end of the summer, the leaders of American astronomy were undoubtedly fully aware of Hale's and Adams's determination to get rid of Ritchey and were ready to support them. Certainly Campbell, in an informal letter to Hale reporting on the meeting, wrote, "On many subjects which I have not touched Dr. Adams will be able to enlighten you."[55]

In June 1919 Hale unveiled the 100-inch telescope to the astronomical community. The Pacific Division of the American Association for the Advancement of Science met at Throop College in Pasadena, and the Astronomical Society of the Pacific met with them. After two sessions for scientific papers and a garden party at the Hales' home on Saturday, June 21, the group toured the observatory shops and laboratories in Pasadena and then went up Mount Wilson. After lunch the astronomers inspected the 100-inch, and Hale gave a lecture in the great dome. Two hundred visitors were present. He talked on the work of the observatory in astrophysics and told that its goal was to study not only the sun but the stars, and that the 60-inch and now the 100-inch were necessary to get high-dispersion spectrograms of the stars, to compare them with the sun. Hale was at his best in this inspiring talk; he had given it hundreds of times, each version updated and improved over the previous one, beginning with the dedication of the Kenwood Astrophysical Observatory twenty-seven years before. In the evening the guests who stayed saw Saturn and other objects through the 100-inch. These visiting astronomers, "several of them men familiar with the best that existing telescopes can do, were filled with admiration for the magnificent definition given by this instrument. In their judgment the definition given by the 100-inch is not only satisfactory, but of the very best." A few nights later the telescope went into regularly scheduled operation.[56] As the Mount Wilson astronomers used the 100-inch to take their data, they could see that the figure of the mirror was excellent and that it gave very fine images. Hale was pleased indeed with the way the telescope had come out, but he never mentioned the name of the optician who had produced the large, beautifully figured mirror. Ritchey was never assigned a single night to use the telescope himself.[57]

By now Ritchey could see the writing on the wall. He believed that the 100-inch mirror and its secondaries, which he had made, were "decidedly more perfect" than the earlier 60-inch but that Adams and Pease had made "atrocious mistakes" in the design and construction of the mounting. Hale he depicted as a weakling, manipulated by Adams, who was determined to pay Ritchey back for his promotion to a position of authority at the observatory during the war. Ritchey could see that he might not be able to continue at Mount Wilson where he was beginning to try to measure proper motions of rotation of spiral nebulae by comparing his earlier and more recent direct photographs, a program he had suggested to young Adriaan van Maanen.

> I do not like to contemplate giving up my astronomical work and retiring to my lemon and orange groves,—a defeated man, my life-work appropriated by Mr. Hale, a millionaire who already has more scientific honors and distinction than he knows what to do with. He has reached a place where scientific work and honors are not enough; he must have vast *power* also; power to dictate the welfare, the making or unmaking, the *positions* even, of scientific men both in the observatory and outside of it,—as far as his influences can possibly reach.[58]

An overdramatic indictment, perhaps, but within six weeks Ritchey was to be retired to his lemon and orange groves without pay.

Soon the first results with the 100-inch were coming in. Pease obtained some very good photographs of globular clusters and the moon at the f/16 Cassegrain focus, with its long focal length and large scale, as Ritchey had planned to do. Other observers, including Shapley, obtained spectra of faint stars that confirmed that the 100-inch, with its fine optical figure, *did* collect much more light than the 60-inch, as it should. Hale spread the word quickly to his friends, the powerful directors and opinion-makers of astronomy in America and England.[59] The Mount Wilson director produced a very quick paper on the comparative tests of the 60-inch and 100-inch, giving these results in quantitative form, and had it published by October 1919.[60] They meant that the 100-inch was a success and that Hale could get along without Ritchey.

The optician realized even more clearly that he was never going to get to use the 100-inch to photograph nebulae, and that Hale, Adams, and Pease had won out. Ritchey could do nothing but continue measuring his photographs of spirals and go back to work on his perennial plan of a book of photographs of nebulae, clusters, and comets. He even dreamt of writing another book or long paper on the work on the 100-inch mirrors, but it was not to be.[61]

Unknown to Ritchey, in August President Woodward consented to Hale's and Adams's demands that Ritchey be fired. On August 19 Hale wrote a formal letter to Woodward recommending that Ritchey be dismissed. The director gave as his reasons Ritchey's "incompatibility with other members of the staff, . . . deliberate dishonesty on several occasions, . . . failure to take his astronomical photographs in accordance with a carefully considered program, . . . failure to keep complete and reliable records of his astronomical photographs, . . . and willingness to modify recorded dates of astronomical photographs." Hale went into detail of what he saw as Ritchey's attempts to prejudice John D. Hooker against the director's ideas for the 100-inch telescope, the optician's "unreasonable attitude regarding the design of the 100-inch telescope, and his unwillingness to accept suggestions from other members of the staff (because of an inordinate desire for personal credit)." Hale and Ritchey each saw the same faults in the other. Hale dismissed Ritchey's measuring the spiral "nebulae" for rotation as "com[ing] rather late" and did not mention his discovery of the novae in spirals in 1917, nor his publication of those results. Ritchey was never given a chance to reply to the charges, as he undoubtedly could have done effectively, or indeed even to see them.

Hale again brought up Ritchey's supposed attacks of epilepsy in 1911 and recommended that because of this illness "and because of the valuable work he has done for the observatory, . . . the advisability of giving him a retiring allowance should be carefully considered, in spite of his seriously objectionable behavior, which would certainly warrant summary dismissal without further pay." However, Hale said, this should only be done *after* Ritchey had turned over to the observatory all his photographs and observing records.[62] A few days later Hale telegraphed Woodward to recommend that the president give as the sole basis of Ritchey's dismissal "personal character and incompatibility with [the] staff."[63] This would mean listing no specific charges to which Ritchey could reply.

Woodward had already made up his mind, and he formally agreed that Ritchey be fired. Hale could do it himself if he wished, or the Carnegie Institution Executive Committee would, if he preferred, retire Ritchey, putting him on leave of absence until the end of the year, and after that "in view of the services he has rendered to the Institution" on half pay for the rest of his life. Woodward believed this action would be simple justice to Ritchey and would also absolve himself and Hale of any blame they might otherwise bear for dismissing such a long-time employee, and also would prevent Ritchey from raising "undue scandal."[64]

Before going ahead, however, Woodward, probably at the suggestion of some of the Carnegie Institution trustees, consulted its legal counsel, Cadwalader, Wickersham, and Taft of 40 Wall Street, New York. The partners included a former attorney general and the brother of a former president of the United States. Woodward described Ritchey as "a man of noteworthy ability [who] has accomplished a number of remarkable feats in optical work, the most remarkable of which are found in the 60-inch and 100-inch mirrors figured for the Observatory" but also as "a cantankerous and generally irreconcilable individual." The Carnegie president asked whether Ritchey would have legal grounds for complaint about being fired.[65]

The lawyers expressed a preliminary opinion that the institution could properly "sever at pleasure" the employment of "a particular person who, however able and worthy, happens to be afflicted with temperamental qualities which render association with him disagreeable," but they wanted to know more about his conditions of employment.[66] This inquiry brought forth the embarrassing revelation that Ritchey had been granted a salary raise from $3,300 to $3,600 effective January 1, 1919, on the recommendation of Hale, no doubt in recognition of his wartime work in the optical shop. There were also glowing letters from Hale in the Carnegie files recommending Ritchey in extravagant terms for his work in earlier days when the observatory was being built.[67] However, the "agreement" (actually Hale's fiat, to which Ritchey had not formally objected) of October 28, 1911, had made Ritchey an employee who could be dismissed by the director at any time, and once the lawyers saw it, all their reservations vanished. They drafted three alternate resolutions in legal form, as Woodward had requested, one "retiring" Ritchey immediately with no pay, one retiring him but with a leave of absence at full pay through the end of 1919, and one granting him a "retiring allowance" at half pay, $1,800 a year.[68]

Adams was implacable. When he read an early draft of the resolutions, one of which the Carnegie Board was to adopt, he strongly urged the words that Ritchey would get a pension "in view of the services he has rendered the Institution" be struck out. This phrase would only strengthen Ritchey's hand, and he would use it as ammunition. Furthermore, Adams questioned "whether it would be entirely just to the loyal and efficient members of the Observatory staff," who had recently come under a retirement scheme to which they made contributions from their salary, to give Ritchey "as much as they will ultimately receive, without contribution on his part and in spite of the acts which have led to a general demand for his dismissal." It might even tempt other staff members to make trouble and court dismissal for the sake of a pension, he argued. Woodward still thought it best to give Ritchey the pension, both to quiet him and "to do him simple justice," but he said that the Carnegie Executive Committee, guided by former Secretary of War Elihu Root, would decide.[69]

They did on October 25, and Ritchey was dismissed without a pension. He was to be given a check for $2,000 as his final and total severance payment on October 31. Even to this, Hale demurred, saying once again that Ritchey should be made to hand in his photographs *before* he got the check. Woodward agreed.[70] On October 31 Hale gave Ritchey a short, hard letter of dismissal, quoting the Executive Committee action and stating that Ritchey had "lost the confidence and respect of many connected with the Institution" and had "destroyed [his] usefulness to the Observatory." He was to receive $2,000 after he had turned over to the observatory all his "photographs, notes, drawings, and other data relating to [his] astronomical, optical and mechanical work."[71]

Ritchey knew that it would do no good to protest. There was no mechanism to do so. All he asked was that he be permitted to keep the enlargements he had, many of them intended as illustrations for the book he still hoped to write. This Hale recommended he be allowed to do, and Woodward approved, but he emphasized in a long stricture that Ritchey should be economical in his publication plans and definitely should not include "a large number" of expensive halftones. Furthermore, Hale stipulated, Ritchey would have to publish his book or books through the regular Mount Wilson Observatory editor, Seares, who could thus retain control and censor any inappropriate statements. Ritchey had to agree to this proviso in writing before he got his $2,000.[72]

Ritchey himself then tried to get just a little even by submitting a bill to the observatory for $140 for "rent" for a 20-inch flat test plate, his own personal property, which he had brought to the observatory shop and used there for two years. Adams indignantly advised Hale to reject the bill, but Woodward advised paying it as a concession rather than a legal responsibility and only on condition that Ritchey make no further demands on the institution.[73] Ritchey accepted the last check on this basis and held his peace.

So in the end Ritchey was fired from the observatory, which he had done so much to build, when he was nearly fifty-five years old and his skills were no longer needed. To his own friends, Hale stated that Ritchey had "bec[o]me simply unbearable and had to be put overboard."[74] His crimes were disloyalty and giving the "wrong" advice on the 100-inch telescope, essentially the charges on which J. Robert Oppenheimer was found guilty in his security hearing thirty-three years later. Hale consistently maintained that Ritchey was a selfish, unprincipled schemer. The director said that everyone on the staff wanted him to get rid of Ritchey and thanked him after he had done so. Probably most of them did, and those who did not knew that if they objected they would be exhibiting their own "disloyalty" and "lack of judgement."

A far different picture of Ritchey was given by Alfred H. Joy forty-five years later. In 1919 he was a young spectroscopist who had joined the Mount Wilson staff only four years previously, very definitely a protégé of Adams and an admirer of Hale. Yet he wrote:

> Although Ritchey was not a trained astronomer he made superlative and lasting contributions to the science by perfecting in this country the optical, photographic, and mechanical equipment of the great reflecting telescopes, so successfully used during the last 60 years in exploring the universe.

Ritchey was "definitely an introvert," but "his devout and kindly spirit was often manifest," Joy recalled. As an example, he remembered a night in June 1918 when Ritchey, assigned to observe with the 60-inch, gave up the telescope to Joy when Nova Aquilae was unexpectedly discovered, so that the younger man could obtain a spectrogram of it as close as possible to maximum light. "[M]odesty was not one of [Ritchey's] dominant characteristics," according to Joy, but "[i]n his optical and photographic work [he] showed extreme patience and perseverance; he was satisfied only with the highest degree of perfection."[75]

Azusa

1919 - 1924

WHEN GEORGE WILLIS RITCHEY lost his job at Mount Wilson Observatory on October 31, 1919, he was banished from American astronomy. George Ellery Hale, who had accused Ritchey of disloyalty and near-sabotage of the 100-inch telescope, was one of the two most powerful men in the astronomical establishment. His fellow director W. W. Campbell of Lick Observatory would not have considered holding out the hand of friendship to anyone whom Hale had fired, no matter how great his skills and experience, and no one below them would have dared make an enemy of a great leader of American science. Only Elihu Thomson, the inventor, engineer, and General Electric research director, commiserated with Ritchey and sought recognition and a job for him. At his suggestion, Ritchey drew up and sent Thomson a statement of his accomplishments at Yerkes and Mount Wilson observatories, an impressive five-page list of achievements, but it was no use. Ritchey was not to get another job in American astronomy for more than a decade, and then only through two other outsiders, themselves both shunned by the establishment.

In 1919 Ritchey had no fortune to fall back on, no money for travel abroad, no continuing salary or pension from the Carnegie Institution, not even any orders for telescope mirrors to make in his shop at his home. He and his wife moved to their ranch near Azusa, and Ritchey threw himself into earning their living by growing lemons, oranges, and avocados. It was hard physical effort for a fifty-five-year-old man who suffered from rheumatism and sciatica and had only one helper, but he thrived on it. Their orchards were on North Vincent Avenue, then a rural area near Azusa but now part of Irwindale, some ten miles east of Pasadena. Ritchey's mother, his son Willis, who was a manual training teacher in Los Angeles, and his daughter Elfleda remained at their home in Pasadena, near the observatory office. Ritchey learned cultivation, irrigation, fertilizing, and controlling insect pests, and pioneered in using orchard heaters to fight frost on cool winter evenings.[1]

Earlier in 1919 Ritchey had reported to the American Academy of Arts and Sciences, which had supported his experimental work on a Ritchey-Chrétien reflector, that its primary mirror was finished and that the secondary needed only slight correcting or retouching. He had not found time to work on it after 1917, and though he kept his shop in Pasadena, he never had the resources to finish the optics and mount this telescope. All his efforts went into his orchard.[2] The first 27-inch set of Ritchey-Chrétien optics that he had started at the observatory shop under John D. Hooker's sponsorship in 1911 had long since been abandoned at Hale and Adams's orders.

After two years, the orchards were thriving, and Ritchey planned to build an optical shop near his Azusa home. He had been experimenting with cements for making "built-up" or cellular disks from relatively thin plates of glass, held together by spacers so that air could circulate through them. Such cellular disks made into mirrors would be much lighter than solid ones but equally strong. Ritchey had made a 30-inch cellular disk in 1912 and figured it into a plane mirror, which had not changed its form at all, he wrote. He dreamt of making a 120-inch or 150-inch disk this way, its front surface softened and annealed ("slumped" in modern terms) to the correct curvature before it was ground and figured. The telescope would be a short Ritchey-Chrétien reflector, devoted entirely to direct photography and specially outfitted for the most precise guiding. The observer would use only the moments of finest seeing, interrupting the exposure with an occulting shutter (operated by a foot pedal or a bulb in his mouth as he guided with both hands). The resulting "superb photographs" would educate the general public in the wonders of astronomy. Ritchey dreamt that Henry Ford, the fabulously wealthy automobile manufacturer, would finance "the Henry Ford Educational Observatory," perhaps even two of them, one in each hemisphere, and put him in charge of building them. But it was not to be. During the bitterly cold winter of 1921–22, the year of the "great freeze" in the San Gabriel Valley, Ritchey lost his crop and $30,000. He could not afford to build even a little shop in Azusa.[3]

In 1922 Henri Chrétien published in France a complete description of the optics of the aplanatic or coma-free reflecting telescope that Ritchey and he had conceived and worked out mathematically at Mount Wilson in 1909–10. This was the first published explanation of just what the "new curves" were, expressed as power series by Chrétien. He showed clearly their relationship to the aplanatic system that Karl Schwarzschild had investigated even earlier, in which the secondary mirror is concave, like the primary.[4] Ritchey, Chrétien, and Schwarzschild had discussed the similarities and differences of their systems at the Solar Union meeting on Mount Wilson in 1910.[5] The two systems are mathematically equivalent, with only the sign of the curvature of the secondary mirror changed, but Schwarzschild had evidently not noticed the possibility of the Ritchey-Chrétien system, and they had not known of the existence of his system. Walter S. Adams, Hale, and other detractors of Ritchey were fond of saying that the Ritchey-Chrétien was "only" a form of a Schwarzschild telescope, but, in fact, in practice the two are quite different.[6] The great advantage of the Ritchey-Chrétien, in addition to the fact that it is coma-free, is that it is short, compact, and therefore relatively cheap to build and enclose in a dome, while the Schwarzschild is necessarily much longer. It also has the disadvantage that its focal point is between the two mirrors and therefore cannot be used for visual observations or guiding.[7] Apparently, no Schwarzschild telescope has ever been built and used successfully for astronomical research.

After his big financial loss in 1922, Ritchey had to work even harder in his orchards. Willis, married, now lived in Hollywood and taught in west Los Angeles, while Elfleda had moved to the ranch with her parents. Ritchey had no shop, but in his drawing room he was planning very large telescopes that he hoped to build someday. He was now convinced that cellular disks could be made with even thinner glass than he had first envisioned. He believed that it would be possible to construct disks with diameters of 25 or even 30 feet (300 or 360 inches). His plan called for a fixed horizontal telescope, much like the Snow telescope, with movable coelostat plane mirrors feeding the light to a fixed primary, which, in turn, worked with any one of a number of interchangeable secondary mirrors, providing focal ratios from f/5 to f/50, as desired. Ritchey was enthusiastic to get back into astronomy, though he feared that Hale would "appropriate" his ideas if he learned of them.[8]

The Mount Wilson director was not feeling nearly so well. After the traumatic experience of firing Ritchey, Hale found himself suffering from lumbago and unable to concentrate. His wife had become severely depressed, but in April 1920 he left her in Pasadena while he went to Washington to push through his plans for the National Research Council and the National Academy of Sciences building. He declined nomination for the presidency of the Massachusetts Institute of Technology but worked vigorously to raise money and transform Throop College into the California Institute of Technology. He declined an offer to become president of the Carnegie Institution, as successor to Robert S. Woodward. Instead, this post went to John C. Merriam, a noted Berkeley geologist and paleontologist, whom Hale felt sure he could manage if he had to. But he never found time to get down to astronomical research.[9] During all of 1920 he nervously watched his health and compared symptoms and cures with Merriam, a great believer in walking and outdoor exercise.[10]

The observatory was thriving and great results were coming out of the 100-inch telescope. Paul W. Merrill and Edwin Hubble both began active work at Mount Wilson in 1919, so either one of them can be considered Ritchey's replacement on its staff. They were beginning their outstanding research careers in stellar spectroscopy and in nebulae and galaxies, which were to take them so far. The observatory had not suffered when Ritchey was "put over the side," as Hale had expressed it. A. A. Michelson and Francis G. Pease succeeded in directly measuring the diameter of a star, the red supergiant Betelgeuse, with an interferometer Pease and John A. Anderson had built for the 100-inch telescope. Hale always passed on the good news of Mount Wilson discoveries to Merriam. In December 1920 Adams went to Washington to represent Hale at the annual conference of the Carnegie Institution trustees with the directors of its various research centers.[11]

Hale had been toying with the idea of resigning the directorship, but by January 1921 he felt well enough again to decide that he wanted to stay at the helm but reduce his hours of work. He embarked on a vigorous campaign to persuade Robert A. Millikan, the famous physicist soon to win the Nobel Prize, to leave the University of Chicago and become head of the California Institute of Technology. Hale made a hurried train trip to Chicago for a series of long conferences with Millikan, and once he got him to Pasadena, worked to convince him to stay. In the end Caltech outbid Chicago and got their man. This whole struggle drew heavily on Hale's nervous energy and physical strength.[12]

At the observatory Pease was proving very helpful. One of Hale's projects was to include a coelostat and a small, solar tower telescope in the National Academy of Sciences building in Washington, and Pease designed them for him. He was getting good photographs of the moon and other objects with the 100-inch, to replace the ones Ritchey had taken with the 60-inch. Under the guidance of Adams, Pease was publishing more papers than Ritchey had and more photographs of nebulae. Hale himself published a paper, in which he reproduced and discussed two lunar photographs Pease had taken with the 100-inch.[13]

By January 1922 Hale had regained some of his strength and was back at work, planning the National Academy of Sciences building and designing a spectroscopic laboratory for Caltech. He felt he had "accomplished very little" in research but jumped into drawing up a city plan for Pasadena. In February he had "a sharp recurrence of my old head trouble," and his physician, J. H. McBride, told him that he must take a long rest or risk a complete breakdown. Soon McBride advised him that he must resign the directorship. Hale did not want to do it, and Merriam told him that the Carnegie Institution would make any arrangement they could to keep him on.[14]

The ailing director decided to take a long leave of absence and travel in Europe, isolating himself completely from astronomy, which was his wife Evelina's favorite cure for him. He would receive half pay while on leave. Adams would take over as acting director.[15] Thus, while Ritchey toiled in his orange and lemon orchards in Azusa, Hale, accompanied by Evelina, sailed to England and toured through France, Switzerland, and Spain before continuing to Egypt and sailing up the Nile to Luxor. There he met his friend, Egyptologist James H. Breasted, who was excavating the tombs of the Pharaohs. Then it was back to Italy and finally England again in July 1923.[16] Hale's restless mind spun off schemes for the National Academy building, the National Research Council, a new engineering foundation, the best way to combat Bolshevism, and almost everything except doing astronomical research.[17]

In March Hale finally decided that he must give up the directorship. He wrote Merriam from Rome, detailing his medical history and problems, from "preliminary nervous attack" in 1908 through his severe breakdown in 1910 that had kept him away from the observatory for sixteen months. He detailed the long, sad litany down to 1923. Hale estimated he had not enjoyed one-third working capacity during the previous fifteen years, but the striking fact to a reader is that in the very stressful but exciting period in which he was engaged in war work, from 1914 to 1919, he reported absolutely no symptoms. Hale actually thrived on crises but was made ill by sustained research effort. He felt he must resign but wished to keep a relationship with the observatory if he could.[18]

Even before the ailing director had left the United States, Merriam had assured him that if he resigned he would be kept on the payroll at his full salary, under the title of honorary director, and that the Institute would help him build a small solar observatory at his home. Hale had made it his business to remind the Carnegie president of this promise frequently. Merriam told Hale how "exceedingly sorry" he was, but that for the sake of his health they would have to accept his resignation. Adams would succeed him, as he had recommended. Hale composed a statement that he had requested a "material reduction of his responsibilities" and had been appointed honorary director, in charge of general policy, while Adams would be director "in charge of operations." Merriam had suggested this arrangement as a possible solution before Hale had departed. Hale kept his salary, and the Mount Wilson staff, which was organized by Adams, signed a cablegram: "regret but recognize the necessity of your decision, and look forward to your return in good health, and the inspiration of your counsel."[19] He retired under very different circumstances from Ritchey's.

Hale did not return to Pasadena until October 1923 after the solar eclipse of September 10, whose track of totality touched southern California, was safely over. Many astronomers were to be there, and an astronomical meeting was to be held in Pasadena. McBride warned Hale not to consider going home until they had cleared out. It might be especially dangerous to his head, Hale believed, as Henri Deslandres, his hated French rival, and "some Germans" might turn up at the eclipse and meeting.[20]

Hale discovered that he had high blood pressure and started receiving treatment for it. He began to write articles for *Scribner's*, the first on Egyptian astronomy and the next on dark nebulae. These articles spurred him into planning a book on the historical development of astronomical instruments. Hale was a good popular writer and found himself enjoying astronomy at this level.[21]

After his return to Pasadena, Hale kept up with his plan to build his own solar observatory near his home. He spent most of his time there, going to the office only occasionally. Adams told him that Carnegie funds were tight and that the observatory budget was stretched to its limit. So, instead of holding Merriam to his promise, Hale decided to build the solar observatory building with his own funds but to "borrow" the instruments for it from the Mount Wilson Observatory. Because the available spectrographs, gratings, and mountings were "in great demand," however, Hale was "compelled to request a special grant" to have the instruments he needed made. Merriam promised him $5,000 for the next fiscal year.[22]

As 1924 dawned, Hale, now fifty-five, was having renewed trouble with his head. He could only endure life by avoiding people. In his reading, his writing, and his personal solar observatory, he was reliving his youth, rebuilding an improved Kenwood Physical Observatory, "revert[ing] in part to an earlier state," as he told Edwin B. Frost. Hale's thoughts were increasingly of his youth and of the past.[23]

Ritchey, on the other hand, at fifty-nine, full of enthusiasm and plans for the future, suddenly found himself going forward. His deliverance came from France. Their country devastated by World War I, French scientists wanted to rebuild their observatories and enlarge their astronomical capabilities. They considered reflecting telescopes a uniquely French contribution to astronomy, which had truly begun with León Foucault in the 1860s. What was more natural to them than that France should have the largest reflector in the world?

In 1920 Benjamin Baillaud, director of the Paris Observatory, had written to Hale, reminding him of their meeting in 1918 in France. Baillaud announced his hope of building a 100-inch telescope, to be erected in France, perhaps at the Pic du Midi, a high mountain site, or perhaps in Algeria. He asked if Ritchey could make the mirror for it. Unfortunately, the French government would not provide the money necessary to build the telescope, which Baillaud estimated as $4 million.

Might it perhaps be possible to raise the money in the United States, from the Rockefeller Foundation or some other source, Baillaud asked with convoluted Gallic politeness. Hale replied in equally flowery English, but the bottom line was that there was no hope of raising this kind of money in the United States to build a French telescope. Baillaud thanked him profusely for the information but reminded Hale that he had also asked about Ritchey's availability. The Mount Wilson director replied that it would probably be better for political reasons to have the optical work done in France by French opticians. He would like to help, but the Mount Wilson shop had been put to other work. "Mr. Ritchey has also given up his connection with the Observatory and gone into other business, but this would involve no difficulty, as our other opticians, who actually figured the 100-inch mirror, are perfectly competent to figure another one of the same aperture equally well," Hale concluded.[24] Without money, the French could not go on, whether they wanted Ritchey, one of his former assistants, or a homegrown optician to make the mirror.

In 1922 their apparent savior appeared. He was Assan Farid Dina, an engineer and wealthy Maecenas (as the French called him), the son of an Indian father and a French mother. He lived in France with his wife, Mary Shillito Dina, born in Cincinnati and the granddaughter of the founder of the largest and most successful department store in that city. Her father, who headed the Paris branch of the store, had brought her to France, and she was educated there. The Dinas' fortune came from her, but under the Napoleonic code of France, where they were married, he controlled it all. This was to cause trouble later, but in 1922 they both were eager to use the money to build a large telescope in France.[25]

The Dinas had a town house in Paris and a country estate on Mont Salève in Haut Savoie near Lake Geneva. There they had founded a meteorological observatory. It brought them into contact with General Gustav-Auguste Ferrié, the Admiral Rickover of post–World War I France. Ferrié had been a key person in the development of wireless telegraphy in France and had increased its range tremendously by installing an antenna on the top of the Eiffel Tower. He was a member of the French Academy of Sciences and "passionately devoted" to astronomy. Under Ferrié's tutelage, the Dinas decided to provide the money to build a large reflecting telescope in France.[26] Ferrié asked André Danjon, then an assistant at the Strasbourg Observatory, to prepare a preliminary plan for the observatory that would operate it. Danjon completed his report in July 1923. It emphasized, in glowing terms, the need for an observatory devoted to astrophysics on a mountaintop site. He skirted the issue of where in France this site would be, for the Dinas envisaged it on Salève, which the astronomers knew was much too cloudy for a big telescope. Danjon recommended it be a 2.65-meter (104-inch), just larger than 100 inches, so that it would be the largest telescope in the world. The St. Gobain Glass Company stated that they could now cast a disk of that size but that going beyond it would lead to "enormous difficulties." Danjon estimated the cost of the mirror alone as Fr 3 million ($600,000) and the total for the telescope in a dome as just over Fr 5 million ($1 million). With additional smaller telescopes and auxiliary instruments, the cost came to Fr 8.5 million, and with the shops and buildings as well, the estimated total was Fr 13 million ($2.6 million).

The telescope could be a conventional paraboloid with several Cassegrain secondaries, but it might be even better to build it as a Ritchey-Chrétien reflector, which, Danjon wrote, had considerable theoretical advantages but had not yet been tested in practice. Careful studies and experiments should be the basis for a decision, and Henri Chrétien should be included in the project. Finally, Danjon considered it obvious that Ritchey, with his unrivalled experience in all phases of making the 60- and 100-inch mirrors, should be brought over to do the French mirror.[27]

Almost immediately after receiving this report, Dina wrote Chrétien and told him of his and his wife's plan to build the observatory on their property on Mont Salève and to endow it for permanent operation. He invited Chrétien to join the project as one of his advisers. The little optical expert, now a professor at the Institute of Optics in Paris, thanked him fulsomely for his intention, "so elevated with respect to Science, and so generous and friendly to our Country." He had "always been very passionate toward Astrophysics" and would be happy to devote himself to Dina's "noble cause."[28]

Dina invited Chrétien and his wife to visit his estate and help him pick out the exact spot where the observatory would be located. Chrétien diplomatically replied that he would be honored to inspect "*possible* sites for your observatory" (author's italics). In fact, it rained the entire time the Chrétiens were at Les Avenières, the Dinas' mansion at Cruseilles on the slopes of Mont Salève. Chrétien made a point of mentioning Ritchey and emphasizing that he had worked with him in California.[29]

Soon after this visit, Danjon consulted Chrétien about the project. The Strasbourg assistant professed himself strongly inclined toward making the new telescope a Ritchey-Chrétien and asked why Ritchey had not done so with the 100-inch. Chrétien explained that Hale had made the decision and that by the time the idea had come up in 1910, it had seemed more important to push ahead and keep John D. Hooker's interest, rather than to halt for tests of the new concept. For the French telescope, Chrétien proposed that they first build an experimental 0.5-meter (20-inch) Ritchey-Chrétien and then, if it was successful, a 1-meter model. Danjon had planned a telescope of this size for the new observatory, and it could be this Ritchey-Chrétien. After it, Chrétien thought, an intermediate-sized 1.6-meter (63-inch) should be built "to gain more confidence in the exactness of the extrapolation." Then they could go ahead with the final 2.6-meter Ritchey-Chrétien.[30] He apparently did not ask himself whether Dina's interest, and fortune, would survive to the end of this program.

Meanwhile, Dina, trained as an engineer, came up with an idea for making the mirror from a composite steel-and-glass disk instead of a thick disk of glass. It was something like Ritchey's cellular mirror idea but with metal spacers rather than glass. Dina's proposed disk would clearly have suffered badly from flexure and thermal distortion. Ferrié referred this concept to Chrétien, who very diplomatically pointed out its flaws but took the opportunity to describe to Dina his proposed series of Ritchey-Chrétiens of increasing size. The optics professor also wanted to carry out a study and series of experiments to test whether it would be better to pierce the central hole in the primary mirror before or after it was figured. Ritchey had avoided this problem in all his previous telescopes by using a flat mirror to reflect the light to a focus at the side of the tube rather than letting it pass directly through a hole in the primary.[31]

All these questions but the last were to plague the large-telescope project throughout its lifetime, up to its dismal conclusion. But no one was aware of that in 1923. Dina, with Ferrié's counsel, set up a group of expert scientific advisers, including several of the top astronomers and optical physicists in France. Among them were Baillaud, Chrétien, Deslandres, Charles Fabry, Alfred Pérot, Aymar de la Baume Pluvinel, and Lucien Delloye. They met in October 1923 and unanimously recommended going ahead with the large reflector, though several of them questioned the need for the smaller and intermediate-sized telescopes of Danjon's plan. Dina authorized Danjon to write Ritchey to invite him to come to France to make the mirror. The Dinas, who had already put Fr 1 million into the project, would pay his salary and expenses. Delloye, whom Ritchey knew well, would provide administrative supervision of the project. Danjon sent the letter on December 7, 1923, and when Ritchey received it on December 28 it must have seemed a wonderful, unexpected Christmas present.[32] One week later he replied in a long, enthusiastic letter that he would be in Paris by April 1, 1924.[33]

Ritchey was on his way to New York before the end of March 1924, and he sailed for France on April 2. His understanding was that he was to act as a consulting engineer for the project, "with the possibility . . . of having charge of the design, construction and optical work of these telescopes, which are to surpass the 100-inch size."[34]

Paris

1924 - 1930

ON APRIL 2, 1924, George Willis Ritchey sailed from New York on the *Berengia*, bound for France. He was fifty-nine years old, "recalled to life" in astronomy, and determined to see his dream come true at last. He intended to build a telescope that would be bigger and better than the 100-inch Mount Wilson reflector. His friend Elihu Thomson was filled with joy that Ritchey had got the job and was certain that there was "no one else . . . so well fitted to carry out the task." The telescope maker went alone, but planned to bring his wife and daughter over soon if all went well.[1]

Ritchey arrived in Paris on April 8. A week later he attended a meeting of the French Academy of Sciences to which he was welcomed by its president. A few months later the Academy awarded the American optician its Jules Janssen gold medal for his "admirable work" on reflecting telescopes.[2] Ritchey set up his laboratory in the Paris Observatory and also had an office at the Institute of Optics on the Boulevard du Montparnasse where Henri Chrétien was now a professor. General Gustav-Auguste Ferrié, head of the Army Signal Service, and Lucien Delloye, director general of the St. Gobain Glass Company, were the American telescope maker's sponsors.[3] André Danjon, the Strasbourg astronomer who had drawn up the preliminary plan for the observatory in 1923 and who translated Ritchey's remarks at the Academy meeting, acted as Ferrié's executive for the telescope project.

Soon after the meeting, Ritchey went to the Bar-sur-Seine country home of Assan Farid Dina, the donor-to-be of the funds for the project, some eighty miles southeast of Paris. There Ritchey worked on the plans for the telescope for nearly a month.[4] Near the end of his stay, Dina met with his board of advisers in Paris to discuss the project but did not invite Ritchey to attend.[5] The scientists soon learned from Ritchey himself, however, that he had no intention of making a 104-inch, slightly scaled-up version of the 100-inch, as Dina had originally imagined, or a Ritchey-Chrétien of that size, as Danjon and Chrétien had planned. Instead, Ritchey wanted to build a much larger telescope, a 5-meter (200-inch) or 6-meter (240-inch) reflector. Rather than a solid glass disk, he proposed to use a built-up, cellular disk, assembled from glass plates,

Ritchey with Yvonne (left) and Madeleine Chrétien, Henri Chrétien's daughter and wife, Paris, probably ca. 1924. Courtesy of Cercle Henri Chrétien.

for the primary mirror. He patented the design and plunged into an experimental program of making successively larger cellular disks, promising to complete a 1.5-meter (60-inch) by June 30, 1925. Some of the members of Dina's board of advisers soon became skeptical. One of them, Charles Fabry, director of the Institute of Optics, began referring to Ritchey as "cuckoo" ("maboul").[6]

The American further antagonized some of the French scientists when he "demanded" a 50 percent salary raise soon after arrival. Actually, Ritchey was quite unrealistic about money; he had first agreed to come at a salary of $4,800 per year, which was less than observatory directors or senior professors were being paid in the United States at that time. It was entirely insufficient for his expenses in Paris and his family's in California, and even at the $7,200 salary to which Dina agreed, Ritchey was hard pressed for cash.[7] Nevertheless, this discussion poisoned the atmosphere. Ritchey put his French assistants to work on the experimental cellular disks. As always, he threw himself so completely into the task that he had little time for other scientists, even his friends Chrétien and Thomson (who visited Paris briefly), except when he needed their help.[8] In making the cellular mirrors, the main problems were to grind all the spacers precisely to the same size and then develop a cement that, when applied in a very thin, uniform layer, would bind the top and bottom plates securely to them. Ritchey and his staff experimented with various forms of the new compound Bakelite, trying to find the perfect cement. Ritchey was happy when he was at work in the laboratory, stayed on good terms with Delloye, and made "every reasonable effort to be on the most cordial terms with Mr. Dina and General Ferrié."[9]

Once they had ground all the pieces of glass to the exact dimensions, the assistants had to assemble the disk carefully on a form that would hold it together while it baked in an oven, just hot enough to activate the Bakelite and make it set. This process lasted several days. Then, after slowly cooling the furnace, they would remove the disk, polish its faces flat, return it to the furnace again for reannealing, then remove and polish it a second time. The disk was then supposed to be ready for use.

Bakelite was a new type of cement, still in its experimental stage. Different batches had slightly different responses to heat, pressure, and tension. This complicated the work, for Ritchey's plan involved using the thinnest possible layers of cement to avoid introducing strain into the top and bottom plates of the disk.[10] By November 1924 he produced his first 40-cm (16-inch) disk of this type. As it cooled, a small crack appeared in its surface, an indication that the spacers were not sufficiently uniform and had strained the top plate of the disk. Nevertheless, Ritchey polished it flat. After a few days a wave pattern appeared in the glass, only 0.02 microns (1/20 of a wavelength of blue light) high. Ritchey declared the mirror a success but one of his French critics (Charles Fehrenbach) later claimed that he had suppressed the truth. Actually, a mirror whose surface is accurate to within 1/8 of a wavelength is generally considered as good as needed; Danjon, the other critic, in his history described this same mirror as "tolerably satisfactory." At the time, Ritchey analyzed the results of this experiment, modified his methods, and started on the next trial, a 75-cm (30-inch) cellular disk. One of the changes in procedure he made was to wait longer after taking the disk from the furnace before beginning the optical work. This mirror did not develop a crack, and Ritchey stated that he considered it "very successful." Privately, however, he was unhappy with the "inefficient help" of the staff in his Dina Laboratory of Optics, now numbering ten, and was worried that he had accomplished so little. Nevertheless, the financial angel for whom the laboratory was named still shared his dream of making a 5- or 6-meter telescope.[11]

This Ritchey hoped to accomplish by constructing a giant "fixed universal telescope." It was the vertical version of the fixed horizontal reflecting telescopes he had so long planned. The primary mirror would be at ground level, the light fed to it by a gigantic two-mirror coelostat at the top of a high tower. Thus it was similar to the solar tower telescopes that Francis G. Pease had designed to George Ellery Hale's specifications at Mount Wilson. They had been relatively small-aperture telescopes, using long focal-length lenses to form the solar image at ground level. In the fixed universal telescope the huge primary mirror would direct the converging light beam back up the structure, and a secondary mirror would send it down again through a hole in the primary to a focus behind it. There an observer, guiding with a double-slide plateholder, could take direct photographs. If desired, various spectrographs or other instruments, all kept at the ready, could be rolled in and used instead of the double-slide plateholder. Ritchey called this system a "universal" telescope to indicate that it was not designed simply for direct photography but could be used with any astrophysical instrument. The design was superior to the earlier fixed horizontal telescope in that the primary mirror lay horizontally on its back, free of flexure, and that the light entered the system far above the ground and was thus less affected by the turbulent surface layers of air that produce most of the bad seeing. Also, the plateholder, spectrographs, bolometer, or other instruments could be rigidly positioned at ground level, after they had been wheeled into the beam. It was an extremely advanced concept.

Ritchey did not publish his plan for a fixed universal telescope until 1927, but when he did so he claimed that he had been working on it since 1919, when he had been fired from the Mount Wilson Observatory staff and had moved to his Azusa citrus ranch. He certainly had referred to it in his first letter to Danjon in January 1924 and had been talking about this design with the French astronomers from very soon after his arrival in Paris. Very probably, he had indeed been working on it long before leaving California. In his first published description of a fixed universal telescope, Ritchey gave the diameter of the primary mirrors as 3 meters (120 inches) and of the coelostat mirrors as 3.7 meters (150 inches). This plan called for *four* primary mirrors, each of which, used with a different secondary mirror, would provide a different focal length. Two were Schwarzschild systems with focal ratios f/3 and f/4.5 for wide-field photography of faint nebulae; the other two were Ritchey-Chrétien systems with focal ratios f/6.5 and f/12 respectively for larger-scale photography and spectroscopy.[12] In another paper, published soon after this one, Ritchey stated that his first plans, made in California before he came to Paris, had called for *three* primaries, one a conventional paraboloid, one a Schwarzschild system, and the other a Ritchey-Chrétien. That plan had envisaged primary mirrors of 4-meter (160 inch) diameter, with 5-meter (200-inch) coelostat mirrors.[13] He alluded to this design in his letter to Danjon without giving any specific dimensions for the mirrors. The first plan he presented to Dina in France was this earlier combination of "only" three primaries, but he gave their apertures as 5 or 6 meters. Each primary mirror was to be mounted on a wheeled carriage, running on rails, so it could be quickly and accurately put in its precise position for use and then afterward equally quickly removed to make way for another. The secondary mirrors were mounted on similar

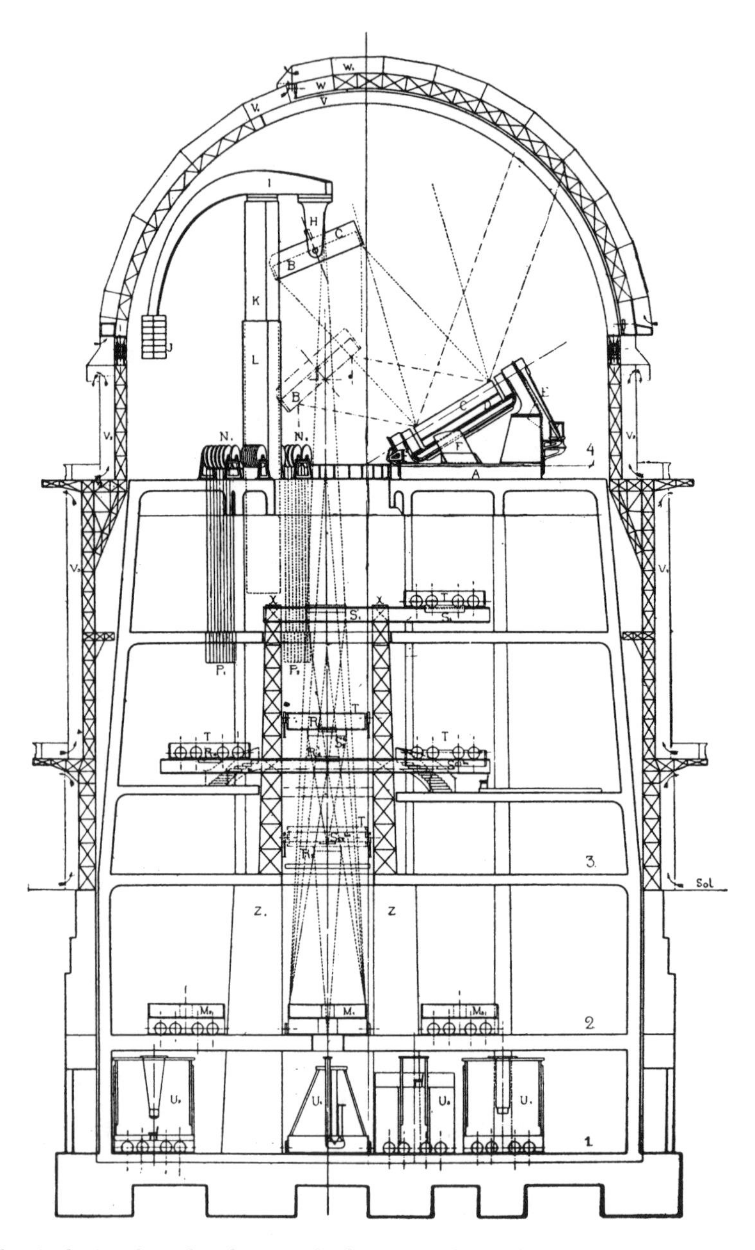

Ritchey's design for a fixed vertical telescope. The coelostat, consisting of two large plane mirrors, is at the top. The ground level is marked "Sol." A 5-meter diameter primary mirror, marked M, is in use; two others, on wheeled carriages on either side of it, may be substituted for it. Various secondary mirrors, marked R and S, may be used with these primaries. The focal plane is at the level 2, and any of several spectrographs and other instruments, marked U, may be used at level 1. From the *Journal of the Royal Astronomical Society of Canada*.

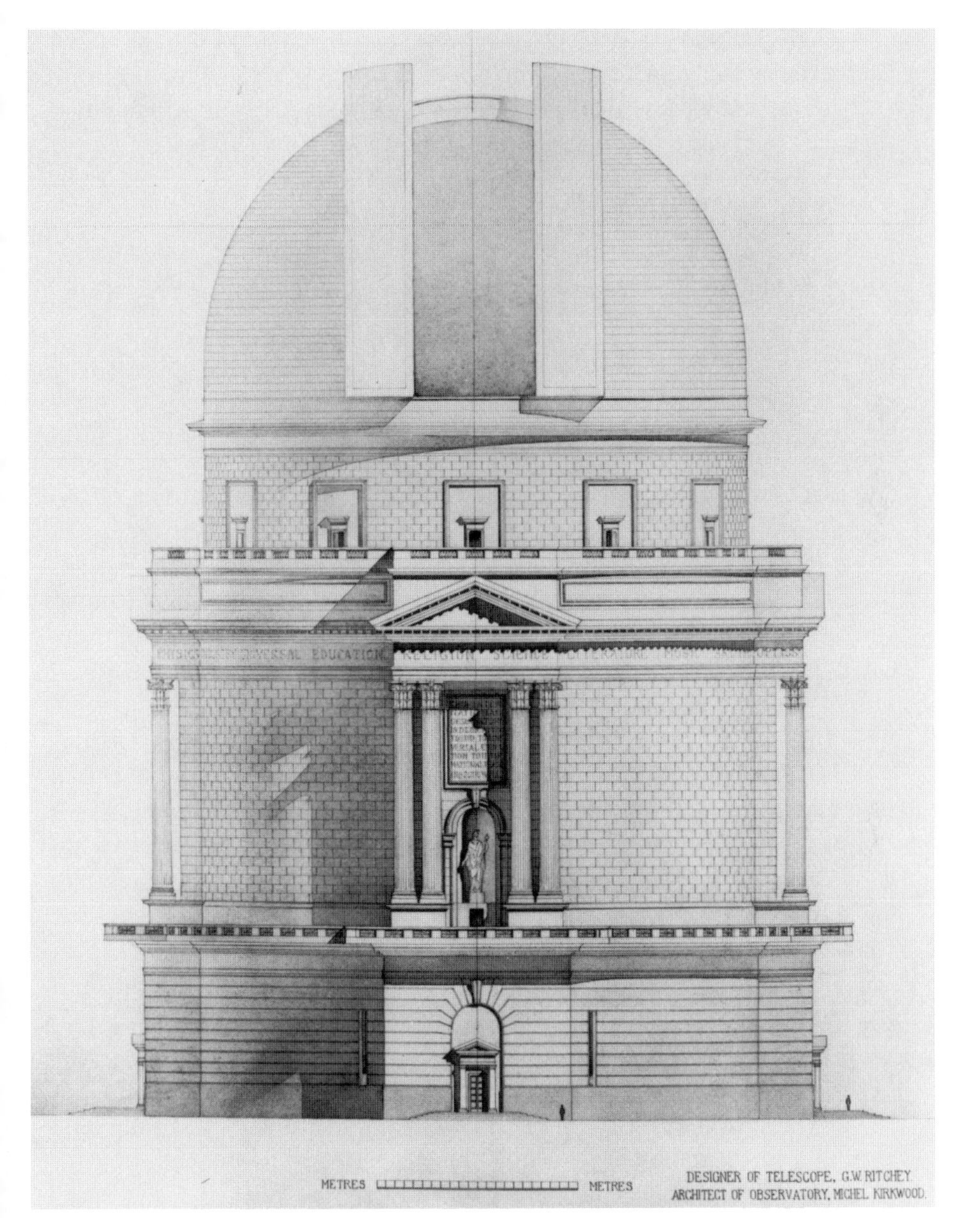

Design of building and dome for a fixed vertical telescope. Courtesy of Richard B. Burgess.

wheeled carriages, each at the correct level within the tower to be rolled in for use with its own primary.

It was a visionary plan, and by 1925 the sixty-year-old Ritchey had become somewhat of a messiah, or at least a prophet, in describing his plans. To William H. Evans, a friend of his son, who wrote from Los Angeles to ask for a job at the Dina Laboratory, Ritchey revealed that he believed that his great project was "surely providentially managed—*divinely* planned." He could see no possibility of failure in "the gigantic work which lies just ahead." The confident American expected to make "not only Mr. Dina's 20 foot [6-meter], but 30 foot [9-meter] and 40 foot [12-meter] telescopes before I die." Ritchey could not offer Evans a job because he had to hire French workmen, but he promised him a position on the future telescope contracts that he knew he would surely get. Ritchey flatly stated that he would "make Mt. Wilson look like 'thirty' cents." He referred to the Los Angeles school superintendents who had fired Evans from his teaching job there as "a bunch of crucifying Scribes and Pharisees," and told him to "[r]emember the ignominy they heaped upon Christ—the ridicule even Lincoln received not long ago; and think of the loving reverence in which they are held now in the hearts of countless good men and women."[14]

Ritchey did not use expressions remotely like these in his dealings with the French, but his relations with them began to deteriorate before the end of 1924. He never learned their language. This was always a problem, making him suspicious of them, and them of him. The exceptions were Delloye, who spoke not a word of English, and Chrétien, who could converse easily with Ritchey in English. Danjon, who was his other frequent interpreter, believed in him at first but soon began to have doubts. Ritchey felt isolated among them while they thought him secretive.

Another problem was the choice of a site for the French observatory. The astronomers constantly chipped away at Dina's idea of putting it at Salève, and by 1925 he gave up and agreed to a site survey. In the long run this was beneficial astronomically. The site eventually chosen, near Forcalquier in Haute Provence, became the original location of the Haute Provence Observatory. But the decision meant that Dina and his wife were no longer as emotionally committed to the observatory as they had been when they had visualized it on their own estate at Salève.[15]

In the spring of 1925 Ritchey's wife Lillie came to Paris to stay with him. She had worried that he was working too hard, but she had remained in Azusa for over a year, trying to sell their citrus ranch. She found no buyer, but in the end came anyway, bringing their daughter Elfleda with her and probably leaving the orchards in the care of a tenant.[16] Ritchey wanted his son to come, too, to work with him, but Dina was not enthusiastic about hiring another American. However, Dina's advisers wanted to put a French physicist to work with Ritchey to learn his methods and to report on what he was doing. Ritchey hated the idea because he recognized that such a top assistant would be both a spy and his own potential successor. They reached a tacit compromise, and Willis G. Ritchey and André Couder both began work in the fall of 1925. Couder had been a twenty-eight-year-old assistant in chemistry at Strasbourg, where Danjon had met him. Couder had a strong interest in astronomy and had participated in the 1924 and 1925 site surveys. Now at the Dina laboratory he began reporting directly to Danjon on the progress of Ritchey's work.

In November 1925 Ritchey lectured to the Astronomical Society of France on his plans for making large, cellular mirrors and exhibited pictures of the recently completed 75-cm disk. Full of optimism, he began work on the 1.5-meter disk, now six months overdue. Immersed in his work at his laboratory in the Paris Observatory, he seldom saw Dina, or even his friend Chrétien, who was equally busy with his own projects at the Institute of Optics.[17]

In December, Danjon submitted a sixteen-page report on the progress-to-date of the project and on plans for the future. Most of it was very positive about Ritchey's work, although he offered a few suggestions on the mounting that Couder had proposed after only three months of experience in making telescopes. Danjon described the advantages of the fixed universal design, as stated above, and also its disadvantages. Ritchey was well aware of these and discussed them freely with other scientists, but he tended to deemphasize or even to omit them from his early published papers and probably from his lectures as well. The main technical problem is that, because the two-mirror coelostat must be above the vertical telescope without blocking the light from entering it, the system can only be used over a limited range in declination, approximately 40 degrees in Ritchey's design. Large parts of the sky surrounding the north pole and the southern horizon are thus always out of reach of the telescope. The other disadvantage is that the coelostat cannot track a star as it crosses the meridian, and long exposures, therefore, could not be centered on the meridian without interrupting them. These were both real disadvantages but by no means fatal in Danjon's opinion. The alternative would be an equatorially mounted telescope. Danjon believed that it should certainly be a Ritchey-Chrétien reflector. Such an equatorial could reach the whole sky and especially with a pierced primary mirror would be very convenient for observing. Ritchey recognized these advantages, and in fact this type of telescope was his second choice, after the fixed universal telescope. The disadvantage of the equatorial was that as a Ritchey-Chrétien with a single primary mirror, it would be free of coma at only a single focal ratio. However, as Ritchey knew and Danjon stated, it could be used with other Cassegrain secondary mirrors, giving other focal ratios as desired. The images would be perfect (free of spherical aberration) at the center of the field and away from the center would probably suffer from coma no worse than in a conventional Cassegrain reflector. (At the time no one had computed the exact size of the aberrations of such a "modified Ritchey-Chrétien" or "hybrid-type" two-mirror system.) Since the longer focal ratios would be used primarily for planetary photography or spectroscopy, which require only very small fields of view, coma would not be a real problem.

As Danjon emphasized in his report, the choice between the fixed universal telescope and the equatorially mounted Ritchey-Chrétien was basically a financial one. The fixed vertical telescope, with its three large, primary mirrors plus two even larger plane coelostat mirrors, its tall building, and the nested rigid mountings within it, would be much more expensive than a single equatorial reflector of the same aperture. He estimated that a fixed universal telescope, complete in its building, would cost four times as much as an equatorial Ritchey-Chrétien with the same diameter. Danjon gave Fr 38 million ($7.6 million) as the cost of a 5-meter fixed universal telescope, compared with Fr 10 million ($2 million) for an equatorially mounted Ritchey-Chrétien of the same aperture. A 6-meter Ritchey-Chrétien would cost Fr 16 million ($3.2 million); Danjon did not even estimate the cost of the corresponding fixed universal telescope, but it would have been about Fr 60 million ($12 million). The real choice, he believed, was between a 4-meter (160-inch) equatorial or a 2.5-meter (100-inch) fixed universal telescope, either of which would cost about Fr 7 million ($1.4 million), which he evidently thought Dina could afford. Between the two, Danjon came down solidly for the 4-meter equatorial. "One must always make the largest, not for the simple pleasure of making it large, but because it is thus that for two centuries Astronomy has progressed," he concluded.[18]

The Indian donor was horrified when he read the report. The sums were far larger than he had imagined. First the astronomers had made him give up Salève, and now they wanted his wife's entire fortune as well! Ritchey, meanwhile, was planning a book with Chrétien on their telescope system, and another volume by himself on astronomical photography, but he was spending little time keeping Dina informed of the progress of the work. The optician seldom even saw Chrétien but wrote to him that the 1.5-meter cellular mirror had finished baking in January and was standing up well under rough grinding, the first stage of making it into an experimental primary mirror.[19]

Dina, however, was fuming. He resented not hearing from Ritchey, and what he had learned from Danjon had not reassured him. Now that he realized how expensive the telescope was going to be, he decided to reduce the scale of the whole project and to get rid of Ritchey. In March 1926 he directed Delloye to stop paying Ritchey's salary and Willis's as well. At last the optician realized the importance of communication with his financial sponsor. He dashed off a four-page list of "[d]ifferent projects carried out at the Dina Optical Laboratory 26 December 1925–26 March 1926," had it translated into French, and rushed it to Dina. To Delloye he sent a long, self-justifying letter, complaining that Danjon should have been reporting his progress to Dina but instead had been working against him. He complained bitterly about the negative points in Danjon's written report and staked his prestige as a telescope maker with thirty-five years of successful experience that *his* plans would work, while those of "the younger men" would, if followed, wreck Dina's project. This was just what had happened, Ritchey claimed, to Andrew Carnegie's project at Mount Wilson. Ninety-five percent of its income had been "*wasted*, squandered, thrown away, by incompetent management and by the fierce internal fighting at cross-purposes of various factions, for supremacy." Ritchey promised to provide frequent written reports in the future and closed by expressing his "profound respect and admiration for [Dina's] great and noble purpose in this project."[20]

It was too late. Probably Ritchey was finished, as far as Dina was concerned, no matter what happened. But, in the event, on April 19 the workmen removed the 1.5-meter cellular disk, which had been reheated

Ritchey, working on a tool to be used in figuring a large 1.5-meter cellular disk, Paris, probably ca. 1926. Courtesy of Richard B. Burgess.

and reannealed after the rough grinding, from the furnace. It cracked. Couder, who was there and knew Dina's frame of mind, immediately telegraphed Danjon: "Cooling again not achieved. Important breaks discovered. Confidential. Couder." Danjon rushed the bad news to Dina. The bitter donor fired Ritchey immediately and put Couder in charge of the laboratory. He had only six months experience working on optics, but he was French and Danjon's friend.

Couder had to let some of the laboratory staff go and reduce its expenses greatly. He concentrated on making three 80-cm (32-inch) mirrors but kept the large-telescope dream alive. Dina was persuaded to pay Ritchey's salary for a few more months rather than discharge him without notice, but he did so only until July 1926. The American continued as director in name only of the Dina Laboratory until the end of 1926, when Couder acquired the title to match the reality of his job. Benjamin

Baillaud, the elderly director of the Paris Observatory, then put Ritchey on its payroll and allowed him to keep working in the laboratory. Ferrié, Delloye, and Baillaud all continued to believe in Ritchey as a telescope maker, though not as a planner or a manager. They still wanted to use his talents. They also felt that in bringing him to France they had become responsible for him and could not cast him adrift without a job.[21]

The remaining history of the Dina Foundation was short and dismal. Ferrié, who was undoubtedly prejudiced against small, dark-skinned men married to rich American women, began to distrust Dina. Various financial irregularities came to light. Ferrié pressed him to make a large cash payment to the Academy of Sciences, as evidence of his good faith and ability to pay for the telescope. Dina refused and Ferrié filed a lawsuit against him. Dina filed a countersuit against Ferrié and Danjon, demanding an accounting of the funds he had given them, which he claimed they had wrongfully wasted.[22] While the case was under advisement, Dina died under mysterious circumstances on June 26, 1928, on a ship between Colombo and Port Said. His widow was far more sympathetic to Ferrié and to astronomy than he had been, but Dina had left most of his assets, the remains of her fortune, to his mistress. This he was entitled to do under French law. The Haute Provence Observatory finally came into existence only years later, with an 80-cm (32-inch) telescope.

In the summer of 1926, Baillaud, his astrophysicist son Jules, and Chrétien tried to find funds to support Ritchey. One possibility was for him to figure a 1.2-meter (48-inch) solid glass disk into a primary mirror for a telescope for the Paris Observatory, but this plan fell through. Then Aymar de la Baume Pluvinel, a wealthy aristocrat (he was known as Count, his family's title in the days of the monarchy) and astronomer, used his influence to get Ritchey a contract from the Ministry of Education to make a 0.5-meter (20-inch) Ritchey-Chrétien telescope.[23] It provided 40,000 francs, enough to pay Ritchey's salary, expenses, and two assistants for the better part of a year. Willis had to return to his schoolteaching job in Los Angeles.[24]

Ritchey had been about to start making the optics for his first Schwarzschild-type telescope but was pleased to get the chance to do the new Ritchey-Chrétien instead. In his respite from managing the cellular-mirror project, he at last finished the optics for his own 0.5-meter Ritchey-Chrétien, begun years previously at his shop in Pasadena. It had an f/2.5 primary mirror and an overall focal ratio of f/6.25. The new reflector that he was to make would have a somewhat larger focal ratio, and its primary mirror would be pierced (unlike his own). It was thus to be a model for the 1.2-meter Paris Observatory job, which Ritchey still hoped to get. It would have an alternate secondary mirror, providing a still longer focal ratio for spectroscopy. He favored f/7.3 and f/11.4 for the two focal ratios but consulted Chrétien and de la Baume Pluvinel as to the advantages and disadvantages of each possible choice.[25]

One hope Baillaud held was to obtain support for Ritchey's laboratory from the International Education Board, an arm of Rockefeller philanthropy. Its scientific representative in France was Augustus Trowbridge, head of the Princeton University Physics Department, who spent his summers in Paris on behalf of the board. Ritchey was anxious to make a good impression on him with a plan for an experimental comparison of a very short Ritchey-Chrétien reflector and a Schwarzschild system of the same focal length.[26]

When Trowbridge visited Ritchey's laboratory at the Paris Observatory, he was very favorably impressed. The American optician was extremely competent. The workshop was spacious and well equipped. Many projects were obviously in progress. However, when the Rockefeller physical and biological sciences "director for Europe" spoke with Fabry, he learned of the "Dina fiasco." The French physicist respected Ritchey's technical competence; Fabry said that he was "unique in the world for construction of large astronomical mirrors." But he had no confidence in the American's estimates of the time or funds required to complete a project. The leaders of the French Academy wanted to salvage some of the investment that Dina had put into the laboratory. They promised to commit some matching funds but hoped to get the bulk of the expenses, including Ritchey's salary of $7,200, from the International Education Board. Trowbridge recommended a grant of $14,000 a year for two years on condition that the academy itself provide the equivalent of $7,000 a year (the franc was falling rapidly in value against the dollar at the time) plus all the facilities and equipment of the laboratory. However, there were many other scientific projects competing for the Rockefeller funds, and the Paris Observatory optical laboratory request was not one of those granted.[27]

Ritchey was seeking other means of support very actively. He knew that Ohio Wesleyan University in the United States and the University of Toronto in Canada both had plans and money to build 60-inch telescopes. Ritchey tried to get contracts to make the mirrors for both these telescopes and to design their mountings. He believed that he could obtain the necessary solid disks for them from the St. Gobain Glass Company, which had been nearly destroyed during the war but was starting to get back into production. With these two orders, Ritchey hoped to keep his Paris laboratory in operation as an international telescope-making center.[28] Edwin B. Frost, the Yerkes Observatory director (who had still not recovered the nebular photographs that Ritchey had taken to Mount Wilson in 1904), thought highly of his work. He recommended that his protégé, Clifford C. Crump, the Ohio Wesleyan astronomer, visit Ritchey's laboratory in Paris. Crump did, but in the end J. W. Fecker (the successor to John A. Brashear) and Warner and Swasey, two established companies that were safer if more pedestrian than Ritchey, got both jobs. As soon as Frost heard of William J. McDonald's bequest to the University of Texas to build a telescope, he recommended Ritchey very strongly for this job also, but again Warner and Swasey and Fecker got the contract.[29]

Meanwhile, Ritchey was working on the optics for the 0.5-meter Ritchey-Chrétien. Baillaud had decided it should have an overall focal ratio f/6.8. As Ritchey proceeded with grinding and figuring its mirrors, he had his assistants start work on an open, rigid, metal tube he had designed for the telescope. The focal surface of such a wide-field telescope is curved, and Ritchey's assistants, therefore, ground concave glass plates to be coated with sensitive emulsion for taking photographs. He could not simply buy and use flat plates. This complication was to prove a real stumbling block for Ritchey's plans to use the telescope as soon as possible. He also began rough grinding the old 1.2-meter mirror that the Paris Observatory had; he planned to convert it into a Ritchey-Chrétien telescope also, "with every possible modern improvement." Chrétien provided the calculations of the dimensions for all the mirrors and plates, as well as the test data necessary to shape them precisely.[30] By October, Ritchey had the 0.5-meter primary perfectly spherical, the first step. Soon afterward he began working toward the final hyperboloidal figure. He planned to compare this telescope directly with a conventional paraboloid mirror with the same diameter and focal ratio to demonstrate experimentally the much smaller off-axis aberrations of the Ritchey-Chrétien system.[31]

In December 1926 Ritchey learned that he would keep his job for another year. The money for his salary came from the Ministry of Education; Dina was still providing funds for the laboratory's operating expenses but would not pay the American. Baillaud was retiring as director of the Paris Observatory, and Henri Deslandres, his successor-to-be, was opposed to everything he had done—including hiring Ritchey. Nevertheless, he agreed to let the optician continue working in the laboratory.[32]

In April 1927 Ritchey learned that Couder, now head of the Dina Laboratory, had himself just completed the primary mirror for an 0.8-meter (32-inch) Ritchey-Chrétien. The American optician did not want to see his hated rival finish the whole system and get the result into print first. Ritchey dashed off a letter to Chrétien, warning him that "you and I should publish a short announcement *very soon* about our type of aplanatic reflecting telescope, if we are to have the priority for it." He had not finished the optics for the f/6.8 Ritchey-Chrétien he was making at the Dina Laboratory, but his own f/6.25 was ready, and they could use it for the demonstration.[33] However, Ritchey managed to finish the second pair of mirrors on May 31, 1927. Mounted in the open steel telescope frame, they formed the first completed Ritchey-Chrétien telescope in existence. Deslandres named a commission, headed by Fabry, to examine it.[34] They judged it acceptable and released Ritchey's payment for making it. He and Chrétien took this 0.5-meter telescope to the meeting of the Academy of Sciences on July 18 and exhibited it there. They published a paper that described the telescope in full and included a photograph of it. It was about twice the size of a chair, a favorite comparison object, which Ritchey included in many of his photographs of mirrors and telescopes to give an idea of their scale.[35] Their paper attracted a fair amount of attention from international press services in Europe, but few if any newspaper articles about the new type of telescope appeared in newspapers or magazines in America.[36]

It proved impossible to set up a camera coupled with a high-power microscope to photograph the images of an artificial star to illustrate the reduction of the off-axis aberrations by the Ritchey-Chrétien system as Ritchey had planned. Instead, he observed the images visually and drew them. For comparison, he observed and drew the off-axis images of a conventional paraboloid reflector of the same focal ratio and aperture, which he had made especially for this demonstration.[37] It exhibited graphically the great advantage of the Ritchey-Chrétien system for wide-field

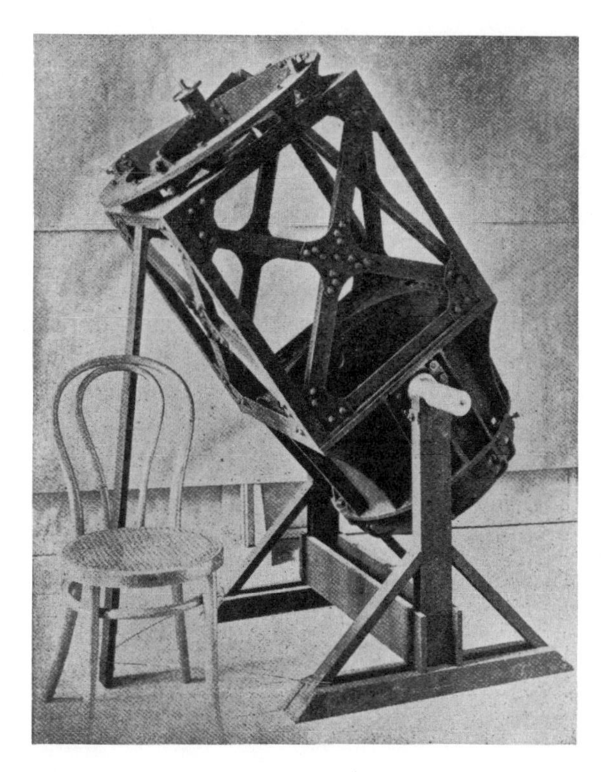

First 0.5-meter Ritchey-Chrétien reflector, displayed on a wooden hurdle, 1927. From the *Journal of the Royal Astronomical Society of Canada.*

photography but also demonstrated, as Ritchey wrote, the need for curved photographic plates to utilize this advantage fully.[38]

He also published a third paper, explaining the other advantages of the Ritchey-Chrétien system. They all flow from its compactness. The very short tube (in comparison with a conventional Newtonian or prime-focus telescope of the same focal ratio and diameter) makes the weight and hence the cost of the Ritchey-Chrétien mounting much smaller. The size of the dome and its cost are correspondingly reduced. Because the weight is smaller, flexure is reduced. Also, a relatively small insulated cover and cooling system can be used to keep the whole telescope at the expected nighttime temperature, reducing thermal distortions of the mirror and mounting. Finally, the observer is much more conveniently located behind the mirror rather than at the upper end of the telescope.[39]

Ritchey was interrupted in making these comparisons and writing these papers by a severe eye infection. His right eye had become ulcerated. An American ophthalmologist in Paris, Dr. Louis Brosch, performed an operation which saved his eye, according to Ritchey. However, his sight was permanently impaired. After the operation he had to rest his eyes and could do nothing for two months.[40] Chrétien sent his student, Vincent Nechvíle, a thirty-seven-year-old Czech, to work with Ritchey and see for him. As a result of this collaboration, Nechvíle, an applied mathematician, became an expert in calculating Ritchey-Chrétien optical surfaces to high precision. Even after he returned to Prague, as a docent at Charles University, he provided many of the detailed optical layouts and test data for Ritchey's later designs.[41]

David B. Pickering, an American amateur astronomer and writer, was traveling in Europe that summer. He visited the Dina Laboratory twice. The first time, before the operation, Pickering met Ritchey, "a good sized, fine looking man." In the large, impressive laboratory, surrounded by giant cranes, motors, chain hoists, and steel tracks, Ritchey showed him two cellular disks, the larger one the rebuilt 1.5-meter disk. There was no reason he could not make such disks up to ten meters in diameter, Ritchey said. He hoped to erect a fixed vertical telescope at a high, dry, even-temperature, low-latitude site, contiguous to the Pacific Ocean. Probably, he was not thinking of Mauna Kea, on the big island of Hawaii, now the home of many of the world's great telescopes, but this is a good description of it. Pickering was obviously inspired by Ritchey. The writer also met Chrétien, who spoke very positively of Ritchey's work.[42]

On Pickering's second visit, after the operation, Ritchey was nowhere in evidence. He was at home following his doctor's orders, but Couder, whom Pickering met, did not tell him this. Instead the ambitious young director confided that Ritchey "was ill, taking a much needed rest at home" . . . "since the expiration of his contract with Mr. Dina." He managed to convey to Pickering that Ritchey had suffered a serious nervous breakdown at the least, and the writer passed on this "news" to the readers of *Popular Astronomy* back in America.[43] He skirted the issue, but his meaning was unmistakable. Pickering even inserted a short note in the magazine that the "Ritchey-Dina project" had been abandoned. This was true, but the note continued with some misstatements about the comparison of the Ritchey-Chrétien and paraboloid that the optician was making. Ritchey corrected the errors, and Pickering subsequently published the corrections, but the whole episode did nothing to help restore Ritchey's image in the United States.[44]

As soon as he could see again, Ritchey began writing a series of articles for *L'Astronomie*, the *Bulletin of the Astronomical Society of France*, giving in detail his ideas on telescope design and astronomical photography. In these articles he described the methods he had used at Yerkes and Mount Wilson, and the plans for future observing methods he had developed on the basis of all his past experience. Ritchey wrote the articles in English, and E. M. Antoniadi, the editor of *L'Astronomie*, translated the first of them into French. Later Chrétien himself did some of the translations, and other French friends of Ritchey helped out with the last-minute insertions.[45]

In this series, Ritchey gave in brief and fragmentary form a synopsis of his entire career, interspersed with his philosophy of telescope making and photographic observing. The first three articles dealt chiefly with the Ritchey-Chrétien design and its advantages.[46] The fourth was devoted to astronomical photography with very-long-focal-length instruments. It included descriptions of his earlier ideas on horizontal fixed telescopes, as worked out in the Snow telescope at Yerkes (before the fire) and at Mount Wilson, and on his current plans for large, fixed vertical telescopes.[47] The last two articles were on "modern astronomical photography" and gave many more of Ritchey's ideas about the fixed vertical instruments.[48] Several of Ritchey's best astronomical photographs from Mount Wilson were used as illustrations for these articles, but the reproductions did not do justice to the originals.

Toronto astronomer Charles A. Chant, long-time editor of the *Journal of the Royal Astronomical Society of Canada*, read the articles and with Ritchey's permission republished them. Actually, Chant did not use translations, but instead published Ritchey's original articles, written in English, slightly modified and rearranged by the author. The first article of the series in the Canadian journal was the English version of Ritchey's first article for *L'Astronomie* and *Comptes Rendus* on the fixed vertical telescope.[49] The rest, pretty much in order but with some changes, followed the *L'Astronomie* series. Two were on the Ritchey-Chrétien reflector,[50] one on astronomical photography with very high magnifying powers[51] and the last two on the new astronomical photography.[52] The Canadian journal used a glossy-paper process for reproducing Ritchey's photographs, which were spectacular.

Thus Ritchey's ideas were published in France and in Canada but not in his own country. No American magazine, such as the *Publications of the Astronomical Society of the Pacific* or *Popular Astronomy*, would dare risk the enmity of the powerful Hale and Adams.

In the articles, Ritchey concentrated on the future, but in his descriptions of his own results he was always the hero. From the description of his first "astronomical laboratory" in his home in Cincinnati, while he was still a student, right up to the Dina Laboratory in Paris, he never mentioned any failure, miscalculation, or mistake. In one moving passage Ritchey named the six "master technicians" whom he had had "the rare privilege of counting among my personal friends," R. Benecke, superintendent of the Cramer Plate Company, "inventor and master of the art of preparing photographic emulsions"; Delloye; Thomson; Edward E. Barnard; Chrétien; "and finally, James Ritchey, my father, one of the great inventors and one of the greatest geniuses of mechanics whom I have ever known." Hale and Adams were conspicuously absent from the list. In a very few remarks scattered through the articles, Ritchey recounted obliquely how Hale had thwarted his plans, especially with Hooker, had for several years denied him the use of the 60-inch telescope he had built, and had never let him observe with the 100-inch. However, Ritchey did not mention Hale's name anywhere in the articles. In his own notebook he was much more scathing about the "wealth and power and egotism" of the "officials of the Mt. Wilson Observatory in 1910."[53]

Ritchey's basic idea, repeated over and over in one form or another in all the articles, was to get the best resolution, to show the finest detail possible, and to record the faintest star images. He included many calculations which demonstrated that concentrating the light from a star in the tiny "diffraction limit," set by the immutable laws of nature, would enable astronomers to record much fainter stars than previously. To do so would mean using the available seeing to its ultimate and improving on it. To use it effectively would require being able to switch rapidly between telescopes of different focal length so that in the rare intervals of best seeing, the most important objects could be photographed in the finest detail. This was the reason for the fixed vertical telescope with its interchangeable primaries and matched secondaries. The observer would be trained to be an expert in guiding the telescope "perfectly." He would be able to interrupt the exposure during instants of bad seeing and continue it during moments of good seeing. To improve the seeing, the telescope would be cooled within an insulated cover to the expected nighttime temperature. These and other similar practices Ritchey advocated seemed visionary if not "crazy" to his contemporary critics but were the early versions of the "new" ideas that are being tested and included in all advanced observatory designs today.

The fixed vertical telescopes themselves, however, are far too expensive to be considered today. They were too expensive in Ritchey's time as well, but he could not accept this conclusion. His first French plan, as stated in the articles he wrote in 1927, was for a system with four different primary mirrors, each five meters (200 inches) in diameter. Two were Schwarzschild systems giving, with their secondaries, focal ratios f/2.75 and f/4 for wide-field photography of faint nebulae. The other two were Ritchey-Chrétiens, working with their secondaries at f/6.8 and f/12. The latter of these could also be used with another secondary, giving an f/20 "modified Ritchey-Chrétien" suitable for stellar spectroscopy.[54]

In 1927 Chrétien had calculated the size of the coma in this type of "hybrid" telescope, whose secondary provides a longer focal length than the Ritchey-Chrétien primary was designed for. He thought that he had proved that its coma is always *smaller* than that of a conventional Cassegrain reflector with the same focal ratio.[55] In fact, he had made an algebraic error; as we know today the coma of such a hybrid telescope is *larger* than in the corresponding conventional Cassegrain. However, for spectroscopy, which uses only the star image on axis, the coma is zero in any case and no harm is done.

Ritchey's later plan for a fixed vertical telescope, in the article written in 1928, called for *six* primary mirrors, each *six* meters (240 inches) in diameter. The Schwarzschild systems' focal ratios remained unchanged, while there were now four Ritchey-Chrétiens with focal ratios f/6.8, f/10.5, f/16, and f/40, and additional secondaries giving modified Ritchey-Chrétiens of f/27, f/60, and f/90. At the same observatory he proposed to have also an equatorially mounted 6-meter reflector with two interchangeable sets of Ritchey-Chrétien optics, f/6.8 and f/12.[56] It would have been wonderful for astronomy but fantastically expensive to build and operate. This plan was completely unrealistic, even for those

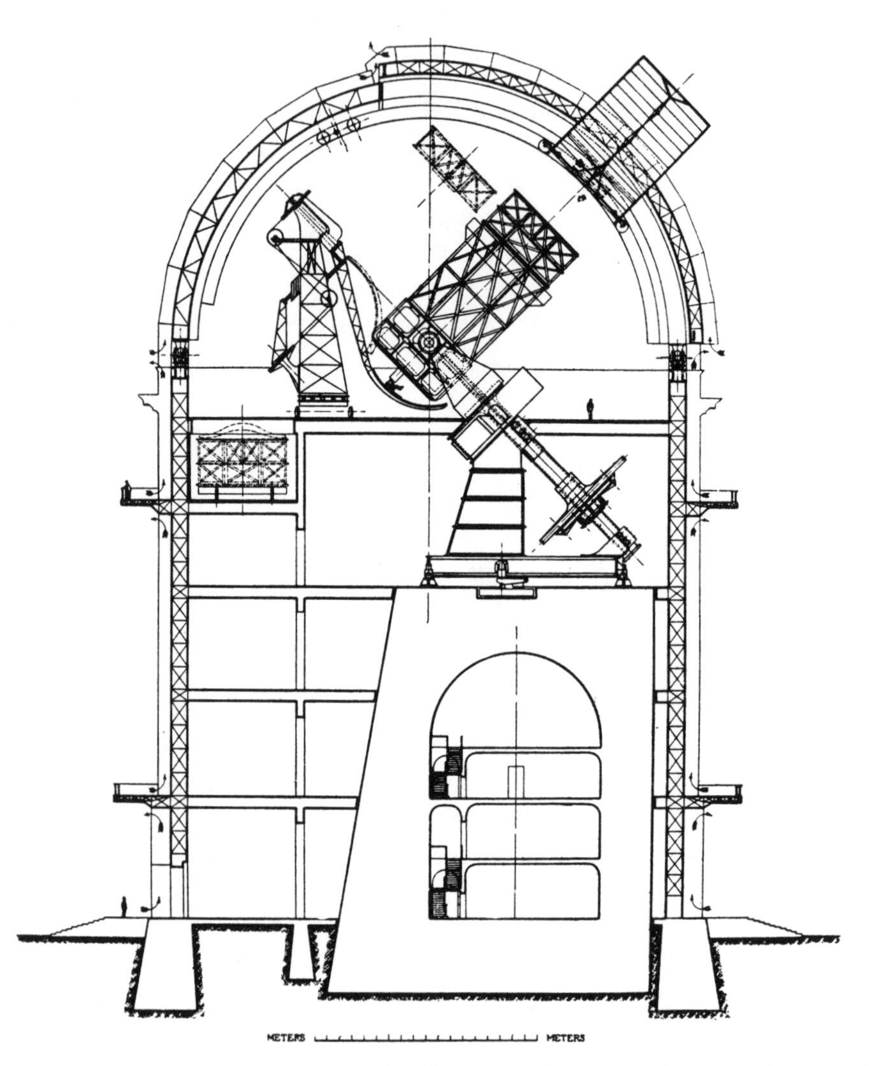

Ritchey's design of a 6-meter equatorially mounted Ritchey-Chrétien telescope. The observer is on the curved movable arm behind the primary mirror. Note the size of the human figures on the floor and the ground. From the *Journal of the Royal Astronomical Society of Canada*.

golden days just before the stock-market crash of 1929 and the Great Depression which followed it.

After Ritchey recovered from his eye operation, he wrote his friend Ralph A. Sampson, longtime astronomer royal for Scotland and professor of astronomy at the University of Edinburgh, to let him know that he would welcome an invitation to come to Britain and describe his plans. Sampson was enthusiastic, for he was an advanced optical thinker and was testing built-up cellular mirrors himself. He strongly recommended that Ritchey be asked to lecture at the Royal Astronomical Society and probably at the Optical Society also.[57]

George Willis Ritchey, Paris, 1927, with a built-up cellular mirror disk. Courtesy of the Astronomical Society of the Pacific.

Invitations soon arrived in Ritchey's laboratory in Paris from both these societies. He delayed accepting them until spring, to complete several of the plans on which he was working so that he could include descriptions of them in his talks.[58] Thus in April 1928 Ritchey crossed the English channel and presented two lectures in London. One, at the Royal Astronomical Society, was on "The New Astronomical Photography," the other, given the following week as the Thomas Young oration before the Optical Society at its meeting at Imperial College in South Kensington, was on "The Modern Reflecting Telescope." Both lectures covered essentially the same material as the six articles published in the French and Canadian magazines.[59] The published version of the Thomas Young oration is an excellent, concise description, in technical terms, of Ritchey's ideas on telescopes.[60]

After his return to France, Ritchey concentrated on preparations for the talk he was to give to an invited audience at a spectacular exhibit of his astronomical photographs at the St. Gobain Glass Company headquarters in Paris. It was a gala event. The exhibit included nearly fifty pictures, all enlarged on glass transparencies, 14 by 17 inches in size and lit from behind. Most were astronomical pictures of the moon, nebulae, and clusters taken at Yerkes and Mount Wilson, but there were also several landscapes Ritchey had taken at the Grand Canyon and Yosemite, as well as a few scenic views of Mount Wilson. He rehearsed the talk with Chrétien several times to make sure that there would be no breakdown in the translation. The exhibit was first displayed at the Institute of Optics in connection with the annual meeting of the Physical Society of France in late May 1928. Then it was taken to the Sorbonne for the meeting of the Astronomical Society of France on June 6. Ritchey gave his talk for the first time there, illustrated by many slides. Afterward the members and guests inspected the mounted transparencies at their leisure. After these two meetings, the St. Gobain Glass Company moved the exhibit to their office building on the Place des Saussaies. There, before an invited audience of scientists and reporters, Ritchey repeated the talk. It was an outstanding success. Among those in the audience were George C. Comstock, longtime professor of astronomy at the University of Wisconsin and secretary of the American Astronomical Society, de la Baume Pluvinel, and the Duc de Gramont.[61]

The duke, actually Armand-Antoine-Auguste-Agénor de Gramont, a distinguished physicist in his own right, was the head of one of the most illustrious noble families in France. Though under the Republic the nobility no longer officially existed, he was often referred to by his title, like de la Baume Pluvinel. The duke had become Ritchey's most important and prestigious sponsor in France. He was the chairman of the board of the Institute of Optics, a director of the St. Gobain company, and the founder and president of a precision optical company. De Gramont pulled strings to get extensive publicity for Ritchey's exhibit and talks.[62] Undoubtedly, he did so as part of a campaign to establish a large French observatory, or at the very least a large telescope *somewhere*, for which the St. Gobain company could supply the glass.

When the exhibit first went on display at the physicists' meeting, Ritchey was named a knight of the Legion of Honor "for his services to science, to optics, to astronomy, and, since his stay in France, to the science of our country."[63] It was a high honor, and the Duc de Gramont was no doubt involved in recommending the American for it.

The exhibit, the publicity for it, and the knighthood did awaken some interest in Ritchey in his own country. The *New York Times* mentioned his large-telescope plans in a column on its editorial page in July 1928 and followed up with a long feature article by Waldemar Kaempffert on September 30. He was a good science reporter who had seen the exhibit and had interviewed Ritchey in Paris. Kaempffert's description of the planned "supertelescope" (fixed vertical telescope) was so vivid that it could easily have given an unsuspecting reader the idea that the instrument already existed.[64]

Ritchey wanted the publicity, for he had been trying to get back to the United States since the collapse of the Dina project. In the summer of 1926, Harlow Shapley, director of Harvard College Observatory since 1921, had met Ritchey in Paris and seen his laboratory. Privately, Shapley wrote to Adams that "Ritchey is temporarily stranded. He had two years of fun but no very concrete results of his experimenting with cellular mirrors and plate glass facing."[65] Shapley did not tell Adams that he had discussed the idea of bringing Ritchey back to the United States to head an optical laboratory at Harvard. Shapley was the one American director courageous (or foolhardy) enough to oppose Hale and Adams, at least covertly. He wrote Ritchey in October of that year to ask the conditions under which he would come. Harvard had funds for a 60-inch reflector to be put in the southern hemisphere and plans for a very large reflector, Shapley said. Ritchey was overjoyed. Shapley's letter seemed to imply so much. The optician could see before him the opportunity to close out his career doing the research and development on telescopes he loved, with a capable staff of assistants, at a laboratory of his own in the United States. He outlined his ideas in a long letter to the Harvard director and followed it up with another a month later.[66]

Actually, as Ritchey soon learned, Shapley was trying to raise the money for the large telescope from the International Education Board. Ritchey's letter, with its budget estimates, was part of the proposal the Harvard director submitted, but it was not accepted. The Rockefeller Foundation granted no funds for the telescope or the laboratory. Shapley had to tell Ritchey he did not have the money to go ahead and could offer him only vague hopes for the future. Ritchey was disappointed, but by this time he had received the French grant to build and test the 0.5-meter Ritchey-Chrétien reflector.[67]

Shapley kept in touch with Ritchey and in September 1927 asked if the optician could refigure a 60-inch disk that Harvard owned. At that time Ritchey had no assurance that his job in France would last beyond the end of 1927. He told the Harvard director that he planned to return to the United States in November and was seeking contracts. He strongly recommended starting with a new 60-inch (or even, if the funds were very limited, a 48-inch) high-quality St. Gobain disk, rather than trying to refigure the old, much thinner mirror. A few weeks later Ritchey postponed his departure date to January 1928.[68]

Evidently, Delloye, de Gramont, and de la Baume Pluvinel, between them, arranged for Ritchey's salary to continue, perhaps on a month-to-month basis. In November he planned to remain in Europe until mid-March 1928 and to begin his return to America with a lecture tour. A few weeks later he postponed it still further to work on his book, probably under a salary or honorarium from the St. Gobain company.[69]

Earlier that fall Shapley had also written to John C. Merriam, president of the Carnegie Institution of Washington, telling of Ritchey's work in Paris. He informed Merriam that Ritchey had prepared the manuscript and illustrations for a book, and he suggested that the Carnegie Institution provide the money necessary to publish it. Merriam replied that it was "difficult" to grant Carnegie funds for publications from "outside sources," but he suggested that Shapley discuss the idea with Adams.[70] Shapley knew very well what Adams's reaction would be and quietly dropped the idea.

In December 1927 Ritchey himself wrote to Henry S. Pritchett, the head of the Carnegie Foundation, appealing for support for this book, because he had done much of the work at Mount Wilson and also because he had been unjustly dismissed eight years before. Pritchett forwarded this letter to Merriam, describing Ritchey as "a man who had an enlarged egotism but who also apparently had a great deal of ability in certain narrow fields." Walter M. Gilbert, the Carnegie Institution administrative secretary whom Adams carefully cultivated on every visit to Washington, prepared an analysis showing that Ritchey had been fired legally and had no basis for a claim. Merriam admitted to Pritchett that Ritchey had undoubtedly done good work and that there would be "advantages" in the Carnegie Institution publishing his book. However, in the past a "serious situation" had developed between Ritchey and "certain members of our staff" (Hale and Adams), so it would be "unwise" to do so.[71]

Ritchey, who did not know of this rebuff, wrote directly to Merriam in February 1928, stating the astronomical importance of the development work he was doing and requesting funds from the Carnegie Institution to set up a laboratory in the United States. The optician outlined in immodest, but basically correct terms, his past contributions and the results he hoped to achieve in the future. Ritchey had already sent Merriam reprints of the French series of articles and appealed to him to check on his technical statements by consulting "some special expert, such as my friend Elihu Thomson (*not*, naturally, my *enemies*)." He also emphasized that he was requesting support for a laboratory "absolutely independent of Mount Wilson Observatory."[72]

This letter arrived at the Carnegie Institution headquarters in Washington just as Merriam, a geologist and paleontologist, was departing for a long field trip in Mexico. He had already sent Ritchey's reprints to Adams, asking whether he felt "that there [was] anything [they] could or should do in furtherance of Dr. Ritchey's work or in assistance to him personally." Now, when Ritchey's letter to Merriam came, Gilbert, his assistant, took it upon himself to send a copy of it to the Mount Wilson director, inviting him to comment on it. Gilbert certainly knew that Adams was Ritchey's enemy. He did *not* send a copy of the letter to Thomson.[73]

Adams, naturally enough, had only negative comments to offer. In a long letter to Merriam, he downplayed all of Ritchey's contributions and provided an extremely slanted critique of the Ritchey-Chrétien system.

He said that its wide field would be "valuable" for photographing large nebulae and "in less degree" for photographic photometry and astrometric work. However, the curved plates would be "very troublesome and expensive." The Ritchey-Chrétien would have no advantage for spectroscopic work. All of this was true, but it over-emphasized the negative. Then Adams went on to write that the Ritchey-Chrétien secondary mirror, since it is necessarily larger than in the conventional design, "would involve a loss of light." Again, this is a factually correct statement but misleading. The loss of light is small in either telescope, while the gain in the size of the field of the Ritchey-Chrétien is a tremendous advantage. Finally, Adams stated that "a very serious, and to my mind fatal, objection" to the Ritchey-Chrétien system was that it "could be used only in a single combination with one focal length" and thus could not be used in the coudé mode, nor in more than one Cassegrain combination. This is absolutely incorrect. Evidently, Adams was not aware of the "hybrid" concept, had not read Ritchey's or Chrétien's papers on the subject, and had not consulted any knowledgeable optical theorist. Yet he made this firm statement as his considered advice. Adams concluded by saying that "a good many years" had passed since Ritchey was at the Observatory and that he thought he could now speak quite impartially regarding him. Adams then poured out all his negative prejudices against the man who had "good technical ability" but "an extraordinary narrowness of mind," whose "conception of astronomy has always been the taking of photographs" but "never" to measure or interpret them. Adams stated that Ritchey was "the most difficult man" that he or Hale had "ever had to deal with." He thought it "extraordinary" that Ritchey "still labor[ed] under the delusion" that Hale had "persecuted" him. Adams doubted that "any competent psychiatrist would have considered Ritchey as of entirely sound mind during his last years at Mount Wilson" but, of course, offered no analysis of Hale's state of mental health. The Mount Wilson director concluded that he was "sorry to have to write in this way regarding one with whom [he] was associated for so many years, but the Institution [was] entitled to as accurate a statement of the conditions as it is in [his] power to give."[74]

This letter ended whatever chance Ritchey might have had for financial support from the Carnegie Institution, the Carnegie Foundation, or indeed from *any* American philanthropic organization.[75] However, in 1928 Ritchey revealed his plan to build a large, fixed vertical telescope at the edge of the Grand Canyon in Arizona. Toward the end of his series of articles he had alluded to a "long-continued series of rigorous comparative tests" he had carried out with a 16-inch telescope "in many different locations" on his vacations. From these tests he had concluded that the best observatory sites were in Arizona, at heights around 7,500 feet. Now he published in *L'Astronomie* a short note on his plans for a "future observatory of great power," with a composite drawing and photograph, showing a proposed large, fixed vertical telescope at the edge of the Grand Canyon.[76]

At the same time he published this plan, Ritchey wrote the director of the National Park Service in Washington requesting permission to use the site, which he described as Comanche Point, overlooking the Grand Canyon and the Painted Desert. Previously, it had been called Desert View. Ritchey also wrote Merriam asking him at least to endorse the project.[77] Though Ritchey did not know it, Merriam was the main adviser to the Park Service on educational activities in the national parks. Like many geologists and naturalists of his time, he envisioned the national parks as centers in which the general population would be gently indoctrinated in the wonders of nature and the beauties of science. He was deeply interested in turning the Grand Canyon into an outdoor display of the forces of nature but had no desire whatever to contaminate its scenery with an astronomical observatory. Though some of the park service personnel, particularly those at the canyon, were initially favorable to Ritchey's dream, Merriam was not. He promised to study the plan, but his view "that the National Park areas should be preserved as nearly as possible in their primitive form and should be used for their scenic values" doomed it from the start. Most informed astronomical opinion today would suggest that Ritchey's site, on a fairly level, high plateau, immediately adjacent to a deep and presumably turbulent canyon, could not have had very good seeing. Whatever the facts, in spite of additional material that Ritchey sent Merriam from Paris, the plan died.[78]

In the winter of 1928–29 a young American student, Clinton B. Ford, met Ritchey at his laboratory at the Paris Observatory. Ford, then fifteen years old, was in Europe for a year; his father, a professor of mathematics at the University of Michigan, was spending a sabbatical leave there. Young Ford, already interested in astronomy, was inspired by Ritchey's plans, designs, photographs, and vivid descriptions of his projects. He could see that Ritchey was bitter at the treatment he had received from Hale but otherwise in good spirits, though he seemed to be living close to the edge of poverty.[79] He desperately wanted to return to the United States but had no prospects to do so. He was reduced to hoping that testimonials his friends wrote for him, such as one by Antoniadi that was published in England, would bring him a job.[80]

In 1929 Ritchey at long last published his book, *The Development of Astro-photography and the Great Telescopes of the Future.*[81] Relatively short but handsomely printed in large format, in both French and English in parallel columns, with many illustrations, it is a spectacular volume. The picture of the planned observatory at the edge of the Grand Canyon adorns its cover. The book is a summary of all of Ritchey's ideas on telescopes, along with his own descriptions of his astronomical photographs taken with the Yerkes 24-inch reflector and 40-inch refractor, and the Mount Wilson 60-inch reflector. In retrospect, he was able to find

Ritchey's composite drawing and photograph of a fixed vertical telescope at the edge of the Grand Canyon. Courtesy of the Carnegie Institution of Washington.

abundant evidence in his papers that spiral "nebulae" are really galaxies of stars, a concept that he *had* advanced in his papers on novae dating back to 1917, but which Edwin Hubble had done much more to confirm systematically.

Ritchey repeated all the material from the six articles in shorter, better organized form in the book. The telescope designs, however, had increased in scale. He showed a plan for a 7-meter (280-inch) equatorial Ritchey-Chrétien and an 8-meter (320-inch) fixed vertical telescope, with five different primary mirrors, one a Schwarzschild and the other four Ritchey-Chrétiens. Furthermore, he proposed building a chain of *five* such observatories, at latitudes 0 degrees, 18 degrees north and south, and 36 degrees north and south, so that one or another of them could photograph any region of the sky outside the polar caps. All five observatories were to be "under one democratic management" (meaning not Hale's). Ritchey emphasized that this group of observatories would be "*international*, . . . its photographs . . . available for competent workers desiring them, *of every nation*," and that in his view they should "be used *equally* for *popular education* and for *scientific research*." The work would necessarily become "a cooperative work of nations," and the five supertelescopes would thus become "the mighty guns of Peace." He concluded in Biblical terms:

> This is the Great Adventure. These telescopes will reveal such mysteries and such riches of the Universe as it has not entered the heart of man to conceive.
>
> The heavens will declare *anew* the Glory of God.

It was an inspiring vision, if quite impractical. Ritchey's "international committee" at the time consisted of David B. Pickering, John Hindle, who was an English optician, and a few French scientists. Their chance of raising even a small fraction of the money necessary to build a single, fixed vertical telescope was essentially nil.

One of the illustrations in Ritchey's book was a striking photograph of a workman kneeling on a huge assembly of plate glass panels on a turntable ten meters in diameter. It suggested the huge cellular mirrors that Ritchey still hoped someday to build. The book was published by the St. Gobain company and dedicated to Delloye, its director and Ritchey's longtime sponsor. From the company's point of view, the book

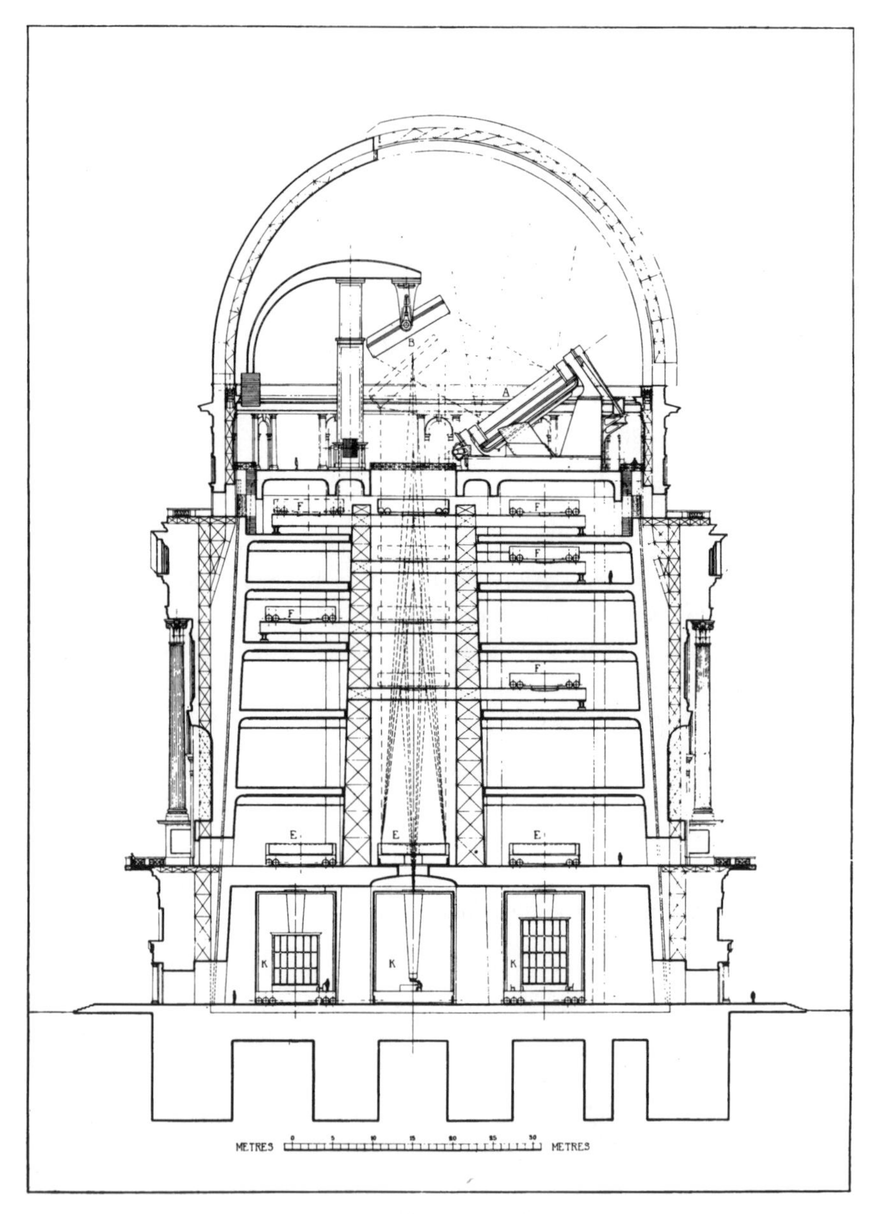

Ritchey's design for an 8-meter fixed vertical telescope. Note the human figures that show the gigantic scale of this proposed instrument. From Ritchey's book *L'Evolution de l'Astrographie et les Grands Télescopes de l'Avenir.*

was a prospectus for large telescopes for which it could supply the glass. Unfortunately, the Great Depression was close upon it, and the direct result of the book on its telescope business was negligible, though it may have helped promote its image in France. Ritchey's book was very favorably reviewed in France and Great Britain but ignored in his native land.[82]

Ritchey himself knew that it was absolutely essential to take a series of spectacular photographs of nebulae with a Ritchey-Chrétien reflector to demonstrate its superiority and gain acceptance for it.[83] To do so was almost impossible with the slender resources at his command. There was only one Ritchey-Chrétien telescope in the world, the 0.5-meter

Ritchey at the guiding eyepiece of the first 0.5-meter Ritchey-Chrétien reflector, mounted at Vallière, with Chrétien looking on, 1930. Courtesy of the Cercle Henri Chrétien.

(20-inch) reflector that they had described and exhibited to the Academy of Sciences. It had a simple mounting, not suitable at all for photography, and it required concave glass photographic plates, which Ritchey had to coat with the sensitive emulsion himself. He could have only a few ready for use at any one time.

Nevertheless, Chrétien was able to borrow an old equatorial mounting from the Nice Observatory, and the Duc de Gramont provided funds to make the double-slide plateholder and other auxiliary equipment required to use it for photography.[84] Ritchey set the telescope up on a pier at the duke's estate, the Chateau de Vallière, some twenty miles north-

Direct photograph of region in the Pleiades, taken with first 0.5-meter Ritchey-Chrétien reflector at Vallière. The brightest star, above the center, is Merope. The circle is approximately 1.5 degrees in diameter. Courtesy of Richard B. Burgess.

east of Paris. A roll-off shed protected the telescope from the weather when it was not in use. There in June and July 1930 Ritchey took the first photographs ever obtained with a Ritchey-Chrétien telescope.[85]

They are not very impressive. The mounting was old and rickety, precluding long exposures. Ritchey himself was old and had no assistant. To strive all night for the best possible photographic results with an inadequate mounting requires drive and energy that he no longer had. Apparently, it was cloudy most of the time, and the exposures that he did get were taken through and between clouds. One, which he marked "field of faint stars in the Milky Way in Cygnus" is a 35-minute exposure, stopped by clouds, that shows only relatively few stars and no nebula. Very probably, he was trying to get a photograph of NGC 6960 or NGC 6992–6995, the "Veil Nebula," parts of the Cygnus Loop, of which he had taken such spectacular pictures at Mount Wilson. If so, he either missed the field or the exposure was too light to show any of the nebulosity. Another 20-minute "net" exposure (suggesting it also was interrupted by clouds) of the Pleiades shows stars but practically no nebulosity. The star images in both pictures are satisfactorily small all over the 1½ degrees–diameter field but not equally round in all parts of it (suggesting that the optics were not perfectly aligned). A picture of the moon is adequate but not spectacular; it is far inferior to the lunar photographs Ritchey had taken with the longer focal-length 60-inch at the much superior Mount Wilson site.[86]

Ritchey published a description of these first photographic results in the *Comptes Rendus*, the journal of the Academy of Sciences, but did not include reproductions of the plates.[87] The paper aroused little attention. Ritchey was hard pressed for money and in his report to the Duc de Gramont had appealed for more generous financial support. The American continued his work in his laboratory in the Paris Observatory but accomplished little more in the way of new designs or ideas. He dreamed of developing more sensitive photographic plates.[88] As always, he remained uncommunicative and concentrated almost entirely on his own plans and ideas.[89]

Ritchey's last article published in France was a popular piece for *L'Illustration*, a glossy weekly. It was entitled "The Great Adventure" and headed by a beautiful reproduction of the design of the observatory at the rim of the Grand Canyon. In the article he described his plan for an 8-meter, fixed vertical telescope, with four primaries, but stated it would be only "the first stage of perfecting the great instruments of the future." He envisioned achieving in the next stage a supertelescope of the same type but with primary mirrors 24 meters (960 inches or 80 feet) in diameter, in a building as high as the Eiffel Tower.[90]

This paper was a fitting conclusion to Ritchey's seven years of spectacular plans, extravagant hopes, and few concrete results in France. Near the end of 1930 his opportunity finally came to return to the United States to stay. After spending November winding up his affairs in Paris, Ritchey, his wife, and their daughter left the continent, stopped for a few days in London, and embarked for America. They reached their country with great hopes just before 1931 dawned, as the Depression deepened and strengthened.[91]

Pasadena and Palomar

1919-1931

PROGRESS TOWARD THE NEXT BIG TELESCOPE did not halt in Pasadena in the 1920s, while George Willis Ritchey scratched out his living in Azusa and dreamed of super-reflectors in France. Success begets success, and no one realized this better than George Ellery Hale. Wealthy men and the foundations they endow prefer to give their money to famous observatories and their directors who have demonstrated that they know how to get results and capture the headlines, not to aging prophets who make great promises but have little to show in the way of photographs, telescopes, or honors. As soon as the 100-inch reflector was completed and in partial operation in 1919, Hale pronounced it a success and trumpeted its accomplishments throughout the astronomical world, just as he had done for the Yerkes 40-inch refractor in 1897 and the Mount Wilson 60-inch reflector in 1908. Immediate validation of the 100-inch was even more essential to Hale than for these earlier telescopes because of Ritchey's claims that it was a failure. Though not published, his criticisms were widely known among the Mount Wilson Observatory staff and Carnegie Institution of Washington administrators. That he had been unceremoniously fired no doubt brought word-of-mouth attention to his cause. Hale had to prove him wrong.

Less than a month after the 100-inch went into regular operation, Francis G. Pease published a short paper based on direct photographs he had obtained with it during its initial trials, of an object then generally known as "Campbell's hydrogen-envelope star." Pease's photographs showed that in fact it was a small ring nebula, and demonstrated that the new reflector was capable of taking good sharp images, revealing very fine detail.[1] Large-scale enlargements of two pictures of the moon that Pease had taken, published in the *Publications of the Astronomical Society of the Pacific*, provided graphic evidence of the 100-inch's excellence.[2] That it was Ritchey's former assistant who had superintended the completion of the 100-inch mounting, and then used the telescope to outperform his former mentor's prize 60-inch, was an irony that could not have escaped knowledgeable observers.

In early September 1919, even before the 100-inch was in regular use, Hale had rushed a paper to the American Astronomical Society meeting at the University of Michigan, on the "preliminary results of a comparative test" of the two telescopes. Needless to say, they "proved" that the 100-inch performed fully up to its expected specifications. Neither Hale nor any of his staff was at Ann Arbor, but he had sent eight slides, pictures of the telescope, and reproductions of spectrograms and direct photographs taken with it. They were the best kind of evidence that the telescope was working.[3]

Hale made sure that his staff produced spectroscopic results with the new telescope quickly, too. Before the end of 1919 Paul W. Merrill published a short paper on the spectrum of a variable star, and Walter S. Adams and his young new colleague Alfred H. Joy published two more on the spectrum of a nova (or "new star") and on the spectra of two eclipsing variable stars. All three of these papers made it clear that observers could obtain spectrograms of fainter stars with the 100-inch than with the 60-inch, though none of them boasted of it.[4]

By the end of 1919 Hale had published a short article on the 100-inch in *Popular Astronomy*, the magazine most amateur and professional astronomers of the time read. It was basically his American Astronomical Society paper rewritten in less technical terms, describing the first results with the telescope. He said that tests of the 100-inch against the 60-inch in use showed that "the full gain in light gathering power to be expected from the increased aperture [a factor of three], has actually been obtained." The 100-inch was thus "found to be a complete success." The frontispiece of the issue of *Popular Astronomy* containing this article was an enlargement of one of Pease's photographs of the moon.[5]

Hale followed up with another paper, sent to the next meeting of the Society, held on the Mount Holyoke and Smith College campuses in August 1920, on the excellence of the 100-inch. By this time some of the astronomers must have wondered why it was necessary to keep proving it. In fact, the audience was different each time, and Hale had more evidence now, too. Pease and physicist A. A. Michelson had used the 100-inch telescope with an interferometer to measure the distance between the two stars that form the binary system Capella, a double too close to be resolved with the 60-inch.[6]

A year later Adams and Frederick H. Seares, now his chief lieutenant in running the observatory during Hale's absences, finally published quantitative details of the comparative tests of the two telescopes. They showed that in conditions of excellent seeing the gain of the 100-inch was nearly, but not quite, as large as expected from its increased aperture. However, in poorer seeing the advantage of the 100-inch was considerably smaller. This was an illustration of a point that Ritchey constantly emphasized. In a larger telescope, with its longer focal length, the apparent disk or "seeing image" of a star is magnified and the light is thus spread out. Adams and Seares deemphasized this issue, as many proud possessors of large, new telescopes have done since, by overstating the goodness of the seeing conditions at their observatory.[7]

These papers and articles were intended to convince typical astronomers and educated amateurs of how good the 100-inch was. To the power brokers, such as W. W. Campbell and Edwin B. Frost, Hale kept up a barrage of personal letters. Perhaps Campbell did not respond quite as enthusiastically as Hale had expected to the news that "his" hydrogen-envelope star was, in fact, a nebula, for the Lick director knew that William H. Wright of his own staff had already seen it on photographs he had taken with the little Crossley reflector. Wright had described the result in a long paper on nebulae, which had been delayed in publication. Hale would never have let that happen. Wright himself was entranced by enlargements of the 100-inch moon pictures. He opined they would open up "an immense . . . practically new field in the study of the moon" and suggested that Hale send reproductions of them to geologists at Berkeley and Stanford.[8] Writing such letters, and many more like them, cost Hale tremendously in psychic strain and led to his frequent depression and withdrawal. Important visitors whom he had to impress, like John C. Merriam, the new head of the Carnegie Institution, nearly always left Hale in hiding after a few days, under his doctors' care.[9]

These glowing reports on the 100-inch were not part of a campaign for a new, even larger reflector but simply part of the normal self-glorifying activities of a large, successful research institution (at least in the post-Hale era). The first clear call for a much larger telescope came from a man who was not an astronomer at all, but a wealthy Michigan businessman. He was William C. Weber, who lived in Detroit and dealt in land and timber options in northern Wisconsin, the upper peninsula of Michigan, and Ontario. He had been instrumental in raising money for the magnificent Detroit Art Museum. Now semi-retired, he had become interested in astronomy, probably as a result of reading newspaper and magazine articles about the new 100-inch telescope. He began collecting astronomical photographs, whose beauty impressed him. Soon he was buying many more of them as gifts for his friends and for schools in Michigan.[10] Before long, he advanced to a higher stage of interest. One question absorbed much of his energy: how large a telescope it would be possible to construct that would be useful for research. He first asked William J. Hussey, now head of the University of Michigan Astronomy Department but soon turned to Campbell and Hale.[11]

Campbell replied to Weber that he did not see any "moderate limitation" on the size of a practically useful reflecting telescope and thought that any size up to 150 inches would be "merely a matter of available funds." This was not what the enthusiastic Detroit timber merchant wanted to know. He was not interested in trying for "a BIT larger" telescope, but in what the "EXTREME limit in size, FIT for use" would be, "one that will do work to satisfy a professional astronomer," a 20-foot (240-inch), 25-foot (300-inch), or a 30-foot (360-inch), he wrote Campbell. The Lick director had not become any less conservative over the years since he had argued much the same point with Ritchey. Though the bigger the telescope the better it would be from the astronomer's point of view, he replied, "[a] person of sagacity does not like to pass from the region of actual experience by one jump into a region unknown to experience." From a 100-inch to a 240-inch was too long a leap for him. What problems would arise he could not foresee, he said, but "a wise man" could not be sure that he could overcome them. Campbell would be willing to build a 12-foot (144-inch) next, but no larger.[12]

At Mount Wilson Observatory, Joy, with whom Weber corresponded, was initially even more conservative. He had no doubt that a 10-foot (120-inch) could be built but confided that Pease would probably want to make it larger. The main limitation Joy foresaw, from the 100-inch experience, would be getting the necessary large glass disk for the mirror. Weber suggested that the mirror might be made of metal, or of many pieces of glass, annealed together. He insisted that the aim should be to go right to the "ultimate limit of practical results for doing good work. Reach out into space as far as our present knowledge permits." Astronomers should not worry about practical details like cost. "Finance is a minor problem—that is what money is for—to do things worthwhile."[13] Weber had already established himself as a free-handed spender by ordering $50 or $100 worth of photographs at a time (the equivalent of ten times those amounts in present-day currency). As a result, Joy had checked with Hussey as to whether Weber might "have the inclination and means to build a reflector of considerable size" and had learned that he was "a very fine man, . . . and presumably quite wealthy." Now the young Mount Wilson astronomer could see that it was time to move in for the kill.[14]

Thus, when Weber next wrote asking for Pease's opinion on building a 300-inch telescope, what it might cost, and what results it might achieve, he received a long, serious reply from Joy. The astronomer said that he had consulted several other members of the Mount Wilson staff and found that they were generally convinced that "at least a 15-foot [180-inch], and perhaps even larger, could be built and operated with great advantage to astronomy if it had the proper financial backing and a suitable location." Pease was going to look into the design, and would send Weber some sketches. Joy had even spoken to Hale, whose "wisdom and experience in such matters" were "unequalled" and whose "judgment in an undertaking of this kind would command the respect of the whole world of science." Joy hoped the great man might find time to write Weber himself. At the same time Joy sent Hussey a package of photographs taken with the 100-inch, asking him to give them to Weber personally and explain them to him.[15]

Within a few days Hale did write the Detroit timber speculator. Speaking from his nearly thirty years of experience at Yerkes and Mount Wilson, he estimated the cost of duplicating the 100-inch at $1 million and of scaling it up to a 25-foot (300-inch) at $25 million. Possibly a new design, due to Pease, would reduce the cost to $15 million, but further study would be necessary to be sure it would be practicable. Hale described the optical problems of obtaining the glass disk as "most serious." He doubted St. Gobain would contract to make a 25-foot disk, or even a 15-foot, but considered it "barely possible" that some new way of making a glass "or, much better, . . . fused quartz" might be found. Most important of all, Hale wrote, was the site. Mount Wilson was "probably unsurpassed" for good seeing. If Weber cared "to follow up the matter," Hale would be glad "to go into the question more thoroughly" with him. A few days later Joy sent Weber a series of pictures showing the 100-inch in various stages of its construction.[16]

After several months' silence, Weber replied to Joy, giving some naive ideas of his own about how the 25-foot glass disk might be assembled, melted together, and annealed right at the observatory site. He offered the inspirational thought:

> A tool of War, to commit murder, now Costs 56 Million $. Very well let us try for a tool of the most wonderfull of all Sciences—Astronomy, to Cost 1/4 as much only: 14 Million $—and Elevate the IDEALS of Mankind—most important of all. Sure we will win out—if we all pull together from the love of the Cause—*Not for Gain.*

He did not offer to contribute any money himself, however. Instead he enclosed a copy of his plan for raising the ideals of the youth of Michigan and Ontario through astronomy. Hale, who was then going through the trauma of deciding whether or not to remain as director of Mount Wilson, must have been disgusted when he saw Weber's letter. He directed Joy to tell the Michigan "benefactor" to keep all the designs and ideas they had sent him confidential and dismissed his thoughts on how to make the disk for the mirror as "impracticable for several reasons." Joy wrote Weber, thanking him profusely for his report on his "activities in creating an interest in Astronomy in your state," asking him to return the drawing Pease had sent him and to keep all the "preliminary ideas" they had given him to himself. Joy also wrote Hussey, warning him that the Mount Wilson astronomers were not working with Weber any longer and did not want him to start to try raising money for them, as he had implied he might.[17] Hale's method of operation was always to work closely with one big donor or foundation, such as Charles T.

Yerkes, Andrew Carnegie, and the Carnegie Institution of Washington, not to get involved in public fund drives.

A few months later Weber wrote Hale himself, describing his ideals:

> Our present civilization is based on greed as the prime motive power, the main spring of all our actions; this is taught to our children, in our schools and colleges. . . . The study of astronomy and research therein will help bring us to our senses, and to higher ideals; and teach us that our present aims of greedy accumulation, selfishness, for individuals and for families is dead wrong, and not worth the effort and war.

He proposed to raise the money for a 25-foot reflector "in this region" from "[s]everal citizens of Michigan [who] could club together and pay for this instrument to-day, and never miss the money." Nothing could be so antithetical to Hale's thinking as the idea of building a large telescope near Detroit. He thanked Weber but told him to wait with any fund-raising campaign until they had more experience with the 100-inch. From then on Hale ignored him. The Detroit timber speculator, like many prospective donors since, was much more eager to help find other, wealthier prospects, than to put up his own money to build a telescope on the astronomers' terms.[18] He hung on the fringes of astronomy, buying countless slides and pictures, and two or three 4-inch telescopes for schools, until he lost his fortune in 1929. Ritchey sent Weber a copy of his book in 1930 but received only a polite acknowledgment in reply, without even an offer to attempt to raise money for a large new telescope.[19]

Pease longed to build a much larger telescope than the 100-inch. Hale had him working on a large interferometer for measuring stellar diameters, but the optician's thoughts kept turning to a 25-foot (300-inch) reflector. Arthur L. Day, the director of the Geophysical Laboratory in Washington, another branch of the Carnegie Institution, had headed America's efforts to produce optical glass during World War I and remained a consultant for the Corning Glass Company. He told Hale he believed it would be possible to produce a 300-inch disk.[20] When Henry S. Pritchett, the former astronomer, former head of the Massachusetts Institute of Technology, and longtime president of the Carnegie Foundation, visited Mount Wilson, Pease found him highly receptive to the idea. The thought of a telescope three times as large as the 100-inch fascinated the old statesman of science. Pease brought the subject up with every business magnate he showed around the observatory, and reported to Hale that "there is something very attractive when you mention the 25-foot. A 10 or 15-foot draws an 'um' and a 20-foot an ah. Three times this (100-inch): Oh Boy."[21]

Hale, however, showed little interest. It was too soon after the completion of the 100-inch. Even after resigning the Mount Wilson directorship, he was still very ill, and in the summer of 1923, when his wife and son returned to California from Europe, Hale stayed in the English countryside to avoid scientific contacts. Back in the United States in 1924, he nerved himself up to take part in the dedication of the National Academy of Sciences building, which he had done so much to make a reality. However, in the midst of supervising the final preparations he found he could not stand the strain of the multitude of personal contacts. He fled to New York City, returning to Washington just before the ceremony. Hale's doctors diagnosed him as suffering from chronic hypertension (high blood pressure) and a "psychoneurotic state."[22]

Hale found relief in travel and in reliving the triumphs of his youth. Summer in Montecito, alongside the sea in Santa Barbara, occasionally observing with a small visual telescope, driving through the redwoods to visit his daughter in Oregon, a return to Egypt and the Upper Nile, or even just riding the Santa Fe railroad between Chicago and Los Angeles were far more relaxing and attractive to him than research and administration. He and his wife had moved to Hermosa Street in South Pasadena. Two miles away, near the California Institute of Technology campus, Hale occupied himself by building his own solar observatory in which he hoped to try out the ideas he had never had time to work out at Kenwood. Somehow, he never got around to them at the new Hale Solar Observatory either, though he constantly had ideas for new instruments, which required appeals to Frost to borrow lenses and mirrors from Yerkes Observatory, and to Adams to assign Mount Wilson shop personnel to build them. Hale enjoyed rearranging his books in his new library and short bursts of observing with his new equipment, but never could bring himself to write up the results, except in letters to Harry M. Goodwin, his close friend ever since their MIT student days.[23]

Although he frequently protested that he did not have the energy and psychological well-being for research, Hale could never resist involving himself in large fund-raising activities for science and Pasadena.

He was one of the main driving forces behind the plan for a National Science Endowment, to be based on the fortunes of American captains of industry and finance. Planning it involved meetings and conferences with such men as former Secretary of War Elihu Root, Secretary of Commerce Herbert Hoover ("an enigma to me as regards his real attitude toward pure science"), and Secretary of the Treasury Andrew Mellon. The endowment plan never really got off the ground, but it provided Hale with an amusing but cynical story of how he made Sears Roebuck mogul Julius Rosenwald feel that the pledge of $10,000 a year for ten years, which he intended to make, would "look like thirty cents" and persuaded him to up it to $100,000 a year for the same period. Hale was also a leader in "the scheme for the Huntington Library," in which he and his friends did succeed in convincing Southern Pacific organizer Henry Huntington, Jr., to leave his art collection and library, with a $10 million endowment, as his monument in San Marino, only a few blocks from the Hale Solar Observatory. These activities drew terribly on Hale's reserves of nervous energy and strength, and left him depressed and withdrawn.[24]

As Hale alternated between retreating into the past and organizing the future, between relaxation and tension, Mount Wilson Observatory rolled on. In 1925 Merriam, always eager to promote interdisciplinary research, set up a "Committee on Study of Physical Features of the Surface of the Moon," which was to include astronomers, physicists, geologists, volcanologists, geophysicists, and astronomical photographers. He named as its initial members Adams and Pease, Day and Fred E. Wright of the Geophysical Laboratory, John Buwalda and Paul S. Epstein of the California Institute of Technology, and Campbell. Advised by Hale and Adams, Merriam never considered appointing Ritchey to the committee, in spite of the outstanding lunar photographs he had earlier taken with the 40-inch refractor and the 60-inch reflector. The shop teacher turned astronomer had very little knowledge of the physics of the moon, but Adams, Pease, and Campbell had hardly more. Campbell, who had recently become president of the University of California, soon resigned from the committee. He was replaced by Henry Norris Russell, the Princeton astrophysicist who made yearly visits to the Mount Wilson Observatory offices as a research associate of the Carnegie Institution.[25] Thus the committee was very much Hale's creature. Wright, who was the most active member doing actual research on the moon, was named its chairman. He was well aware of Ritchey's "excellent photographs" of the moon and secured copies of many of them directly from Yerkes Observatory to use in this work.[26]

American astronomers knew that the former Mount Wilson optician was established in Paris, doing his best to build a telescope larger than the 100-inch. In 1925 the International Astronomical Union met in England. Hale made it financially possible for Harold D. Babcock to attend and to travel in Europe, visiting several astronomical centers. One of them was Paris. Babcock knew Ritchey well from the many years that they had both worked at the Mount Wilson Solar Observatory, and visited his laboratory. Babcock's son Horace, a twelve-year-old boy at the time but decades later the director of Mount Wilson and Palomar observatories himself, accompanied his father on the trip. Horace Babcock still remembers the vast hall in the Paris Observatory where Ritchey had his laboratory, the cellular disk he was trying to assemble, and the cement he was using at the time, made from egg whites. Obviously, Harold Babcock was not a spy for Hale, but clearly, once he got back to Pasadena, he reported fully to his benefactor on all that he had seen on his travels. Charles Greeley Abbot, who visited Paris in 1926, reported to Adams that he had seen Ritchey's "extraordinary [cellular] mirrors, which I believe will succeed if his Hindoo [Dina] pays the bills."[27] Thus, at least in Hale's mind, pressure was building on Mount Wilson to keep its lead. In 1926 he allowed Pease to publish a paper "On the Design of Large Telescopes," which was actually a call to build one very large telescope.

In this paper Pease gave a brief synopsis of how much more light could be collected by a larger telescope than the 100-inch and correspondingly how much more observational information could be obtained on faint stars and "spiral nebulae." He reacted directly to Ritchey's ideas, without mentioning him by name. Pease stated that it was important to reduce the aberration of coma in such a telescope in order to take images or spectra of many objects in one exposure. Gains in the speed of photographic plates were as important as increases in the mirror diameter. In addition, he emphasized the need for very fast spectrograph lenses. Progress could be made in all these directions, he opined, by continued theoretical and applied research. He outlined no program to do so, but only the pious hope that "the happy combination of a mathematician and a practical optician" would accomplish it. Pease also asked a rhetorical question "as to the relative value of ten 100-inch telescopes scattered about the earth, compared with one [much larger] telescope of the same total cost," and answered that "a canvass of scientists" strongly favored the single large instrument. No doubt, this was the reply he got from his colleagues on the Mount Wilson staff, but he would have probably obtained a very different answer if he had polled "scientists" at Lick, Yerkes, Harvard, and Michigan, all of whom were desperate for 60- or 70-inch telescopes of their own.

Pease went on to say that it would be possible to build a telescope with diameter as large as 100 feet (1,200 inches), provided someone would pay for it, but that he had chosen a 25-foot (300-inch) as a basis for discussion. He estimated its cost as $12 million, "a sum well within the means of a group of interested persons, or even a single wealthy individual" in those heady, pre-Depression days. The cost of simply scaling up the 100-inch to a 300-inch would be $18 million, he explained, but making the primary mirror with a focal ratio f/3.3 (instead of f/5, as in the 60- and 100-inch) would decrease the size of the mounting and dome and thus reduce the cost.

In this paper Pease published a drawing of his design for the 300-inch, dated November 1921. It was the "sketch" that Joy had written Weber that Pease would send him, although apparently he never actually did so. The telescope was to be used only at the Cassegrain focus, directly behind the pierced primary mirror. It would be an f/8 system with auxiliary lenses to convert it for various instruments. Pease claimed that a large coudé spectrograph would be carried directly on the telescope itself, but it is not shown on the drawing and therefore it is impossible to understand what he meant. He stated that obtaining a 300-inch disk would be no more of a problem than the 100-inch had been in its day (conveniently forgetting a good deal of the history of Mount Wilson Observatory from 1909 until 1917). He quoted Day's assurance that there would be no difficulty whatsoever in making a 25-foot disk of low-expansion glass, provided that sufficient funds were available.

Pease did not mention Ritchey by name but alluded to his idea of built-up cellular disks. They would make excellent mirrors, Pease wrote, if only the ideal cement could be found that would fuse the pieces of glass together and survive annealing. There was little real chance of doing so, he implied. Pease also referred to the Ritchey-Chrétien system without giving its name: "Considerable attention has been paid to the problem

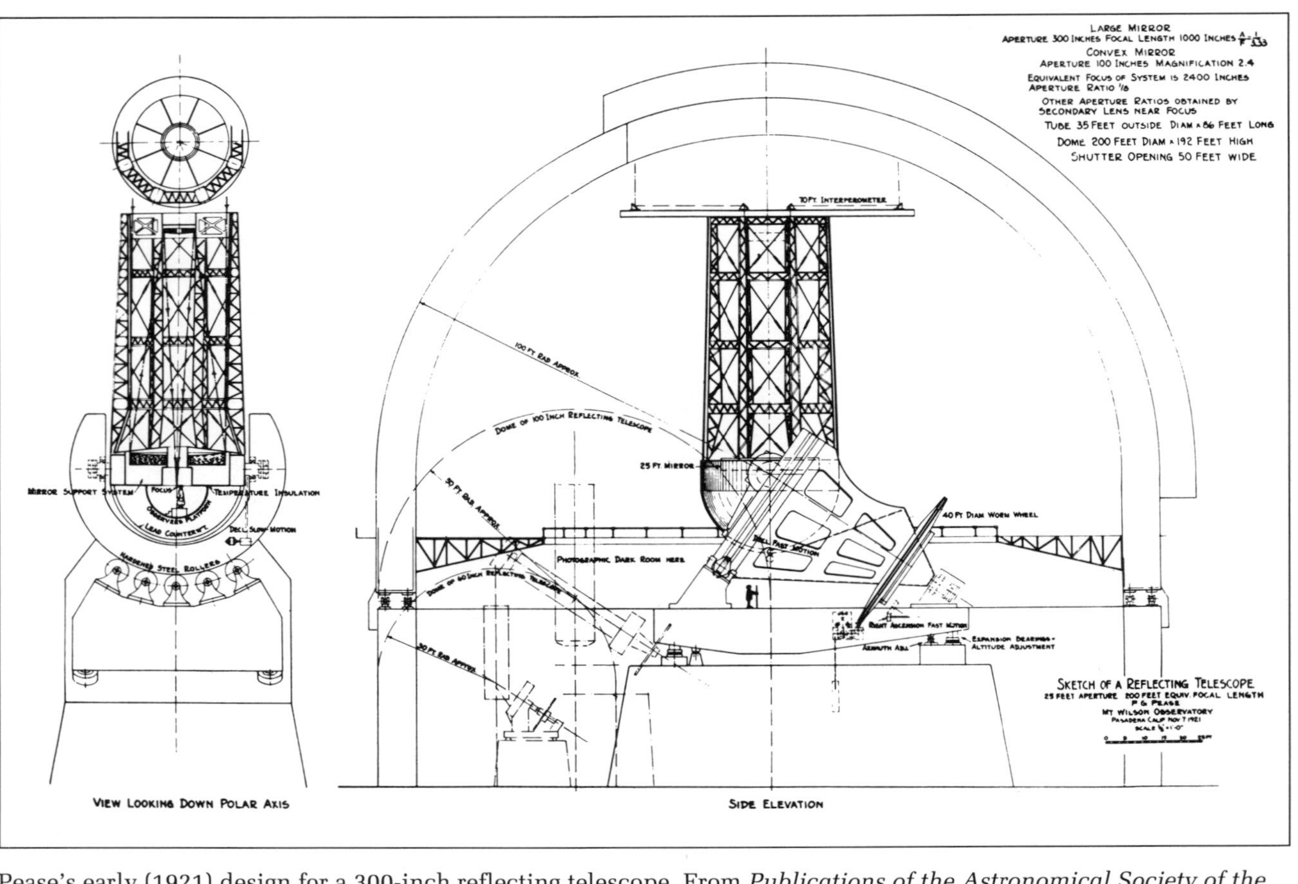

Pease's early (1921) design for a 300-inch reflecting telescope. From *Publications of the Astronomical Society of the Pacific*.

of increasing the size of the field by means of curved surfaces other than the conic sections. For a great telescope, the advantages of these systems are just about balanced by the disadvantages, and the weight of the evidence would favor the retention of the paraboloidal and hyperboloidal curves in existing mirrors."

Pease gave absolutely no reason for this extraordinary conclusion. The paragraph begins with a misstatement, for the primary and secondary mirrors of a Ritchey-Chrétien are essentially hyperboloids, that is, conic sections. Hale, Adams, and even their telescope expert Pease did not understand the Ritchey-Chrétien principle, refused to study it, and simply drew on the immense prestige of Mount Wilson Observatory to label it useless.

Pease concluded with the statement that he had designed the 300-inch reflector for Mount Wilson "where observing conditions are excellent" but that it was conceivable that another, more southern site could be found that would be nearly as good. There was no doubt in his mind about who would use the 300-inch. "Such a telescope is planned for scientific research in the hands of those who have shown themselves particularly fitted for that task."[28] Needless to say, he meant the Mount Wilson staff. The peroration of Ritchey's book, published three years later, demanding that his chain of fixed, vertical telescopes be under "one democratic management," was undoubtedly written in reaction to this arrogant, self-serving statement.

A few months later Russell, who spent the summer of 1926 at Mount Wilson, publicized Pease's plan in his regular monthly article in *Scientific American*. It featured the drawing of the proposed "twenty-five foot 'eye' " and described the design in considerable detail. Russell's article correctly stated that one of the most important questions would be the location of the telescope. He predicted it would be used for a century at least, and although he recognized the excellence of Southern California, he expressed a preference for a southern-hemisphere site where "the Magellanic Clouds, and the great star clouds of the Southern Milky Way would . . . be accessible."[29] Both Pease and Russell mentioned clear skies, excellent seeing, "healthy and tolerable living conditions," accessibility, and "stability against wars and revolution" as important factors in choosing a site, but neither of them mentioned future light pollution from cities. Otto Struve had already noticed its effects at Mount Wilson in 1926. In little more than fifty years, it was to doom this otherwise fine site.[30]

Soon after the publication of Russell's article, General Education Board official H. J. Thorkelson visited Mount Wilson to see and report on the observatory. The university arm of Rockefeller philanthropy, this Board had supported Caltech generously. Hale, who had turned to writing a series of articles on astronomy and astronomical research for *Scribner's Magazine*, presented Thorkelson with copies of the three slim volumes in which he had collected them. Just their titles, *The New Heavens*, *The Depths of the Universe*, and *Beyond the Milky Way*, were evocative.[31] Thorkelson found them fascinating. Adams saw that he got the full tour of the mountain and a good look at the proposed 300-inch design, and presented him with photographs of the moon, taken by Pease with the 100-inch. Soon the Mount Wilson director was surely pleased to learn that Thorkelson had reported to Wickliffe Rose, the president of the General Education Board, that "Mount Wilson of itself was worthy of the trip across the continent." Adams sent Pritchett, who advised Rose on astronomy matters, a copy of Russell's article and additional information, in capsule form, on the 300-inch design.[32]

In 1927 Hale discovered a new sanatorium, in Stockbridge, Massachusetts, a residential center for diagnosis and treatment of psychoneurotic patients. It was operated by Dr. Austen F. Riggs, a psychiatrist. Hale could only stay for a three-week course of diagnosis but learned that his difficulties, though "very deep-seated," could be overcome in time. He did some weaving as therapy and came back home to Pasadena determined to work harder at curing himself. He regarded Riggs as a "thoroughly sane and scientific man" and resolved to follow his advice as consistently as he could. In Santa Barbara Hale heard of a new vegetarian diet to reduce blood pressure. He tried it and reported that although it only brought his systolic pressure down a little, it did wonders for his chronic constipation. His "head," however, "remain[ed] about the same."[33]

Nevertheless, Hale nerved himself for a supreme effort. Somehow, he had decided that the time was ripe to go for the big one. He wrote, for *Harper's Magazine*, what was probably the most important article of his entire life. Its title, "The Possibilities of Large Telescopes," was prosaic, but its text was evocative from its opening lines: "Like buried treasures, the outposts of the universe have beckoned to the adventurous from immemorial times. Princes and potentates, political or industrial, equally with men of science, have felt the lure of the uncharted seas of space, and through their provision of instrumental means, the sphere of exploration has rapidly widened."

Hale skillfully blended the lure of the adventure of building the largest telescope in the world with the confidence based on his own experience with the 40-inch, the 60-inch, and the 100-inch. "[S]ome other donor" might have the opportunity of joining the ranks of Yerkes, Hooker and Carnegie in advancing knowledge and pushing out the frontiers of the universe, secure in the realization that on the basis of the Mount Wilson telescopes Pease had concluded that "an increase of aperture to 20 feet or more would be perfectly safe."[34]

Although Hale wrote the article for publication in *Harper's*, its real target was not the readers of this quality magazine but Rockefeller's General Education Board. As soon as he received the proofs, Hale rushed a copy of them to Rose and Thorkelson in New York. Going east, he followed up with a personal visit to Rose in his office at 61 Broadway. He found the man who controlled the Rockefeller money "so keen about a huge telescope that he [would] talk about nothing but the largest possible, and remark[ed] that there [was] no reason why fifteen millions should not be spent on such an instrument." It was Charles T. Yerkes all over again! Rose and Thorkelson were about to go west to inspect the results of Rockefeller philanthropy at Caltech. Hale easily persuaded them that they must visit Mount Wilson and dashed off a letter to Adams, scripting a tour of the mountain that would impress them with glimpses of the telescopes in operation. Adams should stay with them at all times, Hale wrote, and "if any question about lights in the Valley etc [came] up [he should] show how easily they [could] be met."[35]

Thus even before Hale's article appeared in print, the two General Education Board officials were on Mount Wilson, where Rose got his first view of the 100-inch. Adams followed Hale's instructions closely. He, Pease, and the persuasive Russell, who was in Pasadena on another working visit, showed them around Mount Wilson and Pasadena, and pressed the plan on them for a day and a half. At the end, the ever-cautious Adams could report to Hale: "There is no doubt that Rose is enormously interested in the project for a very large telescope." He believed that Rose "would be strongly in favor of locating a large telescope either on Mount Wilson or in the immediate neighborhood." However, the observatory director was "rather astonished that Rose [favored] putting such an instrument in the hands of a university rather than the [National] Academy [of Sciences] or a research organization [Mount Wilson Observatory]." They discussed this subject at length and Adams believed that Rose had the California Institute of Technology in mind but was not at all sure.[36]

Hale remained in the East until Rose returned. Rose was "in earnest about a big telescope" and only wanted Hale to tell him whether it should be a 200-inch or a 300-inch. The astronomer was uncertain about the bigger size but felt sure the 200-inch could be built. He "unhesitatingly" said so; that is what it became. Hale telegraphed Adams for his best estimates of the cost; the next day the figure came back: $3.2 million plus the disks for the mirrors, which were to be of quartz. Rose was willing to recommend a total gift of $6 million. The only fly in the ointment was that it was to be given to Caltech, not to the Mount Wilson Observatory. Rose was adamant. Rockefeller millions could not go to finance a Carnegie observatory. Hale visualized a plan of close cooperation of the new observatory and Mount Wilson, each with its own director, under a high-level board of policy, perhaps composed of Robert A. Millikan, the head of Caltech, Adams, and himself. He left for Pasadena, stopping to visit his brother in Chicago on the way.[37]

Merriam, the president of the Carnegie Institution, abruptly objected, however. He had only read Hale's *Harper's* article late in the game and had thought of it as an "extremely interesting statement," which they should talk over "[s]ometime when we have the opportunity."[38] When he learned of Rose's plan, Merriam was decidedly unfriendly to the idea of a rival foundation building the world's largest telescope under the aegis of his own observatory's honorary director and with the full cooperation of its staff. Merriam was probably reflecting to some extent the views of Adams, his actual director, who was far less happy than Hale with the thought that Caltech, which had no astronomers on its faculty, would control and operate the 200-inch. Hale hastened back to New York and plunged into a series of conferences with Root, Pritchett, Rose, and other high-powered figures in the world of corporate and foundation science. Hale telegraphed Adams that he had been quoted as opposed to the project and requested him to wire back his endorsement and that of other members of the Mount Wilson staff. Adams complied, and Merriam surrendered and pledged his cooperation and the Carnegie Institution's. It was a tremendously wearing experience for Hale, and he had to draw heavily on Medinol, a drug that had become his refuge, to get through each night. His mission completed, he returned to Pasadena in mid-May.[39]

Years afterward, Hale defended himself in letters to Adams against any thought that he had planted any idea in Rose's mind of giving the largest telescope in the world to Caltech. He did admit, however, that he had hoped that Mount Wilson Observatory could be split away from the Carnegie Institution as an independent research organization with a separate endowment of its own, and he had stated the same in letters to his wife at the time.[40]

Hale's friend Goodwin congratulated him fulsomely on the success of "the scheme" for the 200-inch. He only hoped his old fellow student would not work too hard on it and wear himself out. In reply Hale confided, "This has surely been a strange adventure and seems to show, like my previous experiences, that new observatories are not likely to fall from the heavens." Even then, in mid-June, the Rockefeller board had not made its official decision, and the telescope was not a certainty.[41]

The Rockefeller money was assured, however. It was to come formally from the International Education Board, which had turned down the French proposal to fund Ritchey's laboratory in Paris in 1924. Rose headed both Boards. He had secured formal assent for an appropriation of $6 million to Caltech to build the 200-inch, subject to the conditions that the Institute itself would provide the operating funds for the observatory when it was completed and that the Carnegie Institution of Washington and Mount Wilson Observatory would cooperate with it. Hale had already seen to the latter; only the problem of the assurance of the Caltech funds remained. This was quickly solved. Rose and Thorkelson came to Pasadena in June for a meeting with Merriam (who usually spent much of the summer in California), Hale, Adams, Millikan, and Arthur Noyes, who was Hale and Millikan's partner in the triumvirate that had built Caltech. Henry M. Robinson, a very wealthy Los Angeles oil man and Caltech trustee, was also present. Hale and Millikan had long marked him out for a big gift, and Hale had sold him on the observatory project. As soon as Rose brought up the endowment problem, Robinson said he personally would guarantee $150,000 a year, the figure Hale and Adams had estimated would be needed to operate the observatory. The deal was closed informally then and there, though it remained confidential for several months.

Rose and Hale exchanged memoranda on what the formal proposal should request and what the International Education Board would provide. Caltech would set up an Observatory Council, consisting of Hale, Millikan, Noyes, and Robinson, to build the new observatory and later operate it. An Advisory Committee, headed by Adams, and consisting chiefly of Mount Wilson staff members, together with "other leading astronomers and physicists" would help them as needed. The council would appoint "an experienced astronomer or physicist" as its executive officer to be in immediate charge of the project.[42] Hale at once named John A. Anderson, trained in technical physics at Johns Hopkins University, a member of its faculty from 1908 until 1916, and afterward of the Mount Wilson staff. He had not done much astronomical research but was an expert in optics and instrument design, and he was absolutely loyal to Hale. Anderson and Pease, who worked under him on the 200-inch project, remained on the Mount Wilson Observatory staff, but half their salaries came from Caltech funds.[43]

In the spring and summer of 1928, as these negotiations were going on, Ritchey's articles on the great telescopes of the future were appearing in France and Canada. His Thomas Young oration on "The Modern Reflecting Telescopes and the New Astronomical Photography" was published in England. He was writing to Pritchett and Merriam, trying to get support for his telescope designs. Only a year later Hale was to confide to his close friend H. H. Turner that one of the reasons he had written the *Harper's* article was to head off Ritchey's plans.[44] In 1928 he gave Anderson, as one of his first assignments, the task of studying various focal ratios that might be adopted for the 200-inch, the diameters of the fields over which they could provide direct images, the possible use of curved plates, and the practicability of Pease's original plan of using only the Cassegrain focus. He directed Anderson to report specifically on Ritchey's plans, as described in his recent papers. Hale instructed him to make those studies in collaboration with Adams, Seares, and Pease, and to summarize their joint views in his report. The implicit message was that Anderson should provide ammunition to shoot down Ritchey's claims.[45]

Anderson did so very quickly. Within only three days the group had decided unanimously that although a built-up cellular mirror would be ideal if it could be made, and if it kept its shape and behaved as a single mirror, there was no evidence that either of these conditions could be met. Years of experiments would be necessary to prove or disprove them. Even if a cellular mirror could be constructed, the saving in weight would be minuscule in comparison with the weight of the other moving parts of the telescope. Supporting it would be more difficult than supporting a solid mirror. In any case they expected the 200-inch mirror would be made of fused quartz, which would do away with many of the thermal difficulties the cellular disk was supposed to overcome.[46]

The International Astronomical Union was to meet in the Netherlands late that summer. Hale cabled Charles E. St John, already in Europe, that he and other Mount Wilson staff members should collect what information they could on built-up mirrors. St John reported that no one whom he consulted approved of Ritchey's "cemented mirror." He described a trial by the Parsons Company, in England, to fuse together (by heating very nearly to the melting point) a glass built-up mirror. The experiment was a failure, and St John "felt particularly pleased" that the Parsons officials were willing to admit it and let him examine the deformed disk.[47]

For theoretical studies and design of the optics Anderson hired Frank E. Ross, who was strongly recommended by Adams. A member of the Yerkes Observatory faculty who had previously worked in the Eastman Kodak research laboratory, Ross was the outstanding astronomical lens designer in America. Anderson offered him a half-time position and Ross, who pined to return to his native California, rushed to Pasadena on a "vacation trip" to start work immediately. He later formalized an arrangement by which he could work six months of each year on the 200-inch project.[48] By early September Anderson was able to provide Ross's answers to the other questions Hale had asked. For an f/3.3 200-inch mirror used at the prime focus, the field over which usable star images could be photographed would be very small. However, at the conventional Cassegrain focus, planned as an f/10 by then, the images would be "essentially perfect" over a field much larger than any photographic plate likely to be used. Ross stated that "[w]ithout curved plates the Ritchey-Chrétien combination [was] really much worse than the straight [conventional] Cassegrain of the same relative aperture." This statement is true, but it hides the fact that with curved plates the Ritchey-Chrétien eliminates coma, as no conventional Cassegrain or prime-focus system can do. Hale, Adams, Pease, Anderson, and everyone at Mount Wilson had already dismissed the Ritchey-Chrétien possibility; they "knew" on the basis of its name alone that it would not work. Ross produced the theoretical basis for their decision, for he was practically certain that he could design a correcting lens to be placed near the primary focus, which would remove the coma and thus give good images over a fairly wide field.[49] From this time on the Ritchey-Chrétien option was not considered. Pease's original plan to have only a single, Cassegrain focus was also dropped. The 200-inch was to be usable at the prime focus (with a Ross corrector lens), at the Cassegrain, and at the coudé. Pease had already published a design for a 300-inch of this type before Ross came on board; now scaled down to 200 inches, it became the first working plan for the telescope.[50]

By the end of summer news of the 200-inch project had leaked out not only to Ross but also to many other astronomers. Shapley, the Harvard director, had heard of it and passed the word on to Ritchey. The optician, still in Paris, was just learning from Merriam that there was no chance of support for his large-telescope project. He must have been crushed, but all he replied to Shapley was that he would be glad to learn more about "the big American project."[51] He avoided mentioning Hale's name, as Hale avoided mentioning his.

Another decision which remained to be made was the site for the 200-inch observatory. Hale, Adams, Pease, and the rest of the Mount Wilson staff had originally assumed that the large telescope would be on their mountain, but Rose's ultimatum that it must be a Caltech observatory made this impossible. The astronomers envisaged observing with the 200-inch themselves, commuting easily to it from Pasadena. They only considered nearby sites. From the beginning Hale and Adams favored Palomar Mountain, recommended by Hussey back in 1903 as the best site in the San Diego area, but too inaccessible then. By 1928 automobiles had made it a real possibility. Initially, Anderson sent observers to test five sites, from Horse Flats and Table Mountain near Mount Wilson to Palomar, forty-five miles from San Diego. All were out of range of the bright lights of Los Angeles and San Diego. The observers made visual estimates of the seeing with small visual telescopes and collected weather records of cloud cover. Within a few weeks Hale reported that Palomar seemed to be "the most promising."[52]

Walter S. Adams. Courtesy of the National Academy of Sciences.

In October Hale went east again, this time accompanied by Adams, to the formal meeting of the International Education Board in New York. The $6 million proposal sailed through and was quickly approved, although Hale had feared that Shapley might "thr[o]w the monkey-wrench into the gear train" with a rival proposal of his own.[53] Actually, the International Education Board had already conditionally funded a much smaller Harvard proposal for a 60-inch in the southern hemisphere, and Shapley was still trying to get the required contribution from Harvard University itself. It had no Henry M. Robinson on its board—and Shapley was no Hale.[54] The Mount Wilson honorary director who was equally the guiding spirit of Caltech had won, but he had nearly exhausted his mental reserves on the struggle. "It was hard to go, but there was no escape . . . ," he wrote his wife.[55]

One of Ritchey's claims was that his proposed site on the edge of the Grand Canyon was superior to any in California. Hale and Adams had to combat it. That Ritchey was taken seriously they knew well; Waldemar Kaempffert, preparing a feature article for the *New York Times*, mistakenly believed that Hale was collaborating with him.[56] In fact, Hale had already written Charles F. Marvin, chief of the Weather Bureau in Washington, asking for long-term comparative weather data related to the cloud cover and seeing at Palomar, Mount Wilson, and Table Mountain. As soon as he learned of Ritchey's letters to Merriam, he added the Grand Canyon to the list. Hale telegraphed Marvin to be sure he would have the data at the crucial meeting in New York in October.[57]

Hale and Adams soon dispatched Edwin Hubble to Arizona to have a member of their own staff measure the seeing in the Grand Canyon area. Hubble took a 2-inch visual telescope and made measurements for one night at the edge of the Grand Canyon, then two nights at Cameron, forty miles east of it, then one night at Flagstaff, sixty miles to the south. To V. M. Slipher, the director of Lowell Observatory, it seemed obvious that in reality the decision had already been made to put "the new big Bertha" in southern California, and that Hubble's tests were "evidently [intended] for publicity material."[58] Later, when Anderson hired an Arizona resident to make a longer series of seeing measurements, Hale insisted that "[l]ocal observers are almost certain to be biased in favor of their own regions," and he directed that one of "our own men" be sent to check up on him thoroughly. As late as 1930 Hale was still reemphasizing this point. He directed Anderson to send Milton L. Humason, the Mount Wilson night assistant who had become a trusted staff member, or Ferdinand Ellerman, his assistant ever since Kenwood days, to measure the seeing at Flagstaff and the Grand Canyon. Hale knew he could count on their results.[59]

One new person joined the team, Russell W. Porter, the artist, designer, and popularizer of amateur telescope making. Ten years before, he had invented and published the concept of the split-ring mounting, which now was adopted for the 200-inch design. Pease's original 300-inch plans had shown the telescope with a fork-type mounting as used in the 60-inch. Its disadvantage is that the tube is supported near its bottom end, which is far from its center of mass. Balancing the telescope requires a great deal of extra weight, which complicates the whole design. These problems increase with the size of the telescope. The 100-inch had, therefore, been made with a yoke-type mounting, rotating on bearings at its upper and lower ends. This improves the balance but makes it impossible to point the telescope at stars near the North Pole. With the 100-inch, the whole area of the sky north of about 60 degrees declination (similar to latitude) is inaccessible. Porter's idea was to make the upper bearing very large and split the resulting ring so that the telescope could swing down into it and point right to the North Pole.[60] Though Porter was hired initially for only six months, he came back to the 200-inch project several times, and his drawings of the big telescope from every angle and in every detail made it known to a generation of professional and amateur astronomers.[61]

In 1929 Hale followed up his original *Harper's Magazine* article with a second one, "Building the 200-inch Telescope." He confided that he "had to write [it] . . . to save time in answering questions and disposing of false rumors about the 200-inch."[62] In the article Hale described the split-ring design and Ross's newly designed field-correcting lenses, soon to be tested on the 100-inch. Hale barely mentioned the Ritchey-Chrétien possibility, dismissing it along with the Schwarzschild, "neither of which," he said, "is applicable in our case." He gave no reason for this statement. He described the seeing tests in progress at the various sites, writing that the 200-inch could be used very effectively at Mount Wilson but that "[t]he probabilities now are that we can find a still better site within a short distance of Pasadena." Nowhere in the article did Hale mention the possible problem of light pollution. He described graphically "the most promising means" of doing away with thermal distortion of the 200-inch mirror by making it of fused quartz, but he did not mention the cellular-mirror concept.[63]

Hale, Anderson, and Pease all hoped that fused quartz could be used for the mirror but realized that a large development program would be required. Elihu Thomson, now seventy-six years old, would head it. Putting quartz mirrors into big telescopes was his life's dream. He had stopped exchanging letters with Ritchey not long after receiving a query from Pease about quartz mirrors in 1926. Hale had consulted Thomson as soon as he began his discussions with the Rockefeller officials, and the old man was eager to start. Robinson, the generous donor, was also a member of the General Electric board of directors. With his help, Hale easily persuaded Gerard Swope, the company president, to agree that Thomson and his laboratory could work on the 200-inch project on a cost basis, with no profit or overhead charges.[64]

As soon as Hale knew that the Rockefeller money would come through, he had briefed Thomson on the entire 200-inch project. A. L. Ellis headed the work force. They were to complete a 22-inch quartz disk in the laboratory at West Lynn, Massachusetts, and then work up through successively larger disks to the 200-inch. The smaller mirrors produced in the course of the development would be used at Mount Wilson, or as secondary mirrors for the 200-inch. Ellis had estimated the total cost as $252,000, back when the disk was still planned as a 300-inch. In the fall of 1928 Anderson visited Thomson and Ellis at West Lynn and saw the work in progress. After four months of trials, they had spent $13,500 but had not produced an acceptable 22-inch disk. Fused quartz, made by melting pure silica, requires much higher temperatures than ordinary glass. Perfect disks are correspondingly more difficult to produce. Thomson's general procedure was to build a high-current electric furnace in which to fuse and anneal the quartz. It formed with air bubbles throughout its interior, and his idea was to grind the resulting disk to the approximate figure needed for the mirror, then glaze or spray coat it with molten quartz to produce a final smooth surface. Anderson and Hale conceived the idea of casting the 200-inch disk with a ribbed structure on the back, to save weight. Thomson had earlier suggested molding holes in the back for the same purpose.[65]

In January 1929, when Hale went abroad, he warned Thomson that progress on the fused quartz project was too slow. If his laboratory could not produce a disk soon, the 200-inch mirror would be made of Pyrex (low-expansion glass), not fused quartz. Thomson was traveling in Africa to get away from the cold New England winter, but he assured Hale that the work was progressing "as rapidly as could be expected." Ellis and his men in West Lynn did finish the 22-inch fused quartz disk, but work on the next step, the 60-inch, dragged on and on.[66]

By now the Great Depression was very bad. If Hale had not gotten the International Education Board commitment in 1928, he would never have been able to do it in 1930 or later. The 1929 stock-market crash forced even the Rockefellers to retrench. There was no possibility that their foundation would increase its gift for the telescope. As time went on, General Electric continued running up costs but seemed no closer to being able to furnish a 200-inch quartz disk. By the end of March 1930, the expenditures on the quartz project totaled $174,000, but not even a 60-inch had been finished. To Hale it must have seemed the story of the 100-inch disk all over again, with an open-ended development project, a fixed amount of money, continual "difficulties and disappointments," and no mirror in sight.[67]

In 1930, Hale sent Mount Wilson staff member Theodore Dunham, Jr., to Thomson's laboratory in West Lynn to see for himself what was going on, just as he had sent Seares to St. Gobain in 1912. Ellis still had not produced a 60-inch quartz disk. As soon as Dunham left, Thomson once again reported rapid progress and forecast success very soon. It was too late.[68] Max Mason, the former president of the University of Chicago who had succeeded Rose as head of the Rockefeller Foundation (formed from the General and International Education Boards) was rapidly losing faith in the quartz-mirror concept. He demanded a realistic estimate of the cost to complete the 200-inch quartz disk. It came in at $1.6 million, plus what had already been spent. It was an impossible sum of money. Hale sent Thomson an ultimatum, just as Hooker had sent him one in 1910. All expenditures must cease on June 1, 1931. After that, Hale and the Observatory Council would decide whether or not to go on with the quartz disk.[69] Thomson protested that it was impossible to estimate the costs of the 200-inch before the first 60-inch disk had been finished. A few weeks later he had to report that it had cracked while cooling, and they would have to start over. He claimed that he and Ellis saw "more clearly than ever that a final desirable result will be obtained," but the fused quartz mirror project was dead. Hale politely but firmly notified Thomson that the experiment was over, that all the Observatory Council wanted was the 60-inch disk, and that it would pay a maximum of $40,000 more for it.[70] Work dragged on until October 1931, but the decision was firm to go ahead with a Pyrex mirror rather than quartz.[71]

Thus before the end of 1931, the main outlines of the 200-inch were clear. The basic design for the mounting would be the split ring, the site would be Palomar, and the mirror would be Pyrex. A large team of engineers and technicians, headed by Anderson, would build it. The Mount Wilson astronomers, directed by Adams, would provide advice at all stages and be poised to use it on completion. Finally, Hale would preside over the whole enterprise.

Miami and Washington

1930 - 1936

WHEN GEORGE WILLIS RITCHEY came home to America at the end of December 1930, he had high hopes for a new big telescope project in which he was involved. They were to be dissipated in less than a month, but his presence in his own country did trigger another project—the last telescope he was to build in his life.

Ritchey's return to the United States was engineered by George H. Lutz, an amateur astronomer, promoter, and true believer in metal mirrors for telescopes.[1] He lived in Philadelphia but had his eye on the main chance, wherever it might be. Through his contacts, Lutz learned of Robert Henkel, another, much older amateur astronomer who was the president of Commercial Milling, a flour company in Detroit. He also had wide holdings in banks, manufacturing companies, and Florida real estate. Henkel was interested in large telescopes, and Lutz planted the idea in his mind of getting Ritchey to build one. As a result in April 1930 the American optician had hurriedly left Paris to meet Lutz and accompany him to Detroit to confer with Henkel. Though he went on very short notice, Ritchey apparently was optimistic enough to think that the project was a certainty, for in getting his passport he told

George H. Lutz, Ritchey's friend and promoter, with a small metal mirror disk. Courtesy of Richard B. Burgess.

the State Department he intended to reside permanently in the United States.[2] Undoubtedly, Henkel financed Ritchey's quick transatlantic trip.

In Detroit Ritchey met the flour company president and his friend and fellow astronomer Gar Wood, world-famous speedboat racer and owner of an automobile parts firm. Both Henkel and Wood had large winter homes in Miami, Florida, each with its own astronomical observatory. They had several conferences at the Detroit Athletic Club, where the rich of the city were accustomed to meet and deal. Henkel's immediate plan was not to build a 25-foot (300-inch) fixed vertical telescope at the edge of the Grand Canyon, as Ritchey still hoped to do, but rather a much smaller observatory in Miami to attract other wealthy men to give their money for an eventual large telescope somewhere. George Ellery

Hale would have brushed off Henkel, as he had William C. Weber, but Ritchey had no alternatives except this project, starvation in Paris, or raising oranges in Azusa. He agreed to return to the United States in June. Meanwhile he hurried back to France, in a last effort to take some spectacular photographs of nebulae with the 20-inch Ritchey-Chrétien telescope.[3]

His return was postponed several times, and Ritchey never got any striking pictures. By October he had obtained permission to bring the 20-inch to Miami and mount it there. Henkel said he would have Ritchey build him a 36-inch Ritchey-Chrétien for his own observatory in the Florida city. Lutz had persuaded Henkel to order a metal mirror for it, instead of glass, and Ritchey had agreed, no doubt grudgingly, to "experiment" with grinding the "stellite" (a hard, shiny, steel alloy) disk. He had hopes for the big telescope for which Henkel and Wood still said they would raise money, specifying that it be erected in Miami. In November Ritchey's son Willis got a news story into the *Los Angeles Times*, headlined "New Type Telescope Will be Used—Great Strides Made in Celestial Photographic Science," announcing the plan to mount the existing Ritchey-Chrétien in Miami. No doubt Ritchey's nephew, William M. Ritchey, an editor and film writer, was instrumental in placing the article. *The New York Times* picked up the story, but its rewrite man thought that Ritchey was still on the Mount Wilson staff and that the telescope would be erected there. Both newspapers described the Ritchey-Chrétien concept as "revolutionary" but gave little more information than that. Only the *Journal of the Royal Astronomical Society of Canada* correctly related it to his earlier articles, expressed polite skepticism about the observing conditions in Miami, and wished him well.[4]

As soon as Ritchey, his wife Lillie, and their daughter (who now preferred to be called Lila) finally arrived in New York near the end of December 1930, they hurried south with Lutz. The Depression had hit Florida hard, especially land values, and Henkel had no doubt been hurt financially, but they still had high hopes. On Saturday December 27 Ritchey was the center of attention at a luncheon, arranged by Lutz and attended by many members of the Southern Cross Astronomical Society, at which he expansively described his plans to the press. Ominously, Henkel was not present. Three days later he died. The seventy-year-old executive and financier had suffered from heart disease for several months; he had arrived from Detroit hoping to live through another winter in Miami. At the memorial service, organized by the astronomical society, Ritchey paid tribute to his would-be benefactor's plans and described the "Henkel Memorial Observatory," soon to be erected. But the rich man's widow, daughter, and sons were not interested in a telescope, and the project soon expired.[5]

Less than a month later, however, Ritchey headed north for Washington, D. C., where his real chance to build a telescope had at last arisen. He was going to the United States Naval Observatory, where he would build the world's first Ritchey-Chrétien telescope actually used for research. The Naval Observatory dated back a century, to a Depot of Charts and Instruments, founded in 1830. It had become an observatory, run by the Navy, devoted entirely to positional astronomy and a time service. Astrophysics was outside its purview. Its largest telescope was a 26-inch refractor, built by Alvan Clark & Sons, with which Asaph Hall had discovered Deimos and Phobos, the two tiny moons of Mars, in 1877. Its observers did much of their work of determining accurate positions of the sun, moon, planets, satellites, and stars, using specialized meridian circles and transit instruments. The Naval Observatory staff included a few older commissioned professors of mathematics, with the status of naval officers (no new appointments were being made in this grade), while the rest were government civil service employees. Typically, they had started work at the observatory after undergraduate training in astronomy and mathematics, and once they were hired they seldom left. There was almost no back and forth movement between jobs at the Naval Observatory and positions at universities or astronomical research institutions such as Mount Wilson Observatory. The superintendent of the Naval Observatory was always a commissioned regular navy officer, who, in the 1920s and 1930s was almost invariably on a two- to four-year tour of duty ashore between commands in the fleet that had nothing to do with astronomy or even science. The superintendents were not astronomers, but executives who were expected to be able to carry out any assignment they were given in ship-shape fashion.[6] Most outside astronomers believed that the Naval Observatory should be under civilian control; all of them were positive that its director should be a professional astronomer (like themselves). They had taken this issue to Congress in 1901, with Hale as one of their leaders, but had lost their fight.[7]

In 1924 the astronomers tried again. Herbert Hoover, an engineer and graduate of Stanford University, was Secretary of Commerce under

President Calvin Coolidge. Yale astronomer Frank Schlesinger took the lead in proposing a bill to move the Naval Observatory to the Commerce Department, where it would be renamed the U. S. National Observatory. Hale, W. W. Campbell, who was president of the University of California, and other astronomical leaders joined the campaign, but the Navy easily repelled these would-be boarders and kept the observatory.[8]

Its superintendent from 1923 to 1927, Captain Edwin T. Pollock was more scientifically oriented than most of his predecessors. He had served three tours of duty as an instructor in mathematics and in electrical engineering at Annapolis, and after retirement he became head of the mathematics and astronomy department at the Cranbrook School near Detroit.[9] He recognized that the Naval Observatory instrumentation was badly out of date. Practically all its observing was done visually, though photography had long since supplanted this method in other American observatories. Pollock hoped merely for a somewhat larger telescope than the 26-inch, but one public-spirited citizen, W. D. Uptegraff of Niagara Falls, New York, reading about Francis G. Pease's 300-inch telescope plan in 1926, wrote his Congressman and Senator David Reed of Pennsylvania, arguing that the government should build it (at a cost of $15 to $20 million), and that the Naval Observatory or the Smithsonian Institution should operate it.[10] Pollock knew there was no chance of this idea becoming a reality, but he encouraged his successor, Captain Charles S. Freeman, a destroyer, cruiser, and battleship commander, to use his access to the top admirals to get the new telescopes the observatory needed.[11]

Freeman, a conscientious organization man, invited his astronomical staff to suggest needed equipment. Some of their ideas were pathetic. Asaph Hall, Jr., son of the discoverer of the satellites of Mars, requested a quotation for the price of a new lens for the 26-inch refractor, a technology outmoded since 1900.[12] The superintendent, however, also sought advice from outside astronomers. There was no money for scientific advisory committees in the Navy's budget, but Freeman had his staff write the leaders of American astronomy, soliciting their recommendations, and followed up himself, inviting them to a conference while they were in Washington for a meeting of the National Academy of Sciences.[13] Freeman was aware that "photography of precision" was the great need of the Naval Observatory, and his questions centered on it. Most of the astronomers he consulted gave intelligent, well-thought-out advice, though Edwin B. Frost replied with the tired old ideas of a superannuated director. Schlesinger, who had pioneered in photographic astrometry (measurements of accurate positions of stars), was especially helpful but also quite critical of the Naval Observatory astronomers' professional qualifications.[14] At the conference held in April 1928, Freeman's advisers agreed that the Naval Observatory should "expand" its photographic work and recommended as the first priority a 6-inch, wide-angle camera based on the newly developed Ross lens (invented by Frank E. Ross). They also recommended obtaining a larger, but narrower-field, 20-inch photographic refractor, for measuring the positions of comets and asteroids.[15]

The final Navy Department plan for a $160,000 appropriation for "modernization of Naval Observatory" included the 6-inch wide-angle camera but expanded the 20-inch to a pair of 24-inch photographic refractors, mounted in tandem.[16] Freeman, assisted by James Robertson, presented this plan to a subcommittee of the House of Representatives Committee on Appropriations during the hearings on the Navy Department budget request in January 1929. Robertson, who was to become Ritchey's most effective supporter, was the assistant director of the Nautical Almanac Office, the part of the Naval Observatory responsible for calculations of the positions of the sun, moon, and planets.[17] He was an absolute nonentity as an astronomer but a world expert at pulling wires and flattering his superiors. Starting with a bachelor's degree from the University of Michigan in 1891, he had moved to Washington, done some part-time computing (all with paper, pencil, and logarithm tables in those pre-electronic days), and worked briefly as a lowly assistant to Simon Newcomb near the end of this great astronomer's life. Referring continually to this experience, Robertson presented himself as a masterful theoretician of celestial mechanics, although in fact he never had an original scientific idea and had not published his first paper, a short account of the orbit he had calculated (by standard methods) for Jupiter's fifth satellite, until 1924.[18] Nevertheless, by careful attention to his political connections, Robertson has risen nearly to the top of the tree by 1928 and was soon to go higher.

In the summer of that year, he had taken part in the International Astronomical Union meeting in the Netherlands. In the course of this assignment, he had traveled around Europe, visiting observatories and

James Robertson, Ritchey's sponsor at the U. S. Naval Observatory. Courtesy of the late John S. Hall.

collecting information and ideas on photographic telescopes. In France Robertson had met Ritchey, visited him at his laboratory in the Paris Observatory, and seen and heard his telescope plans. Robertson recognized Ritchey as another outsider like himself, shunned by the American astronomical establishment. The Nautical Almanac computer decided almost instantly that what the Naval Observatory really needed was a Ritchey-Chrétien reflector, which he recommended on his return to Washington in 1928. His stated scientific "reasons" were decidedly confused, and they show that he did not understand telescopes at all. Freeman did not include the recommended Ritchey-Chrétien in his modernization plan, but Robertson did not give up. He kept in touch with Ritchey and predicted that he would eventually get it into the budget.[19]

At the hearing before the House subcommittee in January 1929, Robertson, who was an expert in adopting the image of a somewhat rustic, extremely knowledgeable, and very economical scientist, spoke up loyally for Freeman's recommendations but also managed to express his own ideas about a Ritchey-Chrétien reflector. He made several monumental gaffes, including referring to such a telescope's "lens" instead of its mirror, a mistake he was to repeat frequently, but the Congressmen never noticed it. The House left the Naval Observatory modernization plan out of the appropriation bill that year, but the stage was set for the following session of Congress.[20]

In 1929 the director of the Nautical Almanac Office, William S. Eichelberger, reached the mandatory retirement age. He was a competent scientist, though hardly inspired. His successor, in the normal course of events, would be Robertson. All the leaders of American astronomy were appalled. Some of them had never heard of him before—those who had considered him a bad joke. They mounted a campaign to prevent the appointment, launching a barrage of letters to the secretary of the navy and trying to use what little political influence they thought they had. Campbell, Hale, Harlow Shapley, and several others joined the movement, orchestrated by Schlesinger. Walter S. Adams, who had read Robertson's testimony that Ritchey was "the man who has made the three greatest telescopes in the world today, and he predicted that the one at Mount Wilson would not be entirely satisfactory, for the reason that it weighs five tons," and that "the question of temperature" was "one defect of that 100-inch reflector," was especially vituperative.[21]

They had no chance to defeat Robertson. He had spent much of his life organizing the situation so that he could not be displaced. Years before, his political sponsor, Senator Wesley Jones of Washington, had denounced Ross, who had been Newcomb's real chief assistant and who many astronomers thought would someday become director of the Nautical Almanac Office, driving him out of government service. As a result there was no potential rival to Robertson within the organization, and his political connections, which he openly boasted the Naval Observatory needed, made it impossible even to think of bringing in an outsider.[22] In September 1929 Eichelberger retired, and Robertson became the Nautical Almanac director.[23]

The following year Freeman again proposed the same modernization program for the Naval Observatory. Once again Robertson accompanied him to the congressional hearings and supported his testimony. This time the Ritchey-Chrétien telescope did not come up. Most of the questions were about a proposed "service building" (a garage) and whether its design need be approved by the Fine Arts Commission. The appropriation went through and was adopted. The Naval Observatory was allotted $160,000 for new photographic instruments, its first appropriation for new, major telescopes since 1894.[24]

In May 1930, just before the bill was passed and became law, Freeman went back to sea as skipper of the USS *Arkansas*. The new superintendent who replaced him at the Naval Observatory was Captain J. F. ("Fritz") Hellweg. He had served mostly in engineering and ordnance assignments but had commanded the battleship USS *Oklahoma* for two years immediately before his assignment to Washington. Hellweg prided himself on being able to solve any technical problem on the basis of his Annapolis training and paid far less attention to the advice of experts than Freeman had. Robertson gave the captain a pamphlet describing Ritchey's telescopes almost as soon as he arrived at the Naval Observatory, and Hellweg studied it with interest. He quickly decided to change Freeman's plan and build a single 28-inch or 30-inch photographic refractor instead of the pair of 24-inch ones, but initially he did not consider a reflector.[25] However, soon after Henkel's death, Robertson brought up the idea of building an even larger Ritchey-Chrétien reflector instead of the photographic refractor. Hellweg was definitely interested. He liked the idea of getting "the very latest and the very best equipment possible for the money," especially if it could be a 40-inch reflector instead of a 30-inch refractor. Although the whole concept of a reflecting telescope was new to him, and he did not know what questions to ask, he was eager for more information.[26]

Robertson, who with Lutz had sponsored Ritchey's return to the United States, wrote the optician of the opportunity. Ritchey replied with a glowing description of how much better "*one* highly efficient, highly refined modern photographic telescope," a 40-inch Ritchey-Chrétien, would be than either a 30-inch photographic refractor or a conventional

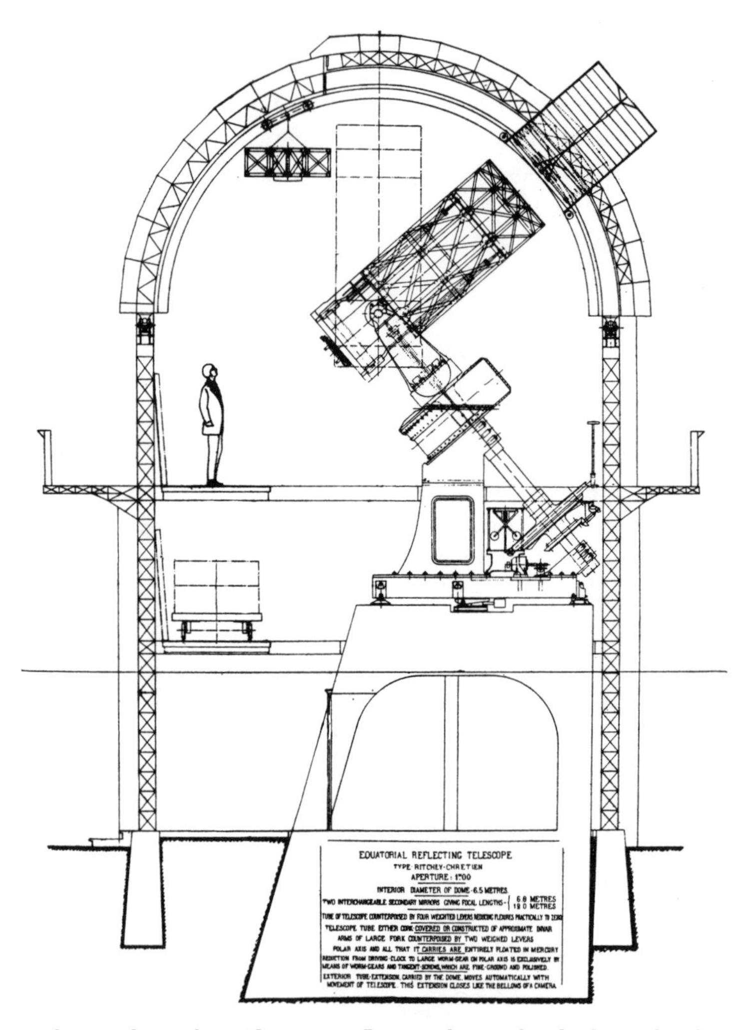

Design of 40-inch Ritchey-Chrétien reflector, drawn for the latitude of Paris but otherwise essentially as built at the U. S. Naval Observatory in Washington. From the *Journal of the Royal Astronomical Society of Canada*.

40-inch reflector ("of the Mt. Wilson type"). He had a preliminary design for a 40-inch Ritchey-Chrétien, prepared in France, which he could build instead in Washington. It would be "the *initiation*—the *inauguration* of the Modern Reflecting Telescope and The New Astronomical Photography of which I wrote, . . . a new epoch in Astronomy. . . . And this will be done where it *should* be done, in the U. S. Government Ob[servator]y, and under your administration." He sent Robertson one of his photographs taken with the 20-inch Ritchey-Chrétien in France and also arranged for a complete set of one hundred of his best pictures, taken with the 40-inch, the 24-inch and the 60-inch, to be delivered to him.[27]

To Hellweg, Ritchey was even more positive. He knew the captain would read the letter he had written Robertson the previous day, and now added that in his opinion, and those of the French experts who had examined the world's only existing Ritchey-Chrétien telescope, "it *must* soon render obsolete all other known types of photographic telescopes excepting only the very wide-angle type of photographic *refractor*" (the Ross lens). The money available in the appropriation for a photographic telescope was $76,000; Ritchey said he *thought* he could build the 40-inch for this amount. He made this estimate within no more than two or three days and with practically no knowledge of current financial conditions and manufacturing costs in the United States, where he had not lived for the past seven years. He promised Hellweg that "[i]f I build your photographic telescope I shall demonstrate the *use* of it in photography of the sky; we shall make photographs with it which CORRESPOND TO 90% OF THE FULL THEORETICAL RESOLVING POWER DUE TO THE APERTURE, INSTEAD OF THE 5% TO 10% REPRESENTED BY THE BEST CELESTIAL PHOTOGRAPHS AT PRESENT; we shall do this in the U. S. Naval Observatory, where, of all places, it *should* be done."[28] Ritchey, who had been trying to find a local donor in Miami, was soon on his way to Washington and met Hellweg at the Naval Observatory on February 6, 1931.[29]

They hit it off perfectly. Ritchey, with his sweeping, positive statements, and his promises of quick results, was just the type of person to appeal to Hellweg. Before long the captain came to believe, or at least to write to his predecessor and Naval Academy classmate, "Hoss" Freeman, that Ritchey had taken the hundred "magnificent" astronomical photographs with a *Ritchey-Chrétien* telescope, though the astronomer had in fact told him that they were from Yerkes and Mount Wilson. Hellweg had remembered this correctly the previous week. He had decided to go the Ritchey-Chrétien route and demanded to know why Freeman had favored building a 24-inch refractor rather than a 30-inch or 40-inch reflector. Hellweg parroted in distorted form some of the arguments Ritchey had been giving him about the superior definition and light transmission of a reflector. He claimed that most members of the Astronomical Council, the senior astronomers of the Naval Observatory staff, agreed with him that a Ritchey-Chrétien reflector would be best. Freeman was unable to contradict him. He had followed the consensus of the experts in his time, he said. They had hardly mentioned reflectors. Now they had changed their minds, Hellweg told him. Freeman was familiar with that, "just as many opinions practically as there were astronomers consulted, . . . the old Navy line stuff, all over again." His conclusion was that if a majority of the Astronomical Council favored it, and "you can get a 40″ or 45″ reflector for less money [than a 24-inch photographic refractor], and are assured of a satisfactory piece of workmanship on the glass itself, there would appear to be little room for doubt as to the correct decision." Still Freeman could not help but wonder "[i]f the reflector type of lens is fundamentally better . . . than the refractor type, why haven't we gone in for reflector type cameras for ordinary photography?" Yet his arguments were, on the whole, *more* cogent than Hellweg's![30]

One of the reasons Hellweg wanted to build the Ritchey-Chrétien reflector was that the price quotations he was getting for photographic refractors were higher than he had expected. With a fixed appropriation, the only solution would be to make but one of them, and to make it small; on the other hand Ritchey promised the bigger 40-inch. Furthermore, the lens for the refractor would require special types of glass that might take years to produce. Hellweg wanted a telescope in operation before his tour of duty ended, and Ritchey was positive that he could get a mirror blank quickly from St. Gobain.[31] Hellweg did not seek outside advice, and when he consulted his Astronomical Council, he made it very clear that he wanted support for the Ritchey-Chrétien, not debate. Harry E. Burton, the elderly head of the Equatorial Division, responsible for the regular telescopes, strongly favored it, while Chester B. Watts of the Transit Circle Division realized he could not vote against it but recommended a smaller photographic refractor to supplement "the work of the reflector which you have in mind."[32]

By making his decision without all the facts, Hellweg never learned of the main drawback to building a Ritchey-Chrétien reflector for the Naval Observatory—light pollution from the city of Washington. Heber D. Curtis had discovered just how effective the street lights and advertising signs of the 1920s were at fogging astronomical plates when he moved from Lick Observatory on isolated Mount Hamilton to Allegheny Observatory in Pittsburgh. He had warned Freeman that although focal ratios of f/18, f/15, and possibly f/10 would be safe, a telescope as fast as f/8 might be a problem. He had advised making experiments with small cameras before reaching a decision. Schlesinger and Henry Norris Russell backed him in this recommendation, and Freeman acknowledged it, but no one at the Naval Observatory seems to have taken it seriously. Curtis had specifically warned the superintendent that he doubted that a telescope as fast as f/6.5 "could be used to its full effect in a locality surrounded by the bright lights of a city like Washington." The Ritchey-Chrétien reflector would almost certainly be an f/6.8, but Hellweg apparently never considered this problem. A letter he had signed, no doubt prepared for him by one of the Naval Observatory astronomers, advised a maker of small telescopes about the fogging of astronomical plates by "sky light . . . in, or near, a city" and recommended focal ratios f/7 or longer, but clearly Hellweg did not understand it himself.[33] In fact, even f/7 would have been too fast.

By mid-February 1931, less than a month after meeting Ritchey, the superintendent had decided the new telescope should definitely be a Ritchey-Chrétien reflector. The optician had imagined that he would be a consultant to the Naval Observatory on the project but learned that he would instead get a fixed-price contract to build it. Hellweg described the Ritchey-Chrétien concept to his superiors as a proprietary secret and Ritchey as the sole supplier of this type of telescope. On this basis he could be given a contract to make it for the navy without any public advertisement or bidding.[34] In late February Ritchey returned to Miami to unpack and assemble the 20-inch Ritchey-Chrétien telescope and to prepare his family for the move to Washington. On March 20, 1931, Hellweg sent him the good news that the contract was a certainty, and before the end of the month he was back at the Naval Observatory, ready to start work.[35] A few months later Ritchey had the 20-inch repacked and returned to France; it was never mounted or used in Miami.[36]

On June 5, 1931, Ritchey signed the contract to provide the 40-inch reflector for the Naval Observatory at a fixed cost of $76,000. He agreed to complete it in twenty-seven months. He was to be paid in monthly installments, but he would not get the final 10 percent of the price until the telescope had been adjusted and tested to the complete satisfaction of Hellweg.[37] Ritchey received his first payment of $3,788.41 on July 30. He had already ordered the 40-inch glass disk for the mirror from St. Gobain, although a "Buy American" law was in effect on government contracts. The navy's legal minds ruled that the telescope was to be part of a building, to which the act did not apply, and hence the disk could be bought abroad.[38]

Hellweg detailed W. Malcolm Browne of the Naval Observatory staff to work full-time with Ritchey, familiarizing himself with every piece of the telescope as it was built. Then twenty-four years old, Browne had completed three-and-a-half years study toward an electrical engineering degree at George Washington University in Washington before getting a job at the Naval Observatory in 1927. With his electronics background, his first assignment had been in the Time Division. In 1928 Freeman had designated him for future photographic work with the telescopes to be built under the modernization plan. Browne's preparation was to visit several observatories, including Yerkes, Mount Wilson, and Lick, on his summer vacation that year. He had done some visual observing with the 12-inch and 26-inch refractors but had no direct experience in astronomical photography when he began working with Ritchey. However, he had signed a statement for Hellweg, recommending a Ritchey-Chrétien reflector as the large photographic telescope of the modernization plan.[39]

Ritchey, now an independent contractor, hired a crew of men to work with him. By the end of the summer he had six or seven men on his team. Jobs were hard to find in the Depression, and Ritchey could get skilled craftsmen at low wages. Draftsman D. O. Landis received $225 per month, and head machinists Edward W. Pearson and (later) R. F. Webber, $200 each. Often they needed advances on their salaries to get through the month; at other times Ritchey could only pay them part of their salary on account for a month or two. Young Lawrence Braymer, many years later the president of Questar Corporation, a telescope-

W. Malcolm Browne, Ritchey, C. A. Fleming of the Baldwin-Southwark Corporation, Mrs. Betty Browne, Mrs. Lillie Gray Ritchey, and Mrs. Harry E. Burton standing next to the lower part of the mounting of the 40-inch Ritchey-Chrétien reflector. Courtesy of W. Malcolm Browne.

making company, worked as a volunteer assistant for a few months at the start of the project. Willis Ritchey came east during his summer vacations to work with his father. William H. Evans, the Californian who had wanted a job in France, came to Washington as an optician in September 1931, and stayed until the project ended.

Ritchey faced high start-up expenses for materials and tools. He had to borrow money to meet them from a bank, from his sponsor, Lutz, and from Burton, whose job would include advising the superintendent of the Naval Observatory whether or not the completed telescope measured up to specifications. Ritchey had no accountant, purchasing agent, or business organization of any kind. He kept his own financial books in which he mingled checks for his family's household expenses with payments for materials and his workmen's salaries. His wife was the secretary who typed his business letters, except the most important ones, which Hellweg had Naval Observatory personnel do.[40]

The captain turned a large area on the second floor of the main building of the Naval Observatory over to Ritchey for his office, a drafting room, and the optical laboratory where all the fine grinding, polishing, and testing of the telescope mirrors would be carried out. He also allowed Ritchey to use an old sheet-metal building next to the main building for his woodworking, pattern and instrument shops, for the rough grinding of the mirrors, and for the erecting shop where the telescope would be assembled. Ritchey's men partitioned off the rooms, and he paid for the electric power to the shops, but the Navy provided the electricity in the main building and the heat in both buildings, as well as the space.[41]

The new telescope to be built for the government observatory was news as soon as the contract was signed. Hellweg, Robertson, and Ritchey were only too eager to publicize it, long before construction had begun. An initial article in the *New York Times* in June 1931 was relatively reasonable, though the reporter enjoyed the concept that it would penetrate "at least 60,000,000,000,000,000,000,000 miles into space." However, an Associated Press article, based on a follow-up interview with Ritchey a few days later, was devoted to explaining how the new Ritchey-Chrétien 40-inch reflector would work at 90 percent of the theoretical maximum efficiency, and far surpass the Mount Wilson 100-inch "only five per cent efficient in this respect." This was too much for a citizen of Wytheville, Virginia, who wrote the Naval Observatory to ask how this could possibly be true. Hellweg assured him "that basically the article is correct." The great advances were based on "professional secrets" belonging to Ritchey and Chrétien, "which those gentlemen have not patented in order to prevent premature publicity." To preserve security the captain could not reveal them, he said. "I have seen some photographs made by a telescope of this type and they were very startling," he closed. This was completely false. The few photographs Ritchey had taken with the 20-inch Ritchey-Chrétien in France were far from startling; all the spectacular nebular pictures Hellweg had seen had been taken with the Yerkes 24-inch or the Mount Wilson 60-inch. Most probably, the captain was not consciously lying but simply scientifically confused and naively optimistic.[42]

Then in October 1931 the *Washington Post* published a feature article in its Sunday magazine section on "the new telescope now building at the Naval Observatory [which] . . . will surpass all existing instruments." Work had not begun on the mirror, indeed the glass disk had not even arrived in the United States, but the article flatly stated that the 40-inch Ritchey-Chrétien "is the most modern instrument today." It quoted Ritchey as saying that the new type of telescope would give about 90 percent efficiency, while the 60-inch which he had built, "itself of proven revolutionary value," was about 10 percent efficient. "The 100-inch telescope at Mount Wilson is far below this in efficiency—below 5 percent," the article went on. Yet it was illustrated with a spectacular picture of M51, the "Whirlpool Nebula," taken by Ritchey, with the 60-inch (although this was not stated), and with photographs of the Mount Wilson 100-inch telescope dome and the 60-foot solar tower. This article was apparently too extreme for the astronomers at the Naval Observatory; at any rate Hellweg learned it had made him a laughingstock among scientists. He apologized formally to Adams, the Mount Wilson director, saying that the article was in very bad taste and assuring him "that the Observatory and its staff had nothing to do with the writing of that article." Adams accepted the apology in good grace. He had had enough experiences with reporters and writers so that he was "accustomed to taking their articles not too seriously." Hellweg also complained to the editor of the *Post*, but in fact the writer of this article had taken all the comparative figures from Ritchey's published statements and had read the article to the astronomer and to the captain, who had both approved it. When confronted with this fact, Hellweg backed down weakly but did not retract his statement to Adams.[43]

The first 40-inch glass disk arrived from France at the end of December 1931. Ritchey's assistants had completed building the optical grinding machine, and started working on the glass just after Christmas. It was to be ground and polished as a plane test mirror. The second disk, to be used for the primary mirror of the telescope, came in April 1932. Each disk cost Ritchey approximately $1,300, plus $500 for the duty and shipping charges. Grinding the primary began within a week after the arrival of the second disk.[44]

To figure the primary mirror to its proper concave hyperboloid form, Ritchey needed precisely calculated test data for the knife-edge measurements of the center of curvature of each zone. Vincent Nechvíle, Henri Chrétien's former student who had returned to his native Prague as a faculty member at its Charles University, provided these calculations. Chrétien, very busy and successful as a professor and a designer of commercial optical systems, had no time to do them himself, but he checked a few of Nechvíle's formulae and first computations. Simply developing the power series and equations to be used in numerical form required long and painstaking effort; the calculations themselves, done to many significant figures with logarithm tables and a mechanical hand computer, represented an enormous piece of work.[45] Apparently, Ritchey did not pay Nechvíle any fee for performing these calculations; he did them for their scientific interest. He later published some of these results in a paper that established him as an expert on Ritchey-Chrétien reflectors.[46] Browne recalculated all the test data and confirmed Nechvíle's numerical results to very high accuracy.[47]

All the work on the primary mirror went well until the time came to enlarge the hole through the center of it. The St. Gobain Glass Company had bored a hole in the disk, but it was too small. In July 1932, as Ritchey's opticians enlarged the hole, the disk cracked. It was ruined. The disk must have had internal stresses, but it probably would have been satisfactory if it had not been necessary to make the hole larger. Ritchey immediately ordered a replacement disk from St. Gobain, which the company agreed to provide at half price. This time he ordered it pierced with a hole big enough so that no later enlargement would be

Vincent Nechvíle, who first calculated the test data for the 40-inch Ritchey-Chrétien reflector. Courtesy of the History Section of the Czech Astronomical Society.

necessary. But he was worried, for he knew that it would be months before the second disk could be produced and shipped to Washington.

For years Hellweg's dream had been to prove that metal mirrors would be better to use in telescopes than glass mirrors. What better time than now? Lutz was certain the 40-inch could be made this way. The superintendent persuaded Ritchey to request a change in his contract to permit him to substitute a "Lutz-metal" mirror for the glass one originally specified. In his heart, the old optician surely felt the alloy mirror could not be made satisfactorily, but he was willing to try—at least for a time. Hellweg shepherded this request for the modification through the Navy Department bureaucracy, specifying it would not delay completion of the telescope and that Ritchey would substitute a glass mirror if the metal one did not prove superior. Everyone, including the secretary of the navy, agreed, and Ritchey began work on the metal disk. At least it was much less expensive than the glass disk, costing only $382.[48]

Simultaneously, work proceeded on fine grinding and polishing the 40-inch flat test mirror. Ritchey pronounced it finished in October 1932. How he tested it is not clear; his usual method was to use a spherical mirror of the same diameter, but he did not have one this large. Probably, he tested it in overlapping sections with a smaller spherical mirror.

Ritchey had not informed Hellweg that he had ordered a replacement glass disk from France. Now suddenly one morning the metal mirror, on the grinding machine, was discovered marred with many deep scratches. They had mysteriously appeared overnight. Since Ritchey kept his shop locked and jealously guarded access to it, it is fairly clear how the "accident" must have occurred. Ritchey now let it be known that he intended to complete two primaries, one of glass and one of a metal, so that whichever was better could be used. The metal disk, however, thin to begin with, would probably be too flexible if it were cut down further to eliminate the scratches. Hellweg was stunned. He was convinced that Ritchey should make the metal primary, but the optician was obstinate. Hellweg came up with a secret plan to let Ritchey finish the glass mirror and then afterward get Lutz to make a replacement metal mirror for it, to the captain's own design. However, when Lutz estimated the cost as at least $5,000, Hellweg had to drop that idea. That was "an awfully big amount of money to the Naval Observatory, particularly as the government [had] cut [their] appropriation so heavily."[49]

That same month, February 1933, the new glass disk for the primary mirror arrived from France. It went on the grinding machine almost at once. This time there was no question of enlarging its central hole. Ritchey and his men worked rapidly and by April they were ready to begin grinding the preliminary spherical surface into the disk. Nevertheless, the time lost in working on the first, broken disk and on the aborted metal disk, and then the delay in waiting for the second disk had cost Ritchey dearly. He had had to pay his men's salaries and his family's living expenses. He had spent nearly all the $76,000 fixed price, except the final 10 percent, which the Navy would not release until the telescope had been completed and accepted. He knew that it would be later than the delivery date he had promised in the contract, and that under its terms he would be penalized financially. To top it all, his bank failed in March 1933 and he lost nearly $900. Ritchey became very depressed and stopped keeping his financial records. His wife probably took over this duty, but her books no longer survive. Hellweg, concerned that the optician would break down and never finish the telescope, tried to get Ritchey released from the penalty clause on the grounds that the cracking of the mirror had been an act of God. In his peculiar Navy language he worried about the "psychological effect" of the "drain on [Ritchey's] resources . . . All astronomers are apt to be extremely temperamental and it is hoped to remove as far as possible the effect of this temperamental condition from the final result" [the telescope]. Though the chief of the Bureau of Navigation approved this request for an extension, the hard-boiled judge advocate general turned it down. If the accident had happened, it was not "unforeseeable," and if Ritchey was familiar with his work he should have foreseen it, this sea lawyer argued. Ritchey kept up a bold front and maintained that the telescope was very nearly completed, but inwardly he must have trembled. He tried to lock Browne out of the optical shop, so he would not be able to check on the real state of progress on the telescope, but Hellweg insisted that his inspector to be allowed to enter.[50]

Somehow, Ritchey kept up the work. By the end of May his men had the disk rough ground to spherical form, in July fine ground, and in August polished. Then he began figuring it to the final hyperboloidal form. It was a long, slow process, but in January 1934 he finished it. His assistants had rough ground the 16-inch secondary mirror for the telescope in intervals between work on the primary. Now he had them begin fine grinding it to the preliminary spherical form, and in February he began testing it with the primary and the flat, which they had silvered under his direction. By March he had the secondary fine polished as a sphere and started figuring it to the final hyperboloidal surface. In April he declared it finished and silvered it, too.[51]

On April 27, 1934, Hellweg announced to the Astronomical Council, the senior astronomers of the Naval Observatory who by law were his advisers, that the 40-inch Ritchey-Chrétien was completed. He directed them to inspect "and determine its readiness for operation." As soon as they reported it ready for use, the contract would be completed. He made it very clear that he expected a positive report. Yet Ritchey had not taken a single photograph with it, and it was impossible for them to state its optical condition. The weather was almost continuously cloudy, and they could not even observe visually with it. Nevertheless, Hellweg reported the telescope completed and tentatively accepted on April 28. Final acceptance would have to wait for clear weather, but Ritchey would not be penalized $10 a day for late delivery past this date. On June 16 a majority of five of the seven members of the council, including Robertson and Burton, signed a document stating that they had "inspected the 40-inch reflector telescope and [found] it complete in mechanical details," carefully expressing no opinions on its optical quality. Only H. R. Morgan and Chester B. Watts refused to certify the new telescope. No one attempted to challenge Hellweg's tentative acceptance, and on its basis the navy considered the world's first real research Ritchey-Chrétien reflector as finished and ready for action. Ritchey received his final payment.[52]

At Hellweg's advice, he filed a claim for $2,300, the penalty for the 230 days overrun of the delivery date set by the contract. This amount had been deducted from his final payment. Ritchey stated that the entire delay was due to the breaking of the first disk, a completely unexpected event. Using language that the navy lawyers had suggested, he stated that he had used St. Gobain disks for forty years, at Yerkes, Mount Wilson, Paris, the Naval Observatory, and in his own laboratory, but that "not one of these discs, large or small, broke or gave any trouble or delay, except this one." He had, very conveniently for himself, forgotten the four-year delay with the Mount Wilson 100-inch disk. This time the entire chain of command accepted his argument that the breaking of the first 40-inch disk "should be regarded as something beyond [his] control, and that consequently [he] should be held guiltless of negligence or blame." He received the last $2,300 also.[53]

In 1933 Ritchey had phased out most of his employees, one by one, since the mounting for the telescope was essentially completed and only the final figuring of the mirrors remained. He had only two men still working with him in 1934—Evans, his optical assistant, who stayed on the job until April, and Webber, his head machinist, who lasted until May. Clinton B. Ford, who had met Ritchey in Paris in 1928, now an undergraduate at Carleton College, hoped to work with him in the summer of 1934, but the optician had to tell him that no money was available.[54] Hellweg was desperate to keep Ritchey himself at the observatory. No one on the staff had any experience at astronomical photography (George H. Peters, the one person who had done a little of it with his own small cameras, had retired in 1931), and no one but Browne even knew how the new 40-inch was supposed to operate. Hence in the fall of 1933 Hellweg had begun a campaign to have Ritchey appointed to the Naval Observatory staff. There were tremendous bureaucratic obstacles to bringing a senior person into the Civil Service, and there was very little money in the navy budget. Nevertheless, the captain persisted. He stated that "Ritchey [had] made all the finest and best known astronomical telescopes in the world during the past 35 years, and even his most bitter enemies acknowledge[d] that he [was] the foremost photographer in the world." Hale and Adams, no doubt, would have disagreed with that statement; it had been true until 1919 when the Mount Wilson team had completed the 100-inch and Francis G. Pease and John C. Duncan had begun using it to surpass Ritchey's 60-inch photographs. But Hale and Adams were not consulted, and Hellweg put all his navy influence and connections on the line to push the appointment. It worked, and in the end Ritchey became an "Astrographic Consultant," an "absolutely essential position," at the highest Civil Service Grade, P-7. The salary was $6,500 per year, but his appointment was a temporary one for only six months. He could not begin it until after the telescope was judged completed and it belonged to the government. Thus he went on pay status at the Naval Observatory on April 30, 1934.[55]

Hellweg now ordered Browne, who had previously spent one night a week showing visitors through the Naval Observatory, to give up this duty and work full-time with Ritchey at the telescope. That spring and summer the astronomer did succeed in obtaining some photographs of stars with the instrument.[56] They were far from impressive, however. One problem was that Ritchey had only a few plates, probably twenty at most, available for use. The focal surface of the telescope is concave, curved inward toward the incoming light rays. Thus it was impossible to use flat photographic plates. Ritchey had thick glass plates, ground to the correct curvature, which he had to send to Eastman Kodak to be coated with photographic emulsion each time they were to be used. Because the plates were bowl-shaped, the layer of emulsion was not uniform; in addition the coating process left many flaws in the resulting photographs. To bring a new telescope into perfect alignment and operating

Captain J. F. Hellweg (left) and Ritchey at the Cassegrain focus of the 40-inch Ritchey-Chrétien reflector, U. S. Naval Observatory, Washington, probably ca. 1934. Courtesy of Richard B. Burgess.

condition requires taking many test exposures, but they were impossible with the limited number of plates.[57]

Another problem was light pollution at the Naval Observatory site in northwest Washington. Hellweg finally became aware of the "light glare" around the observatory in 1932. Probably Ritchey, who knew it would be bad for long exposures with the f/6.8 Ritchey-Chrétien, showed it to him. There was no home rule in Washington, and the superintendent suggested to the Navy Department that Congress pass a law to prevent building more apartment houses, which he saw as the main offenders, close to the observatory. The Navy lobbyists, however, realized that there was no chance that congressmen would halt building in the capital for the sake of a few astronomers. Then Hellweg learned that the commissioners of the District of Columbia, appointed by Congress to administer the city, could use their zoning power to restrict the heights of buildings. One of the commissioners, Major John C. Gotwals of the Army Engineers, was friendly. Through him Hellweg arranged that the Naval Observatory would have veto power over any changes in zoning that might otherwise allow higher buildings near the observatory.[58] This was no solution, however, for actually the light pollution came from the whole city of Washington, not the local area. Thus, though it was possible to take short exposures on the few nights when the skies were actually clear, the long exposures that Ritchey needed to obtain the spectacular photographs of faint nebulae which would validate his new telescope were impossible. Hellweg, impatient for results, complained of lack of progress.[59]

Ritchey, however, decided that the secondary mirror he had made was not perfect. At the end of August he took it out of the telescope and began refiguring it, even though the Navy had accepted the telescope months previously. Ritchey completed refiguring the secondary in early October, resilvered it, and reinstalled it in the telescope. This was his last month of employment by the Naval Observatory. Hellweg tried to get his temporary appointment extended for an additional three months but was unsuccessful. The stated reason was that no funds were available, but probably the fact that Ritchey would be seventy years old before those three months were over was also relevant. On October 29, 1934, he went off the Naval Observatory payroll.[60]

Before Ritchey left, however, he briefly held the center of the stage at the unveiling of the new telescope to a scientific group. This occurred at a meeting of the Optical Society of America in Washington in October 1934. Ritchey, a loner and outsider, was basically secretive. He long had rigorously guarded access to his workshop, especially in Washington. With his background in manual training and telescope making, and ostracized in America since 1919, he had practically no friends among American astronomers. The only recorded visitors who got into Ritchey's laboratory and saw the 40-inch Ritchey-Chrétien under construction were the Duc de Gramont in 1931, Henri Chrétien in 1932, three minor figures whom Robertson shepherded through in 1933, and one whom Shapley sent in August 1934. Only one of these six was an American. Perhaps there were more who saw Ritchey's telescope before the Optical Society meeting but not many.[61]

Hellweg was equally secretive, except for information that he dispensed himself in suitably laundered form. In moments of stress, he had a tendency to "call attention to the lack of loyalty of certain of the employees," by which he meant their reporting failures as well as successes. Such frankness could only cause trouble, he warned, perhaps even the closing of the observatory and sending its instruments elsewhere.[62] In fact this is just what several of the leaders of American astronomy were trying to do. Campbell, the former Lick Observatory director, had retired from the presidency of the University of California in 1930, but soon after he became president of the National Academy of Sciences. In this post he lived in Washington several months each year, and he considered it part of his duty to provide the government with scientific advice. Naturally, he was especially interested in astronomy. By the end of 1932 Campbell had become convinced that Washington would be a very poor site for the 40-inch reflector Ritchey was building because of the dust, smoke, and light pollution of the city, and the large daily temperature range. To him, the whole problem was that the superintendent of the observatory was a naval officer with no astronomical training, experience, or knowledge. As the new administration of President Franklin D. Roosevelt came into office in 1933, Campbell lobbied vigorously for a change at the Naval Observatory. He criticized Hellweg, the site, and above all the system by which the Navy ran its astronomical institution. He never criticized Ritchey or the Ritchey-Chrétien concept. Campbell did not see the president himself, as he had hoped, but he was assured that Budget Director Lewis Douglas would look into his charges. Hellweg was worried enough to come to Campbell's office at the academy building and appeal for his assistance to save the observatory, but the strong-willed old astronomer denounced him to his face. Once again, however, the Navy Department's power was stronger than a scientist's advice, and nothing changed.[63]

Thus Hellweg dreamed of showing the astronomers the completed 40-inch reflector and turning their censure to praise. Ritchey had told him that no improvements had been made on the Mount Wilson telescopes since he had left its staff in 1919, that the 100-inch was no good at all and the 200-inch would be worse, and that the new 40-inch with "all the latest refinements" would give "far superior pictures" to those he had taken with his prized 60-inch. The credulous captain had visualized as early as 1931 that once "our big telescope" was completed, the formerly doubting astronomers would come clamoring to the doors of the Naval Observatory to learn the "secrets" of the Ritchey-Chrétien reflector.[64] Thus he had welcomed the opportunity to show it off to the Optical Society meeting in Washington in October 1934. Probably he had made or approved the invitation at least a year in advance, when he thought the telescope would surely be in operation months before the event. However, as October drew near, Ritchey had practically no results to show and was working on the secondary mirror almost to the last moment.

Furthermore, Hellweg and Ritchey learned the old optician would not have the center of the stage to himself. His rivals, Francis G. Pease and Ross, would also be on the program. Since starting to work half-time on the 200-inch project, Ross had become an especially virulent critic of Ritchey. The lens system Ross had designed to correct conventional reflectors for coma had made him the favorite of Hale and the confidant of Adams. It made their plan for the 200-inch possible, although modern optical designers consider the Ross type of field corrector a forced, unnatural solution and no longer use it. Ross received his reward in almost immediate election to the National Academy of Sciences, with the strong expressed support of Hale, its high priest, and the solid vote of the Mount Wilson members.[65] (John A. Anderson had similarly been elected in 1928, just before he took charge of the 200-inch project; Hale knew how to reward his helpers.) In 1932 Ross had protested violently when Charles Fabry, the president of the International Astronomical Union Commission on Astronomical Instruments, appointed Ritchey a member of this body. Ross claimed that "[w]e in America are strongly of the opinion that the activities of Mr. Ritchey in scientific publications, on the lecture platform, and in the daily press are not in the best interests of astronomy, in that he disseminates false information, [and] makes preposterous claims." He finished his letter by comparing Ritchey to an inmate of an insane asylum.[66]

Now, at Adams's suggestion, Ross prepared a paper on "Optics of Reflecting Telescopes" to be read at the Optical Society meeting. It was a survey of the possibilities for large telescopes, emphasizing the 200-inch, stressing the disadvantages of the Ritchey-Chrétien system and the advantages of the conventional paraboloid prime focus equipped with a Ross field corrector. Ross would not go to Washington himself to give the paper but planned to have George W. Moffitt, "a cool cutting debater," present it for him, and, he said, "if friend Ritchey gets peevish,

Moffitt will tear him to shreds." Adams, however, had a better idea. He told Ross to let Pease "read" the paper, thus giving Ritchey's former assistant two shots at his former boss. Pease was also scheduled to present an invited lecture on "Modern Large Telescope Design," that is, on the 200-inch, at the dinner at the end of the meeting. The idea for this invitation may well have originated from Adams also.[67]

Thus the meeting was a poor anticlimax to Hellweg's and Ritchey's hopes. Pease spoke with confidence, but only in the most general terms, of the plans for the largest telescope in the world and how the Ross corrector would give a coma-free field large enough for any photographic work that was likely to be useful. Ritchey, on the other hand, could only show a slide of one last-minute photograph on which but a few stars were dimly visible. The Optical Society members were allowed to look through the 40-inch, but it made hardly any impression on them. The general attitude of those who were there was that the new Ritchey-Chrétien telescope was a failure.[68] After the meeting, as he traveled around the East on 200-inch business, Pease collected negative remarks on Ritchey's work and gleefully relayed them to Adams. However, the less biased Fred E. Wright reported to the Mount Wilson director that "the new telescope is certainly a good one and the optical and mechanical details are excellent." For his part, Adams made sure that the papers by Ross and Pease were speedily published where astronomers and physicists would read them.[69]

Ritchey now had no job and no income. He had hoped to get other telescope projects on the basis of the 40-inch, but none had materialized. His best chance had been for the 80-inch reflector the University of Texas was planning, to be funded with money left in the will of W. J. McDonald. All the astronomical advice for it came from Yerkes Observatory, which was to operate the observatory, and its young director, Otto Struve. Early in 1932 John D. Phenix, a Naval Observatory astronomer who was a Texas alumnus, had written its president, Harry Y. Benedict, to recommend Ritchey be given the contract to build the new telescope as a Ritchey-Chrétien. The following year H. J. Lutcher Stark, a wealthy, expansive Texan and an influential member of its Board of Trustees, had come East, met Ritchey, seen the 40-inch under construction, and had been converted. Stark came back to Austin convinced that Ritchey "is far preferable to any other maker of astronomical instruments in the world." Struve was horrified. He wanted a big, reliable telescope soon, optimized for spectroscopy, not a development project to build a Ritchey-Chrétien. Struve wrote Benedict that Ritchey was known as a "slow worker" and had had "the ill-fortune" of breaking the first 40-inch mirror while in Washington, and "a similar unfortunate accident" in France a few years earlier. The Yerkes director, who had heard all the arguments from Ross, described in detail for Benedict and Stark the disadvantages of the Ritchey-Chrétien system. Like Ross and Adams, he did not mention Ritchey's concept of using alternate secondary mirrors for spectroscopy at the Cassegrain and coudé foci, which would eliminate many of these problems. Struve met Stark in Chicago to convince him that a conventional reflector would be best for McDonald Observatory.[70] He also obtained letters of support from Adams and J. S. Plaskett, director of the Dominion Astrophysical Observatory in Canada, which had a new 72-inch conventional reflecting telescope. Both obliged, but only Adams included in his letter a blast against the Ritchey-Chrétien. He admitted that a "moderate-sized instrument" of this type "may well prove to be of value for some special field of work," but he considered it "highly undesirable" for "a large reflector where much of the work would be in spectroscopic lines." The majority of the Texas Board of Regents accepted the experts' testimony, and Warner and Swasey, not Ritchey, built what became the McDonald Observatory 82-inch reflector.[71]

In October 1934, just before the Optical Society meeting, Ritchey had filed a claim for the money he had lost as a result of the "unforeseeable" breaking of the first 40-inch disk. Hellweg, who no doubt helped Ritchey prepare the claim, supported it strongly. The total amount, $8,283.69, included the wages of Ritchey's men from the date the first disk broke to the date that the second one reached the same stage of work; Ritchey's own salary, which he valued at $6,500 per year (his salary during his brief Naval Observatory appointment), for the same period; the expense of the temporary-replacement metal disk; and the salaries for the time his men had spent working on it. The Navy Department recommended Ritchey's claim be paid, and eventually it was, in one of the many special bills Congress passed for this purpose. It nearly went down to defeat by objections from a few economy-minded congressmen, but Representative Sol Bloom of New York, a friend of science and the Naval Observatory, succeeded in calling the bill to the floor and getting it passed by unanimous consent while the objectors were absent from the chamber. The president signed the bill in July 1935, and Ritchey received his final payment, which had to last him the rest of his life.[72]

At Hellweg's order, Browne took charge of the 40-inch in November 1934, after Ritchey retired. In best government-scientist fashion, the young astronomer had "respectfully submitted" a long, numbered list of improvements needed in the mounting, dome, shutter, drive clock, mirror cover, clamps, plateholder, and everything else connected with the telescope even before he took over. The observing program on which he was supposed to work was photographic astrometry, especially the measurement of the positions of the planets and their satellites with respect to stars. Browne himself had drawn it up, with full quantitative details back in 1932. At that same time Robertson, the supposed chief scientific adviser to the superintendent, had submitted as his proposal for the program one short paragraph, which simply stated that all planets, satellites, comets, asteroids, occultations, and double stars should be observed, without any details whatsoever.[73] Though Browne's intentions were noble, he was completely inexperienced in using a photographic reflector. There are always problems with a new telescope, and he had never encountered any of them before. Browne was very slow at working them out, particularly under the frequently cloudy conditions at the Naval Observatory. He and his assistant spent much of their time cleaning up and rearranging the dome, and finding and correcting minor defects in the wiring of the telescope. Hellweg was impatient for results and kept urging him to press on.[74]

He did so especially because on January 6, 1935, the *Washington Post* published in its Sunday rotogravure section a page of pictures of Ritchey, the observatory, the telescope, and of two nebulae, supposedly taken with it. They were beautiful photographs of the Veil nebula in Cygnus and the great Andromeda "nebula" (galaxy), but they were Ritchey's 60-inch pictures, as he no doubt had mentioned but the reporter had not understood. Once again Hellweg had to apologize, this time to the editor of the *Post*, and admit that there were no good pictures taken with the 40-inch yet. The captain himself wrote a paper, "The Naval Observatory's New Telescope," for publication in the *U. S. Naval Institute Proceedings*, a journal read by other professional officers like himself. His paper is shallow and completely lacking in technical detail. No doubt most of its readers favored that omission, but even they must have noted that Hellweg's paper, though full of extravagant "hopes" for the future, showed not a single picture taken with the 40-inch, nor did it describe a single new result. The superintendent desperately wanted some.[75]

One of the serious defects Browne noted in the telescope was astigmatism, or elongation of the images of stars on the photographs. It had become worse over the months, a fact that indicated to him that it was not an inherent property of the mirror. He suspected the problem was in the support system that held it in place. Browne removed the mirror and support system from the telescope to try to track down the problem, but he could not find it. When he replaced the mirror, with new felt pads to cushion it, the astigmatism became worse, not better. Hellweg, frustrated and angry, called in Ritchey, who was still living in Washington. For ten days the old man, with no official post at the observatory, was in effective charge of the telescope, modifying the support system and testing the mirror for astigmatism daily. Then, on May 4, 1935, as he was hoisting the mirror, it slipped from the clamps which were holding it and it fell to the steel "deck" of the observatory. As the mirror fell, Browne, who was closest to it, tried to catch and protect it. Several of his fingers were badly mangled as the heavy glass disk struck the deck. It fell face down, and for two days no one had the courage to turn it over and see the extent of the damage. Luckily, the glass had been well annealed and did not break; only a few small cracks developed near the edge. News of the accident quickly leaked out to the newspapers, despite Hellweg's efforts to impose secrecy. He soon found himself a leading character in newspaper stories headlined "US Loses Its Biggest Eye." The captain was furious.[76]

It was Ritchey's last hurrah. The superintendent lost all confidence in him. After only brief consideration, Hellweg entrusted the task of grinding out the cracks in the mirror to rival optician C. A. Robert Lundin rather than to Ritchey. This operation left only a very small area of the mirror unusable, which caused no problem. The superintendent banned Ritchey from the observatory, taking away his keys to his shop and to the dome.[77]

Now Browne had complete charge of the telescope. Where Ritchey had insisted on using ground concave glass photographic plates, and Hellweg had wanted him to use films, cemented to curved forms, Browne boldly changed to ordinary flat photographic plates, cut round with a diamond cutter and held curved against a concave metal form by a vacuum pump. Ritchey had been well aware of this possibility, for it had been used in astronomy with wide-field photographic cameras as early as 1900. Both Chrétien and Ross had emphasized that plates taken this way can be accurately measured and are just as useful and accurate as plates taken with a telescope with a flat focal plane. Nevertheless, Ritchey had resisted this solution. Browne used it, and it worked.[78]

Hellweg continued to harry Browne for results, who in turn continued to submit lists of "repairs and improvements requested for 40-inch reflector." Light pollution remained a problem. Toward the end of 1935 a majority of the senior astronomers at the Naval Observatory recommended to Hellweg that the 40-inch be used to show the stars to public visitors, so that the old 26-inch visual refractor could be relieved of that duty, which had "interfere[d] with [its] . . . scientific work." A few months later, in February 1936, Burton, the elderly visual observer, reported that he could see Triton, the satellite of Neptune, better and resolve closer double stars with the 26-inch than with the 40-inch. Clearly, the new Ritchey-Chrétien telescope had not impressed the old guard at the Naval Observatory.[79]

Browne did succeed in getting some useful photographs to measure, if none of spectacular nebulae. In 1937 he published the first paper based on photographs taken with the 40-inch, giving accurate measured positions of Comet Peltier 1936 II.[80] A few months later, feeling he had no future at the Naval Observatory under Hellweg, he left to take a job as an engineer with the Bendix Radio Corporation.

Ritchey, seventy years old in January 1935, hung on in Washington, and though he was not permitted to work in his shop or with the reflector, he still hoped for another mirror or telescope contract. That summer he received orders from Captain A. S. Hickey, the assistant superintendent, to remove his desk and other belongings from his former optical shop at the Naval Observatory, which was to be reconverted to a museum. Ritchey procrastinated, but in September Hellweg sent him an ultimatum to get his property out within a week or it would be removed. This time the optician complied. Two months later Hellweg made him clear his tools out of the building that had been his machine shop. Ritchey still had a storage area in yet another building on the observatory grounds, where he was keeping his mirrors, disks, large optical grinding machine, and the other equipment of his shop. In February 1936 Hellweg gave him his final notice that it, too, must go within a month. On the appointed day, Ritchey appeared with a truck and took it all away for storage elsewhere in the Washington area. The one thing he left was the broken 40-inch glass disk, originally intended for the primary mirror. The government owned it as a result of settling his claim. To Hellweg it was an all too solid and obvious reminder of what had been a nightmare for him. He certainly did not want it in the museum, and he had no way of getting rid of it without attracting attention to what the Navy chain of command would certainly consider a failure. Hellweg had the broken disk buried on the Naval Observatory grounds at night.[81]

Now Ritchey had to face the reality that he would never get another telescope job. He could afford neither to ship his optical equipment to the ranch he still owned in Azusa nor to store it in Washington. His only recourse was to sell it. Potential buyers were few, but Ritchey knew them all, and they knew that his tools and materials were first-class. He managed to sell the 42-inch glass flat disk, which he had used to test the Ritchey-Chrétien primary, to the Warner and Swasey Company for nearly $2,000. He sold his optical machine to John E. Willis, an astronomer on the Naval Observatory staff, who stored it in a barn in Maryland. Ritchey's asking price had been $1,000, one third what it had cost him, but he probably let it go to Willis for less in the end. The Depression was very bad in 1936. Ritchey had cashed in the capital equipment of his professional existence for less than $3,000.[82]

Deeply disappointed by his failure to obtain spectacular photographs that would prove the Ritchey-Chrétien concept, he avoided corresponding even with his friend and faithful supporter, Chrétien. Ritchey was now displeased with Browne, whom he believed had displaced him as Hellweg's "favorite," just as Adams had become Hale's right-hand man years before. Ritchey still wanted to see the Ritchey-Chrétien telescope gain recognition by its results; he did not understand that Hellweg and Browne, both of whom shared his aim, were at loggerheads.[83]

In their apartment in Washington, Ritchey devoted himself to helping his wife and to working on another planned book. He tried half-heartedly to raise money for a new big telescope, but he knew there was little hope until the Depression ended. As he worked and reworked his manuscript, he collected astronomical photographs to illustrate it from the Royal Astronomical Society in England and from the American astronomers who still were on good terms with him. His best friends were the astronomers at Lowell Observatory, who were in the 1930s still near-outsiders, in spite of their discovery of Pluto, and therefore closest to him.[84] In contrast, Struve, the young director who was pulling Yerkes Observatory out of its lethargy, hectored Ritchey time after time to return the original photographic plates of nebulae and clusters he had taken in the days of Hale. Struve knew from reproductions that Ritchey's plates "excel[ed] all those now in [their] possession in definition," and he wanted the originals back. Ritchey simply ignored him.[85]

Finally, sometime toward the end of 1936, or in early 1937, Ritchey pulled up stakes in Washington and moved back to Azusa with his wife and daughter. His interaction with astronomers was nearly at an end, though he was to remain vitally interested in astronomy and continue studying and writing on it to the end of his life, nine years later.[86]

Southern California

1932 - 1945

WHEN GEORGE WILLIS RITCHEY left Washington and returned to California for the last time, probably in 1937, he moved directly to his orange and lemon ranch near Azusa. There he owned twenty acres of land, with a home at 1345 North Vincent Avenue, just north of Bonita (now Arrow) Avenue. Then a rural area, its location is now in Irwindale, about halfway between Azusa and Baldwin Park, in a flat area bordered by the San Gabriel River and the Little Dalton Wash.[1] Ritchey had kept the property all the years he was in France and in Washington. His son, Willis Gray Ritchey, a manual training teacher at Lafayette Junior High School in Los Angeles, had supervised and operated the ranch in his father's absence. It was a good orchard that produced large crops of oranges. Minoru "Paul" Nakada, a high-school student who lived nearby, worked part-time as a helper on the ranch, probably in 1937 and 1938. He remembers the Ritcheys' house as a typical middle-class home of the time. It was not palatial, but it was not poor either. Ritchey, always wearing white shirt and tie, worked on his astronomical photographs and manuscripts on the pleasant screened porch. When he went outside, even if only to walk around the orchard, he always put on a black coat. Half a century later, Nakada still remembered how the remote, formal-appearing old man liked to pause and chat with him.[2]

Within the Ritcheys' first year or two back in California, their daughter Lila, not quite fifty, died. She had never married and had lived with them all her life. Evidently, she had been chronically ill for several years, for Ritchey collected recipes for tonics for her in France and described her as an "invalid" toward the end of their stay in Washington. No record of the nature of her illness has been found.[3]

Ritchey devoted the remaining years of his life to writing. He planned a series of five books, to be entitled "Exploring the Universe." All of them were to be "richly illustrated"; they ranged from "Our Nearest Neighbor World, The Moon: Is It in Fact a Child of the Earth?" and "Our Kindly Mother Earth, with a supplementary volume The Grand Canyon and The Great American Plateau" to "Countless Far-Distant Rotating Systems of Stars: The Spirals." He wrote and rewrote hundreds of pages of drafts and continued to collect new photographs to illustrate the volumes. In earlier years, Lila had typed them for him; after her death his wife, Lillie Gray Ritchey, did them, and they planned to dedicate the series to the memory of their daughter. Ritchey's descriptions of the pictorial features of the moon and nebulae are graphic, but he had little understanding of their physical implications. Nevertheless, he filled chapters of the book on the moon with elementary calculations of dimensions and masses of the visible features on its surface. Excluded from the Carnegie Institution's Committee on Study of Physical Features of the Surface of the Moon, Ritchey worked to the end, completely on his own, on his favorite subject. He never found a publisher for these projected books.[4]

In South Pasadena, some ten miles to the west of Azusa, George Ellery Hale had been living a very different kind of life. After he had decided to end the costly, time-consuming experiment to make a quartz disk for the 200-inch mirror, and to settle for Pyrex instead, he could do little more than watch the telescope project. He was apprehensive about his poor health, with nosebleeds, dizziness, and high blood pressure. In 1932 he went east for another course of diagnosis and therapy at Dr. Austen F. Riggs's sanatorium in Stockbridge, Massachusetts. Recovered, Hale chanced a trip to England to preside at a meeting of the International Council of Scientific Unions, which he had founded. He stayed in England a month, revisiting his old haunts in London and Cambridge,

George Ellery Hale at the solar spectrograph at the National Academy of Sciences, Washington. Courtesy of the National Academy of Sciences.

but his blood pressure shot up alarmingly.[5] He returned to California where he continued taking injections of a special enzyme to aid his digestion. The next year, accompanied by his wife, Hale left suddenly for a second trip to England. In London he took yet another "new treatment." Hale felt himself "very much better" as a result of it, and his wife was amazed at his improvement. Always, he avoided people, even old friends, and excitement.[6] Back in California, he soon fell ill again and underwent an operation at the hospital in Santa Barbara in the summer of 1934.[7] His last years were not happy ones.[8]

During his periods of good health, Hale puttered around in his solar laboratory, the refuge in which he could escape from everyone, even his wife. He worked intermittently on his long-continued attempt to detect the elusive general magnetic field of the sun, still without success. The rushed trip to England in 1933 was partly to investigate a new method of measuring spectra. Using it, Hale initially thought he had confirmed the field's existence, but once again it was a false alarm. Best of all he enjoyed reminiscing about the exciting days of his youth at his Kenwood Physical Laboratory and the Massachusetts Institute of Technology, and of his travels in Egypt.[9] Hale devoted himself to writing a long article on how young boys and other amateurs could build spectrohelioscopes and spectroheliographs for solar research, but in that Depression year of 1933 there were few readers indeed who could follow his path of four decades before.[10]

The following year was an exciting one for Hale. He and Walter S. Adams had the second serious difference of their whole long lives, and then afterward Hale was able to help the Mount Wilson director and the observatory greatly. The difference came about when an audit disclosed that James Herbert, the bookkeeper at the Mount Wilson Observatory offices in Pasadena, had embezzled $13,000 by forging Adams's signature to checks and cashing them himself. Adams immediately notified Hale, who then told him for the first time that Herbert was an "untrustworthy thief," who a few years before had stolen securities entrusted to him. At that time Hale had not informed Adams, the director, of the theft. In reporting the incident now, Hale stated that he had shielded Herbert out of consideration for his wife and children, but there must have been at least a suspicion in Adams's mind that the bookkeeper had been blackmailing the honorary director. Adams immediately sent word of the whole affair to John C. Merriam, president of the Carnegie Institution. Merriam summoned Adams to Washington for a fuller report, and Hale accompanied him at his own expense, no doubt to protect himself against possible charges of malfeasance.[11] In the end the whole episode was hushed up. Hale was absolved of any guilt, and Merriam, fearful that the "sensational" newspapers and the "radical elements" in California would attack the Carnegie Institution for paying a poor bookkeeper so little that he had to steal from his rich employers to support his family, did not prosecute Herbert.[12]

Even before this event, Adams and Merriam had not been on close terms. Merriam believed that Hale and Adams were selling out the Carnegie Institution to the California Institute of Technology and the Rockefeller Foundation for the 200-inch telescope. Hale and Adams, on the other hand, felt that the Carnegie president was holding back astronomy for reasons of petty institutional jealousy. The astronomers on the Mount Wilson staff were becoming restive because there was no direct, official link between them and the 200-inch project except Hale, who was an aging, ill recluse. He recognized this himself and recommended that Adams be elected to both the Caltech Board of Trustees and the Observatory Council that supervised the 200-inch project. Robert A. Millikan and the Caltech trustees were glad to oblige, but Merriam forbade Adams to accept the two appointments. Hale, angry but diplomatic, once again summoned up all his persuasive powers and wrote Elihu Root and Henry S. Pritchett, the heads of the Carnegie board and its executive committee, arguing that the Mount Wilson director must be a member of the groups that controlled the 200-inch project.[13]

Simultaneously, Edwin Hubble, who had already made himself famous by using the 100-inch telescope to explore the distant reaches of the universe, stopped in Washington and spoke with Merriam himself. Hubble was on his way to London to give the prestigious Halley Lecture before the Royal Astronomical Society; Merriam had to listen to him. Hale and Hubble, between them, forced the Carnegie Institution president to back down, and Adams became a Caltech trustee and a member of the Observatory council.[14] There he played a very important role in protecting the interests of the Mount Wilson astronomers and in keeping sound astronomical ideas in the forefront of the 200-inch project.

By now Hale had very little energy. He felt that he was accomplishing almost nothing.[15] Yet he did write two long papers, which described the astrophysical laboratory, the instrument shop, and the optical shop on the Caltech campus, and the plans for the 200-inch. Progress had been swift on the Pyrex disk once the Corning Glass Company got the order. In the summer of 1933 the company had successfully cast a 120-inch disk, to be used as the test flat for the 200-inch mirror. In March 1934 the Corning workmen poured a 200-inch disk, but at the end of the process the molten glass broke loose parts of the mold and the disk was a failure. The second attempt, in December of the same year, was a complete success. The giant disk annealed for a year as the temperature in the furnace slowly fell. Then the huge piece of glass, crated in a high box, traveled across the country on a specially modified railroad flatcar. Its route avoided all low bridges and other danger points. The mirror had arrived safely in the Caltech optical shop, and work on it was in progress by the time Hale wrote his second article. The observatory was definitely to be at Palomar, and the mounting was to be the split ring, invented by Russell W. Porter. Captain Clyde S. McDowell, a Navy construction expert, took over as supervising engineer of the project, and the telescope design was well in hand.[16] As Hale dozed comfortably in his laboratory, the new, improved version of the Kenwood Observatory that his father had given him long ago, recollecting with pleasure the enthusiasms and freedom of his youth, he could be secure in the knowledge that for the fourth time in his life the largest telescope in the world, which he had begun, would soon be a reality.[17]

All the builders of this largest telescope, however, were not to be on hand for its completion. On February 7, 1938, Francis G. Pease died in Pasadena at the age of fifty-seven, following an operation for cancer. He had begun as Ritchey's assistant at Yerkes Observatory, had gone with him to Mount Wilson, and had ultimately supplanted him there. Pease had been in charge of completing the design of the 100-inch telescope (his former supervisor had finished the optical work on its mirror), and he had begun demonstrating its success in use as Ritchey was being fired. The *Pasadena Star News* and the *Los Angeles Times* treated the "famous astronomer's" death as a top local story and published long, glowing accounts of his life and work. The *Times* followed up the next day with an editorial on his life.[18] Adams, John A. Anderson, and Gustaf Stromberg of the Mount Wilson staff wrote highly laudatory obituary articles about Pease's contributions to the observatory, optics, and astronomy.[19]

Two weeks later, Hale himself succumbed to the ravages of his illnesses. He was sixty-nine. Newspapers all over the world reported his death. Many astronomical journals published obituary articles about him. Among the most admiring were those by Adams, Harold D. Babcock, and Theodore Dunham, Jr., of the Mount Wilson staff.[20] Adams wrote a long biographical memoir for the National Academy of Sciences, and a quarter of a century later Helen Wright published her full-length biography of the "explorer of the universe."[21]

A year after Hale's death, Anderson could report that the nearly completed dome for the 200-inch was standing on Palomar Mountain, along with several other buildings and shops. Complete water and fuel systems were in place, along with a diesel generator to supply electrical power. The telescope mounting was in the dome, ready for assembly, and the 200-inch disk was on the optical machine in Pasadena, already ground to spherical form and ready for figuring. Anderson thought the telescope might be ready for operation in 1940, but this was not to be.[22]

Soon after World War II began in Europe, the leaders of American science shifted their resources to helping Britain to survive and to preparing this country for war. Astronomy had few immediately useful products, and the 200-inch project went on hold for the duration. Only after the war had ended was the 200-inch completed. By then Adams had retired, replaced as director of Mount Wilson Observatory by Caltech physicist Ira S. Bowen. On June 3, 1948, the 200-inch was dedicated and named the Hale telescope.[23] Ever since, it has been the most productive research telescope in the world.

After Pease's and Hale's deaths in 1938, Ritchey lived for another seven years on his orange ranch in Azusa, studying his photographs and writing his astronomical manuscripts. Finally, he died on November 4, 1945, two months before his eighty-first birthday. His death was reported in three short paragraphs in the *Azusa Herald and Pomotropic*, the local weekly newspaper. No news story appeared in the *Los Angeles Times* or any other daily newspaper. *The Publications of the Astronomical Society of the Pacific* mentioned his death in a one-sentence item, published in the same issue with a nine-page survey of the year's work at Mount Wilson Observatory by Adams, a six-page article on Adams's retirement as director, and a shorter article on Bowen, who would take over as the new director on January 1, 1946. The astronomical community had already forgotten Ritchey.[24]

Early in 1946 *Popular Astronomy* also mentioned his passing in a one-sentence note. However, two months later, *Sky and Telescope*, published

George Willis Ritchey at his office in Paris, probably about 1930. Courtesy of Richard B. Burgess.

under the aegis of Harvard College Observatory, carried an appreciative short account of Ritchey's life by Dorrit Hoffleit.[25] In 1947 British astronomer and optician F. J. Hargreaves wrote a very positive memorial biography of his American colleague for the *Monthly Notices of the Royal Astronomical Society.*[26] Both these articles contain many errors of fact but are perceptive in their overall tenor.

J. F. Hellweg was still superintendent of the Naval Observatory when Ritchey died in 1945. Under naval regulations, Hellweg had been forced to retire at age fifty-six, in 1935, but continued in his post as a retired officer for eleven more years. He was promoted to commodore in 1944 and remained superintendent until 1946, after World War II had ended. He had been in charge of the Naval Observatory far longer than any officer except the first superintendent, Commander Matthew F. Maury, probably at first because of his desire to prove the 40-inch Ritchey-Chrétien reflector a great success and later because of the wartime emergency.[27]

Hellweg had continued to the end to hope for spectacular results from the 40-inch telescope. Soon after W. Malcolm Browne left the observatory, John E. Willis was placed in charge of the reflector. He was an experienced astronomer, and with the Ritchey-Chrétien he began to get good photographic measurements of planets, including the recently discovered Pluto. During all these years Harry E. Burton continued as head of the Equatorial Division, and his sole research effort remained visual observations with the old 26-inch refractor. All his astronomical interests centered on occultations and double stars for his entire career at the Naval Observatory.[28] When Mars came particularly near the earth at a close opposition in 1939, Hellweg "advised" Burton to have Willis obtain a series of photographs of the red planet with the 40-inch for publicity purposes, and to observe it visually himself. Burton and Willis tried, but actually at these close oppositions Mars is far south, close to the horizon as observed from the northern hemisphere, and hence its image is often distorted by the earth's atmosphere. The main requirements for planetary photography are a long-focal-length telescope, a very steady atmosphere, and unlimited amounts of clear observing weather. The 40-inch Ritchey-Chrétien reflector in Washington met none of these conditions, and pictures of Mars taken with it could not match the ones from Mount Wilson. Hellweg was disappointed once again.[29]

Just before Hellweg retired, Albert G. Ingalls, who wrote a regular telescope-making column for *Scientific American*, asked him if the

40-inch Ritchey-Chrétien reflector was not a failure. Certainly, most of the writer's correspondents thought that it was. Hellweg, drawing on a memorandum prepared by someone on the Naval Observatory staff (Willis had left in 1944), gave sound reasons that it was the best type of telescope for the work it did but could point to very few results obtained with it.[30]

Only when Burton retired in 1948 and John S. Hall came in from outside the Naval Observatory staff to replace him did the 40-inch Ritchey-Chrétien reflector begin to make its mark in astronomy. Trained at Yale, Hall was a skilled research astronomer, well versed in modern techniques. He began using the 40-inch for photoelectric photometry and with it got important new results on the polarization of starlight by interstellar dust.[31] By then it was clear to all that Washington was a very poor site for this telescope. The Navy Department at last began planning to move the entire observatory to a darker location in the Blue Ridge Mountains of Virginia. That would have solved the light-pollution problem for a few years but none of the other problems. Hall himself hoped to move the 40-inch Ritchey-Chrétien away from the rest of the observatory, to a really good site in the Southwest.[32] James Robertson had retired from the Nautical Almanac Office in 1939 but retained his political connections and his interest in the Naval Observatory. In 1950, when a bill was introduced in Congress to move the observatory to the Virginia site, he came back to Washington from California and presented himself at a crucial hearing to testify against it. Appearing more senatorial than most of the legislators, and playing the role of a brilliant but down-to-earth scientist to perfection, Robertson spoke in favor of leaving the rest of the observatory in Washington but moving the 40-inch to Flagstaff, Arizona, the site Hall preferred.[33] The bill was killed, and by 1953 it seemed definite that the 40-inch would move west.[34]

In 1955 the telescope was finally relocated to a site outside Flagstaff, five miles from Lowell Observatory and sixty miles south of the Grand Canyon, where Ritchey had hoped to build his first fixed universal telescope.[35] There Arthur A. Hoag obtained excellent long-exposure photographs of nebulae with the 40-inch. Two of the most spectacular of them, showing the spiral galaxy M51 and the planetary nebula M97, appeared in *Sky and Telescope* in 1956. They provided the graphic evidence that the Ritchey-Chrétien system really works, evidence that

Three pioneer users of the Naval Observatory 40-inch Ritchey-Chrétien reflector: Arthur A. Hoag, John S. Hall, and W. Malcolm Browne, 1982. Courtesy of the late John S. Hall.

Ritchey had been trying to obtain in France a quarter of a century earlier and in Washington in the mid 1930s.[36]

Then in 1958, Aden Meinel, the first director of the Kitt Peak National Observatory decided to build its 84-inch reflector as a Ritchey-Chrétien, rather than a conventional reflector as had originally been planned. Meinel, an expert in optics, had first learned of the Ritchey-Chrétien concept in 1940, when as an eighteen-year-old he worked as an optician at the Mount Wilson optical shop. Later, as a graduate student at Berkeley, he had heard optical designer James G. Baker argue for the system. In 1957, while on a site survey in Arizona, Meinel visited Flagstaff and saw the 40-inch Ritchey-Chrétien telescope and photographs taken with it. By then he was responsible for building the 84-inch, and, fully recognizing the advantages of the Ritchey-Chrétien design, he decided to make it that way.[37] This telescope made the principle known to the astronomical world, and since then essentially all large research telescopes, including the University of Arizona 90-inch, the Kitt Peak and Cerro Tololo 4-meter reflectors, the European Southern Observatory 3.5-meter reflector at La Silla, and the 2.4-meter (94-inch) Hubble Space Telescope, have been built as Ritchey-Chrétiens or closely related two-mirror systems. The Keck 10-meter (400-inch) telescope, recently completed on Mauna Kea, Hawaii, is of this type. For wide-field photography all of them (except the Hubble Space Telescope) use a field flattening, correcting-lens system close to the focal plane, similar in concept to the correctors described by Ralph A. Sampson in Ritchey's time.[38] The advantage of the large, coma-free field of the Ritchey-Chrétien reflector is so great that it is very hard to understand today why it took so long for astronomers to adopt it.

Many of Ritchey's other ideas, which seemed extreme, visionary, or even ridiculous to his contemporaries, are now recognized as necessary measures to build and use large telescopes effectively to reach the faintest possible limits of observation. One is the concept of cellular mirrors. Ritchey's scheme of cementing glass plates together has never succeeded, but fusing methods have been used to make large, lightweight, "egg-crate" type disks. One such mirror is the primary mirror for the Hubble Space Telescope. Apparently, others are in use in classified reconnaissance satellites. More recently, Roger Angel, at the University of Arizona, has developed a method of molding the glass for the disk of a "honeycomb" mirror as a single structure, much like a highly curved version of one of Ritchey's cellular disks. The advantages of a lightweight, minimum flexure, and rapid thermal response are exactly those that Ritchey sought to achieve.[39] Control of the temperature of the mirror, the interior of the dome, and of the telescope mounting to reduce thermal distortion of the mirror and to prevent thermally induced air currents that degrade the "seeing" is being built into all large new telescopes. Likewise, most new telescopes are designed to make it possible to change rapidly from one focus to another, to best use the existing atmospheric conditions.

Methods for interrupting the exposure and using only the fleeting moments of best seeing for the most critical observations are coming into use.[40] All these were Ritchey's ideas. He was a visionary, unable to bring his visions into practice with the resources of his time. Astronomers today have rediscovered many of those same ideas, have recognized the value of others of them, and are beginning to use them. They are difficult and expensive, but they produce smaller, brighter star images and resulting increases in observing efficiency. Thus Ritchey, long after his death, has been vindicated as a prophet of modern astronomical telescopes.

When Adams retired in 1946, he moved his office to the Hale Solar Laboratory, which Hale had left to Mount Wilson Observatory. There he continued to work and to write. By the end of his life in 1956, Adams had published more than 270 scientific papers, nearly all of them on solar and stellar spectroscopy. He was an outstanding research scientist.[41]

Looking back at Hale, Adams, and Ritchey from our vantage point today, we can see that, in the words of Genesis, "there were giants on the earth in those days." Hale was an outstanding observatory director but also the archetype of all directors; Adams was an outstanding research astronomer and the archetype of all astronomers; Ritchey was an outstanding telescope maker and the archetype of all telescope makers. Hale was a fund-raiser who has never been equalled in astronomy. He had inspiring visions and the ability to sell them to the hard-headed businessmen whose personal fortunes and foundations provided the money to finance science in his day. He was an excellent writer, whose articles and books caught the interest of wealthy readers all over America. A tremendously persuasive speaker, Hale charmed businessmen and research astronomers alike. They wanted to give him money, or work for him with the wonderful telescopes and spectrographs that he alone could provide. Hale constantly claimed he wanted to get back to doing research himself, but after his earliest years, he never found time to do so until he was burned out, and then it was too late. Yet he inspired loyal assistants like Adams, Ferdinand Ellerman, Harold D. Babcock, and Seth B. Nicholson to work with him to the end.

Ritchey was an outstanding craftsman. He developed and perfected the methods for making large telescope mirrors that were better than any previous ones. His 100-inch mirror, made from a striated, inhomogeneous disk, pocked with internal bubbles, had a figure unequalled in excellence for fifty years. He studied and analyzed every detail of using the reflecting telescopes he had made to take superb photographs of nebulae and galaxies. A perfectionist, Ritchey was unsparing of himself and of those who worked with him. He believed that he alone knew the secrets of making large telescopes; he was contemptuous of rivals like

John A. Brashear and Francis G. Pease. Ritchey was tremendously self-centered, and in the end his failure to accommodate himself to the vision of his director cost him his job at Mount Wilson. His own talents as a fund-raiser were nonexistent, and although he dreamed of increasingly grandiose Ritchey-Chrétien telescopes, and designed them, he never got a chance to build any but two small, underfunded examples at poor sites. He accomplished nothing with them.

Adams had no desire whatsoever to raise money, nor to build telescopes. He wanted to use telescopes to do research and learn the nature of the universe. He conformed to his director's wishes, first observing for him, then taking over increasing amounts of the administrative work as assistant or acting director, finally becoming director himself and allowing Hale to retire at the age of fifty-five with the title of honorary director. Adams did not want to build a perfect telescope, nor invent a new concept in telescopes; he wanted to get his hands on the largest telescope he could. He wanted to use it to take spectrograms of faint stars just as soon as he regarded it as completed. Thus there was bound to be tension between Adams, the research astronomer, and Ritchey, the optical perfectionist. They were bound to struggle, and once the biggest telescope in the world was finished, with no immediate prospect of building a bigger one in sight, the telescope maker was bound to lose, and the research scientist was bound to come into his own.

Their story is as old as astronomy and as new as today; change the names and every incident in it has been reenacted by other directors, telescope makers, and research astronomers in our generation. They were not such outstanding figures as Hale, Ritchey, and Adams, but their struggles were the same. Yet scientists' lives are not composed of science alone. Personal aspects affected our three main characters' actions, and through them, determined the course of American astronomy.

Hale was the oldest son of a wealthy self-made businessman. From the Civil War to World War I, America was expanding and industrializing itself. It was the age of big bosses and predatory capitalists. Hale's father was very successful. Hale could have been, too, but instead he followed a parallel path in science. He was a master of operating a monopolistic observatory, the biggest and most successful in the world, and in organizing combines, in the form of scientific societies and international unions. Through them Hale, and the tight little circle that he led, controlled American astronomy. Mount Wilson staff members worked on research, each in his own well-defined field of specialization, using the largest telescopes and the best auxiliary equipment in the world, with little fear of competition, internal or external.

Hale's parents brought him up to believe in the literal truth of the Puritan Bible and in going to church twice each Sunday. Yet he, and they, did not really believe that it was easier for a camel to pass through the eye of a needle than for a rich man to enter into the gates of heaven. There must have been tremendous internal conflicts deep in Hale's unconscious between the obvious necessity of enforcing his directoral authority on an unruly subordinate and the stated benefits of turning the other cheek. Once he became a student at MIT, his religious beliefs quickly fell away, but his emotional inheritance remained. It, together with his physical frailty, led to frequent and increasingly severe illnesses, particularly after episodes of intense stress.

These were often caused by Ritchey, the son and grandson of immigrant craftsmen. They, and he, knew in their hearts that they were "saved." Ritchey had few inner conflicts; as he grew older he frequently wrote that his telescopes would proclaim the glory of God. At first he and Hale had mutually reinforced each other, the one raising the money and organizing the observatories, the other building the telescopes. As they grew older, they were destined to clash. Hale's aim was research, Ritchey's was photographing the heavens. Each felt he was responsible for their joint successes and resented the acclaim the other received. To have any chance of getting the support necessary to build ever larger telescopes, Hale had to present himself as omniscient. When Ritchey publicly said that his director did not understand telescopes, one of them had to go. It could only be Ritchey.

The real split occurred in 1910, when Ritchey knew he should build the 100-inch telescope as a newly invented Ritchey-Chrétien reflector. Hale knew that he should not. Ritchey more than half convinced John D. Hooker, the putative angel-to-be of the project. To Hale there was no sin in a staff member more awful than coming between him and the source of funds for the next telescope. From there on to Ritchey's dismissal in 1919, it was downhill all the way.

Adams, the child of New England missionaries to the heathen in the Holy Land, did not suffer from the same conflicts as Hale. He could see that the older telescope maker had to go, to be replaced by a younger, more research-oriented telescope user—himself. He was a master at defining the issues for struggles that Ritchey could not win. Hale and Adams, college-educated sons of families that went back to the Mayflower, belonged to a different world from Ritchey, the son and grandson of immigrants from Ireland who worked with their hands. Ellerman, Hale's eternally faithful assistant, exemplified his idea of a subordinate who knew his place. Ritchey did not. Adams made no secret of the fact that he regarded making telescopes as a somewhat crass, lesser art; Ritchey knew it and resented it.

These all too human hostilities had serious scientific repercussions. Because Ritchey was so positive that the 100-inch should be a Ritchey-Chrétien, Hale and Adams were even more adamant that it should not. Furthermore, no other telescope that they had anything to do with should ever be one. As soon as they heard that Ritchey had helped invent it, they knew the concept must be wrong. They never investigated it seriously but denounced it from ignorance. It would have been a great mistake to make the world's largest telescope, the 100-inch reflector, as the world's first Ritchey-Chrétien. But it was equally a mistake to fire Ritchey rather than to keep him on to make an experimental 0.5-meter Ritchey-Chrétien, and then a 1-meter one, at Mount Wilson. If Hale had kept Ritchey on, when the 200-inch was built it would have been possible to make a reasoned, scientific decision as to whether it should have been a conventional reflector or a Ritchey-Chrétien. Certainly, the observational research on galaxies and the universe that Edwin Hubble began, and which Walter Baade, Allan R. Sandage, Halton C. Arp, and other more recent observers carried out with the 200-inch would have benefited greatly if it had been a wide-field Ritchey-Chrétien. Human animosities had foreclosed this possibility in 1919.

Even so, these same human emotions preserved the Ritchey-Chrétien concept. Robertson and Hellweg, two complete outsiders, scorned by the American astronomical establishment, brought Ritchey back to his native land and, against great opposition, made it possible for him to build at last one small example of his and Chrétien's invention. Most American astronomers ignored it; a few actively opposed its sponsors. Yet finally, in the hands of Hall and Hoag it helped convince Meinel to put national resources into a modern Ritchey-Chrétien telescope at an excellent site. From there it swept the world.

Some of Ritchey's work achieved instant success: his outstanding astronomical photographs, his discovery of novae in spiral "nebulae," the 24-inch reflector at Yerkes, the 60-inch reflector, and the 100-inch mirror at Mount Wilson. Some achieved delayed success: the Ritchey-Chrétien concept, cellular mirrors, thermal control of telescope mirrors, mountings and domes, rapid changes of telescope configurations to use the best existing seeing conditions, and interrupting the exposure to use only the moments of excellent seeing for the most crucial observations. Some never achieved success, at least to date: the fixed vertical telescope and a 24-meter (960-inch) mirror. Some of his most specific concepts may even be abandoned: at present it seems likely that large reflectors of the future, intended to be used in both the optical and infrared spectral regions, will be conventional Cassegrains with sophisticated corrector lenses rather than Ritchey-Chrétiens. But George Willis Ritchey was always testing, inventing, creating new concepts at or beyond the limit of the technology of his time, to push out the frontiers of telescopic study of the universe.

Appendix
Reflecting Telescopes

THE FOLLOWING BRIEF EXPLANATION, together with the figure on the next page, will help the nonastronomical reader understand the various terms used to describe reflecting telescopes.

The simplest form of reflecting telescope shown, in the upper left panel, has a *primary mirror* (at the lower left end of the tube) that brings the light rays from a star (shown as dashed lines) to the *prime focus* at the upper end of the tube. The primary mirror has the form of a *paraboloid*; its cross section is a parabola. The 200-inch Hale telescope can be used in this arrangement, and the 60-inch and 100-inch telescopes were occasionally used in this way. However, in smaller reflectors of this type the observer or auxiliary instrumentation (such as a spectrograph) at the prime focus blocks some of the light that would otherwise enter the telescope.

Therefore, the alternate arrangement shown in the upper right panel in which a *flat secondary mirror* (shown at the upper right end of the tube) directs the converging beam of light rays to the *Newtonian focus* at the edge of the tube is often more convenient. This arrangement is called a *Newtonian* telescope or reflector in English, though in French this term includes the prime-focus arrangement as well. The 24-inch telescope was almost always used in this form, and the 60-inch and 100-inch telescopes were often used in this way.

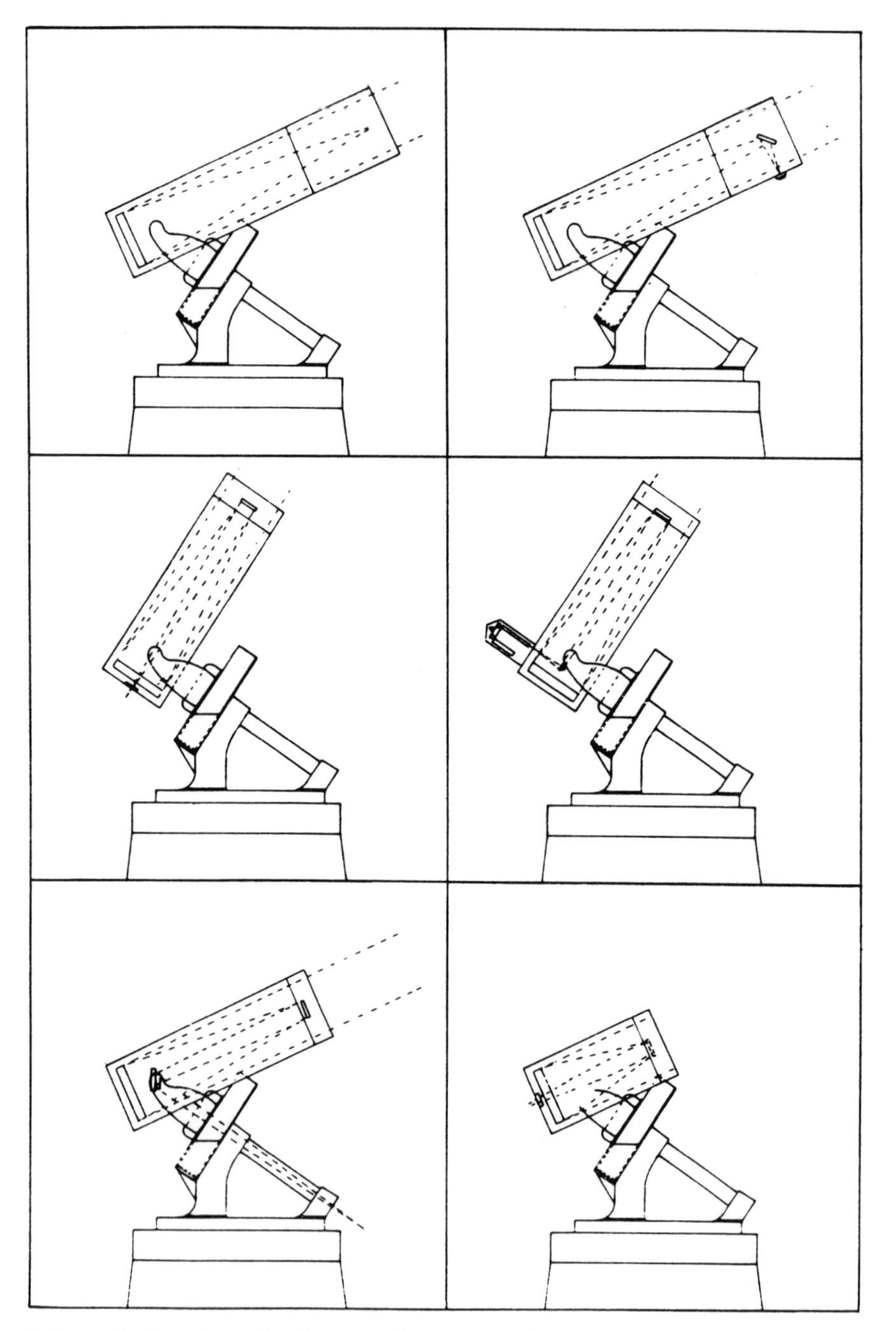

Schematic drawings of reflecting-telescope configurations.

The *Cassegrain* form of reflecting telescope is shown in the middle left panel. It has a *secondary* mirror that is a *hyperboloid* (shown at the upper right end of the tube), which weakens the convergence of the light rays and redirects them back through a hole in the primary mirror to a focus just behind it. Different hyperboloids can be used as secondary mirrors, giving the resulting Cassegrain telescope different focal lengths. The 200-inch Hale telescope is often used in the Cassegrain form today.

Another *Cassegrain* arrangement, shown in the middle right panel, has a tertiary, *flat* mirror in the beam of converging light rays, to redirect them to a focus at the edge of the tube. In this panel a spectrograph is shown mounted on the telescope at this focus. This *folded Cassegrain* arrangement does away with the necessity of cutting a hole through the center of the primary mirror, a dangerous operation with the glass disks available in Ritchey's time. The 24-inch reflector was used experimentally in this folded Cassegrain mode, and the 60-inch and 100-inch telescopes were often used in this form, especially for spectroscopy.

The lower left panel shows the reflector arranged as a *coudé* telescope. The secondary hyperboloid is chosen so that it would bring the light rays to a focus far behind the primary mirror. A tertiary, flat mirror is placed along the prolongation of the polar axis of the telescope. This flat mirror is arranged in such a way that it can be rotated so as to always bring the light rays to a focus far down the polar axis, no matter which direction the telescope points. Thus a large coudé spectrograph, or other instrument, can be mounted in a fixed position, not attached to the telescope, below the fixed focus on the polar axis. The 60-inch, 100-inch, and 200-inch telescopes were all used in the coudé form for high-dispersion spectroscopy.

One of Ritchey's inventions, first introduced on the 24-inch reflector, was to mount the various secondary mirrors and the prime-focus plateholder in different *cages* (shown by solid lines at the upper end of the tube in the different panels). These cages are interchangeable, allowing the telescope configuration to be changed from one form to another rapidly, without the necessity of realigning and adjusting the mirrors each time.

The *Ritchey-Chrétien* system (shown in the lower right panel) uses a different form of primary mirror, essentially a hyperboloid, together with a secondary that is also essentially a hyperboloid. Thus it is another type of Cassegrain system. The primary mirror cannot be used alone; it will not bring the light rays from a star to a single focus. As shown in the figure, a typical Ritchey-Chrétien reflector can be made much shorter than a conventional reflector with a paraboloidal primary, which can be used in the prime-focus, Newtonian, Cassegrain, and coudé modes. The Ritchey-Chrétien primary mirror can be used with other hyperboloidal secondaries to form *hybrid* or *modified Ritchey-Chrétien* Cassegrain or coudé systems. However, these hybrid systems are not free of the aberration of coma, as the basic Ritchey-Chrétien telescope is.

This figure is based on a drawing in Ritchey's 1904 Smithsonian Contribution but has been modified both to show the various systems as they have actually been used and to include the Ritchey-Chrétien system, which was not invented until 1910.

Finally, the focal ratio of any telescope is the numerical value of its overall focal length divided by the diameter of its primary mirror. Thus the 60-inch reflector, used at the prime focus, has a focal length of approximately 300 inches; its focal ratio is therefore f/5. At the Cassegrain or folded Cassegrain focus its focal length is 960 inches, and its focal ratio correspondingly is f/16. At the coudé its focal length is approximately 1800 inches and its focal ratio f/30. As photographers know, the smaller the focal ratio, the faster the telescope is for taking photographs of extended objects like the moon, planets, and nebulae.

Papers, Books, and Articles by George Willis Ritchey

"A Support System for Large Specula," *Ap. J.*, 5, 143–47, 1897.

"Celestial Photography with the 40-inch Visual Telescope of the Yerkes Observatory," *Ap. J.*, 12, 352–60, 1900.

"Nebulosity about Nova Persei," *Ap. J.*, 14, 167–68, 1901.

"The Two-Foot Reflecting Telescope of the Yerkes Observatory," *Ap. J.*, 14, 217–33, 1901.

"Changes in the Nebulosity about Nova Persei," *Ap. J.*, 14, 293–94, 1901.

"Nebulosity about Nova Persei," *Ap. J.*, 15, 129–31, 1902.

"Comet Photography with the Two-Foot Reflector," *Ap. J.*, 16, 178–80, 1902.

"Photographing the Moon," *Harper's Monthly Magazine*, 107, 411–19, 1903.

"Note on the Celestial Photographs made at the Yerkes Observatory by Mr. G. W. Ritchey, and recently presented to the Society," *MNRAS*, 63, 395–96, 1903.

"On Methods of Testing Optical Mirrors During Construction," *Ap. J.*, 19, 53–69, 1904.

"Photographing the Star-Clusters," *Harper's Monthly Magazine*, 109, 508–15, 1904.

"Astronomical Photography with the Forty-Inch Refractor and the Two-Foot Reflector of the Yerkes Observatory," *Pub. Yerkes Obs.*, 2, 387–98, 1904.

"On the Modern Reflecting Telescope and the Making and Testing of Optical Mirrors," *Smithsonian Contributions to Knowledge*, Part of Volume 34, 1–51, Smithsonian Institution, Washington, 1904.

"Note on the Two-Foot Reflecting Telescope of the Solar Observatory," *PASP*, 17, 186–87, 1905.

"The 60-inch Reflector of the Mount Wilson Solar Observatory," *Ap. J.*, 29, 198–210, 1909.

"Astronomical Photography with the Forty-Inch Refractor and the Two-Foot Reflector of the Yerkes Observatory," *Publications of the Astronomical and Astrophysical Society of America*, 1, 176–77, 1909.

"On Some Methods and Results in Direct Photography with the 60-Inch Reflecting Telescope of the Mount Wilson Solar Observatory," *Ap. J.*, 32, 26–35, 1910.

"The Making of a Great Telescope," *Harper's Monthly*, 121, 740–49, 1910.

"Notes on Photographs of Nebulae made with the 60-inch Reflector of the Mount Wilson Observatory," *MNRAS*, 70, 623–27, 1910.

"Notes on Photographs of Nebulae taken with the 60-inch Reflector of the Mount Wilson Solar Observatory," *MNRAS*, 70, 647–49, 1910.

"Novae in Spiral Nebulae," *PASP*, 29, 210–12, 1917.

"A Nebulous Ring about Nova Persei," *PASP*, 29, 256–57, 1917.

"Another Faint Nova in the Andromeda Nebula," *PASP*, 29, 257, 1917.

"Three Additional Novae in the Andromeda Nebula," *PASP*, 30, 162–63, 1918.

"The Small Nebulous Ring about Nova Persei," *PASP*, 30, 163–64, 1918.

"Sur un Nouveau Mode de Construction des Grands Miroirs de Télescope," *CR*, 181, 208–10, 1925.

"Nouveau Mode de Construction des Grands Miroirs de Télescope," *L'Astronomie*, 40, 57–62, 1926.

"Sur le Type d'Instrument le Plus Efficace pour Accroitre nos Connaissances de l'Universe," *L'Astronomie*, 41, 529–41, 1927.

"Le Premier Modèle de Télescope Aplanètique" (with Henri Chrétien), *L'Astronomie*, 41, 541–43, 1927.

"Présentation du Premier Modèle de Télescope Aplanètique" (with Henri Chrétien), *CR*, 185, 265–68, 1927

"Sur Quelques Avantages Mechaniques et Autres Resultant de la Tres Petit Longeur et de la Structure Compacte du Type de Télescope Aplanètique Ritchey-Chrétien," *CR*, 185, 640–42, 1927.

"Sur un Type de Télescope Photographique Verticale Fixe, avec Caelostat, à Rapports Focaux Interchangeables," *CR*, 185, 758–61, 1927.

"Comparaison, en Laboratoire, des Images et des Champs Fournis par un Télescope Newtonien et un Télescope Ritchey-Chrétien," *CR*, 185, 1024–26, 1927.

"The Modern Reflecting Telescope and the New Astronomical Photography—The Thomas Young Oration," *Trans. Opt. Soc.*, 29, 197–224, 1928.

"Le Télescope Ritchey-Chrétien I," *L'Astronomie*, 42, 27–37, 1928.

"Le Télescope Ritchey-Chrétien II," *L'Astronomie*, 42, 60–71, 1928.

"Le Télescope Ritchey-Chrétien III," *L'Astronomie*, 42, 121–27, 1928.

"Photographie Astronomique a Fort Agrandissement," *L'Astronomie*, 42, 169–82, 1928.

"La Photographie Astronomique Moderne," *L'Astronomie*, 42, 225–37, 1928.

"La Photographie Astronomique Moderne (Suite et fin)," *L'Astronomie*, 42, 281–97, 1928.

"La Futur Observatoire a Grande Puissance," *L'Astronomie*, 42, 362–63, 1928.

"The Modern Photographic Telescope and the New Astronomical Photography"

1. "The Fixed Universal Telescope," *JRAS Canada*, 22, 159–77, 1928.
2. "The Ritchey-Chrétien Reflector," *JRAS Canada*, 22, 207–30, 1928.
3. "The Ritchey-Chrétien Aplanatic Reflector," *JRAS Canada*, 22, 303–24, 1928.
4. "Astronomical Photography with Very High Magnifying Powers," *JRAS Canada*, 22, 359–82, 1928.
5. "The New Astronomical Photography," *JRAS Canada*, 23, 15–36, 1929.
6. "The New Astronomical Photography (continued)," *JRAS Canada*, 23, 167–90, 1929.

"Support du Plaque Photographique de Professeur G. W. Ritchey pour Télescope Réflecteur Photographique,"* *Revue d'Optique*, 8, 76–79, 1929.

"L'Evolution de l'Astrophotographie et les Grands Télescopes de l'Avenir," *Publié sous les Auspices de la Société Astronomique de France,* 1929.

"Premiers Résultats de Photographie Céleste Obtenus avec le Télescope Ritchey-Chrétien," *CR*, 191, 22–23, 1930.

"La Grande Aventure," *L'Illustration*, 22 Feb 1930, 255–58, 1930.

*This report was published without any author's name, but the volume index of the journal states that Ritchey wrote it.

Preface to Notes and Bibliography

THE BOOKS AND ARTICLES listed in the bibliography give the general frame of reference of this book. The main primary sources are contemporary letters, memoranda, and other surviving written records, contemporary newspaper and magazine articles, and published scientific papers and news items. In addition, published historical and biographical articles and books have been used to supplement them.

In the chapter notes that follow, individuals for whom there are many citations are referred to by initials only, as listed in the abbreviations of names. The archives in which they were consulted are also mostly given in abbreviated form in the list that follows. Several of these letters are available in two or even more archives, but only one archive is listed for each. Journals are listed by the abbreviations given in the third table. These same abbreviations are also used in the bibliography of Ritchey's published works, which gives the complete titles of all his papers, books, and articles, and their beginning and ending page numbers.

Names or parts of names not explicitly written in a letter, or dates not explicitly written, are enclosed in square brackets in the references. Best estimates for uncertain dates are indicated by a ~. They are based on internal evidence in the letter, the dates of other letters between the correspondents, the dates of letters to other correspondents, and so forth.

ABBREVIATIONS	NAMES
CGA	Charles Greeley Abbot
WSA	Walter S. Adams
RGA	Robert G. Aitken
JAA	John A. Anderson
EEB	Edward E. Barnard
SBB	Storrs B. Barrett
HYB	Harry Y. Benedict
JAB	John A. Brashear
WMB	W. Malcolm Browne
SWB	Sherburne W. Burnham
WWC	W. W. Campbell
HC	Henri Chrétien
LD	Lucien Delloye
ALE	A. L. Ellis
CSF	Charles S. Freeman
EBF	Edwin B. Frost
WMG	Walter M. Gilbert
HMG	Harry M. Goodwin
ECH	Evalina C. Hale
GEH	George Ellery Hale
WRH	William Rainey Harper
JFH	J. F. Hellweg
ESH	Edward S. Holden
JDH	John D. Hooker
WJH	William J. Hussey
AGI	Albert G. Ingalls
LJ	Lucien Jouvaud
AHJ	Alfred H. Joy
JEK	James E. Keeler
SPL	Samuel P. Langley

PL	Percival Lowell
JCM	John C. Merriam
VN	Vincent Nechvíle
SN	Simon Newcomb
FGP	Francis G. Pease
ECP	Edward C. Pickering
HSP	Henry S. Pritchett
GWR	George Willis Ritchey
LGR	Lillie Gray Ritchey
WGR	Willis Gray Ritchey
JR	James Robertson
WR	Wickliffe Rose
FER	Frank E. Ross
HNR	Henry Norris Russell
RAS	Ralph A. Sampson
FS	Frank Schlesinger
FHS	Frederick H. Seares
HS	Harlow Shapley
VMS	V. M. Slipher
OS	Otto Struve
ET	Elihu Thomson
AT	Augustus Trowbridge
HHT	Herbert H. Turner
CDW	Charles D. Walcott
WCW	William C. Weber
RSW	Robert S. Woodward
FEW	Frederick E. Wright
CAY	Charles A. Young

ARCHIVES

AIP Niels Bohr Library, American Institute of Physics, New York
Karl Schwarzschild Papers

APS American Philosophical Society, Philadelphia
    Elihu Thomson Papers
BL Bancroft Library, University of California, Berkeley
    University of California Archives
BHL Bentley Historical Library, University of Michigan, Ann Arbor
    William C. Weber Papers
CHC Cercle Henri Chrétien, Nice, France
    Henri Chrétien Papers
CIT California Institute of Technology Archives, Pasadena
    John A. Anderson Papers
CIW Carnegie Institution of Washington, Washington, D.C.
    Files
DDO University of Toronto Archives
    David Dunlop Observatory Records
HCO Harvard Archives, Pusey Library, Cambridge, Massachusetts
    Observatory Director's Correspondence
    Harlow Shapley Papers
HHL Henry E. Huntington Library, San Marino, California
    George Ellery Hale Collection
HL Archives of Industrial Society, Hillman Library, University of Pittsburgh
    Allegheny Observatory Records
HPM George Ellery Hale Papers, Microfilm Edition, California Institute of Technology, Pasadena, California
LC Library of Congress
    John C. Merriam Papers
    Simon Newcomb Papers
LOA Lowell Observatory Archives, Flagstaff, Arizona
MEL Milton S. Eisenhower Library, Johns Hopkins University, Baltimore
    Special Collections Division
MWO Mount Wilson Observatory Archives, Henry E. Huntington Library, San Marino, California
    George Ellery Hale Letterbooks
    Walter S. Adams Papers
    Alfred H. Joy Papers
    60-inch and 100-inch telescope logbooks

NA	National Archives, Washington, D. C. Civil War Pension Records Records of the Naval Observatory
NAS	National Academy of Sciences Archives, Washington
NO	Library, U. S. Naval Observatory, Washington
NUA	Northwestern University Archives, Evanston, Illinois Henry Crew Papers
PUA	Manuscripts Division, Firestone Library, Princeton University Henry Norris Russell Papers
RAC	Rockefeller Archives Center, Tarrytown, New York International Education Board Records
RBC	Richard B. Burgess Collection, West Covina, California
RL	Special Collections, Joseph Regenstein Library, University of Chicago President's Papers, 1889–1925 William Rainey Harper Papers
RGS	Royal Geographical Society, London David Gill Papers
ROE	Royal Observatory, Edinburgh, Scotland Director's Papers
SGA	St. Gobain Glass Co. Archives, Blois, France
SIA	Smithsonian Institution Archives, Washington, D.C. Office of the Secretary Records Albert G. Ingalls Papers
SLO	Mary Lea Shane Archives of the Lick Observatory, University of California, Santa Cruz
UC	University of Cincinnati Archives
UT	Eugene C. Barker Texas History Center, University of Texas, Austin
VU	Special Collections, Vanderbilt University, Nashville, Tennessee Edward Emerson Barnard Papers
YOA	Yerkes Observatory Archives, Williams Bay, Wisconsin Director's Papers Edwin B. Frost Papers
YUA	Yale University Archives, New Haven, Connecticut Department of Astronomy Archives

JOURNALS

AA	*Astronomy and Astro-Physics*
AJ	*Astronomical Journal*
Ap. J.	*Astrophysical Journal*
ASP Leaflet	*Astronomical Society of the Pacific Leaflet*
BMNAS	*Biographical Memoirs of the National Academy of Sciences*
Bull. AAS	*Bulletin of the American Astronomical Society* or (earlier), *Bulletin of the Astronomical and Astrophysical Society of America*
Cont. MWSO	*Contributions of the Mount Wilson Solar Observatory*
JHA	*Journal for the History of Astronomy*
JOSA	*Journal of the Optical Society of America*
JRAS Canada	*Journal of the Royal Astronomical Society of Canada*
MNRAS	*Monthly Notices of the Royal Astronomical Society*
PA	*Popular Astronomy*
PASP	*Publications of the Astronomical Society of the Pacific*
QJRAS	*Quarterly Journal of the Royal Astronomical Society*
Rep. Dir. MWSO	Report[s] of the Director of the Mount Wilson Solar Observatory
SM	*Sidereal Messenger*

Notes

CHAPTER 1 OHIO AND INDIANA: 1864–1891

1. Clipping from an unknown Cincinnati newspaper, (~June 1928). This article, written soon after GWR was named a Chevalier of the Legion of Honor in France, is clearly based on information provided by his brother, J. Warren Ritchey.

2. GWR to SPL, Jan. 4, [18]88, HL.

3. JAB to GEH, Dec. 6, [19]02, YOA.

4. JAB, *The Autobiography of a Man who Loved the Stars* (Boston: Houghton Mifflin Co., 1925).

5. George Ritchey's year of birth is estimated from the ages he gave in various U. S. Censuses, beginning in 1850. They are mutually contradictory, but the early ones seem most in accord with the ages of his wife and children. His date of marriage is not known but is estimated from the fact that several of his sons and grandsons had their first child born within a year of their marriage; he may be assumed to have done the same. His wife's maiden name is from the death certificate of their daughter, Mary Ritchey, Denison, Texas, June 11, 1931. Their Scottish way of speaking is described by their grandson J. Warren Ritchey in the three-page *Genealogy of the Ritchey Family*, 1947 (hereafter *Genealogy*), a copy of which was provided to me by Catherine Ritchey Miller.

6. James Ritchey's date and place of birth are from a family *Bible*, copied by Belle McD. Ritchey in the 1940s and recopied and provided to me by C. R. Miller; Samuel's month and year of birth are from the 1900 U. S. Census (Meigs Co., Ohio); John's date and place of birth are from his Civil War pension file, NA.

7. Passenger list for Brig *Congress*, arrived Sept. 7, 1841, at New York, NA. The ages of George, Sarah, James, Samuel, and John given in it are consistent with early census records. *New York Journal of Commerce*, Sept. 7, 1841, confirms the number of passengers and gives the departure date from Liverpool.

8. The description of the early history of Meigs County here and elsewhere in this chapter is based on:

Hardesty's Historical and Geographical Encyclopedia Containing . . . Special History of Northwestern Ohio and the Geological History of the State. Outline Map of Meigs County, Ohio (Chicago and Toledo: H. H. Hardesty & Co., 1883).

Edgar Ervin. *Pioneer History of Meigs County, Ohio to 1949.* [Pomeroy], Meigs County Pioneer Society (c. 1951).

Meigs County, Ohio History Book. [Pomeroy], Meigs County Pioneer and Historical Society, 1979.

Agnes C. Hill. *A History of Tuppers Plains, Ohio and Surrounding Area.* [Pomeroy] Meigs County Pioneer and Historical Society, 1985.

9. These and other descriptions of George Ritchey's financial transactions in this chapter are from *Deeds*, Meigs County Court House, Pomeroy, Ohio.

10. George's birthdate is from his Civil War pension file, NA; the other years of birth are from the 1850, 1860, and 1870 U. S. Censuses (Meigs County, Ohio). The younger daughters progressively falsified their ages in later censuses and records.

11. Recorder's Office documents, Meigs County Museum, Pomeroy.

12. *Catalogue of the Officers and Students of Marietta College*, 1851–52. James Ritchie (as his name was misspelled) of Tuppers Plains is not listed for earlier or later years.

13. *Genealogy.*

14. H. Z. Williams & Bro. *History of Washington County, Ohio* (Cleveland: W. W. Williams, 1881).

15. Benjamin A. Gould. *The Family of Zaccheus Gould of Topsfield* (Lynn, Mass.: Thos. P. Nichols, 1895); Ebenezer Alden. *Memorial of the Descendants of the Hon. John Alden, with Supplement* (Randolph, Mass: S. P. Brown, 1867); *Vital Records of Ashfield, Massachusetts to the Year 1850* (Boston: New England Historic Genealogical Society, 1942). All the Gould family genealogy

in this chapter is derived from these three sources.

16. Eliza Gould's childhood, education, and teaching are from *Ritcheys*, a five-page manuscript by Belle McD. Ritchey (c. 1955). A copy of it was given to me by C. R. Miller.

17. John and George Ritchey's respective Civil War service records (Co. B. 116th Ohio Volunteer Infantry) and pension files, NA; "Memorandum of Marches by William H. H. Dye 1862–1865" in Ervin, op. cit.

18. *Genealogy; Bible.*

19. James Ritchey, affidavit, Hamilton Co., Ohio, Sept. 24, 1887, T. Curtis Smith Civil War pension file, NA.

20. *[Pomeroy] Meigs County Telegraph*, Sept. 10, 1868; Oct. 26, 1870; Nov. 24, 1879.

21. U. S. Census (Pomeroy, Meigs Co., Ohio), 1870.

22. *[Pomeroy] Meigs County Telegraph*, Jan. 11, 1871, Jan. 24, 1872; Ohio Secretary of State: *Record of Incorporations*, 9, 503, 1872.

23. *[Pomeroy] Meigs County Telegraph*, Apr. 26, June 7, 1871.

24. *[Pomeroy] Meigs County Telegraph*, Mar. 6, 1872.

25. *[Pomeroy] Meigs County Telegraph*, Apr. 24, July 10, 1872.

26. *[Middleport] Meigs County News*, Jan. 29, Mar. 5, May 7, 1873.

27. *Ritcheys.*

28. *[Pomeroy] Meigs County Telegraph*, Sept. 29, 1875.

29. George Ritchey [Jr.] Civil War pension file, NA; *Evansville City Director[ies]*, Bennett & Co., 1876–1879; *Genealogy; Ritcheys.*

30. *[Middleport] Meigs County News*, Jan. 26, 1876.

31. George Ritchey, Jr., Civil War pension file, NA; *Genealogy;* U. S. Census (Meigs Co., Ohio) 1870, 1880.

32. *Bible.*

33. *Genealogy.* The earliest surviving student records for Hughes High School in the Cincinnati Public Schools Records are for 1912.

34. *University of Cincinnati Catalogue(s) for the Academic Year(s) 1876–77 through 1890–91,* UC Archives. These are also the source for statements later in the chapter about GWR's academic career at UC.

35. Reginald C. McGrane, *The University of Cincinnati: A Success Story in Urban Higher Education* (New York: Harper & Row, 1963). This book is the main source for information on the buildings, courses, and faculty of UC, and on its relations with the Cincinnati Observatory.

36. Lillie May Gray's parents, adopted brother, and so forth, from U. S. Censuses 1860 (Gallia Co., Ohio), 1870, 1880 (Meigs Co., Ohio); *Evansville City Directories*, Bennett & Co. 1877–1880; *Genealogy.* The possible distant relationship is based on Ohio cemetery records. The quotation is from *Ritcheys.*

37. G. Edward Pendray, *Men, Mirrors and Stars* (New York: Funk, Wagnalls and Co., 1935), pp. 266–269. The glowing little biography of Ritchey on these pages is obviously based on an interview with him that probably took place in 1934.

38. Cincinnati Observatory Visitors' Book, April 1879 to May 19, 1893, UC Archives.

39. Jermain G. Porter, *Historical Sketch of the Cincinnati Observatory 1843–1893* (Cincinnati: University of Cincinnati, UC Archives).

40. Joseph S. Stern, Jr. *Cincinnati Historical Society Bulletin*, 39, 231, 1981; John E. Ventre and Edward J. Goodman, *A Brief History of the Cincinnati Astronomical Society*, [Cincinnati], Cincinnati Astronomical Society, 1985; *The Centenary of the Cincinnati Observatory*, [Cincinnati], Historical and Philosophical Society of Ohio and the University of Cincinnati, 1944.

41. GWR, *JRASC*, 22, 303, 1928

42. See reference 2.

43. GWR to [F. W.] Very, Dec. 15, [18]87, HL.

44. See reference 2.

45. *Genealogy*.

46. Charles A. Bennett, *History of Manual and Industrial Education 1870 to 1917* (Peoria, Illinois: Manual Arts Press, 1937).

47. Samuel E. Ritchey, *High School Manual Training Courses in Woodwork* (New York: American Book Co., 1905).

48. *Chicago City Director[ies]*, Chicago City Directory Co., 1887–1896.

49. [WSA] to WMG, Apr. 16, 1914, HPM.

50. *PASP*, 3, 79, 1891.

CHAPTER 2 KENWOOD OBSERVATORY: 1868–1896

1. Helen Wright, *Explorer of the Universe: A Biography of George Ellery Hale* (New York: E. P. Dutton & Co., 1966). Many of the facts of Hale's life and career not otherwise referenced have been taken from this excellent book and from the biography by Adams in the next reference. Both these sources are packed with information but are almost idolatrous in their treatment of Hale, and it is necessary to read between their lines to grasp the failures which he, like other scientists and human beings, sometimes had.

2. WSA, *BMNAS*, 21, 181, 1941.

3. GEH, *Biographical Notes*, HPM.

4. EBF, *Ap. J.*, 54, 1, 1921; EBF, *University [of Chicago] Record*, 7, 117, 1921.

5. GEH to Santa Claus, Nov. 6, [18]81, HHL.

6. GEH to HMG, June 5, [18]87, HHL.

7. W. E. Hale to GEH, Apr. 25, 1888, YOA.

8. GEH to HMG, July 21, 1888, HHL.

9. [GEH] to [HMG], Aug. 5, [1888], HHL.

10. GEH to HMG, [~Aug. 26, 1888], HHL.

11. "Willie" [W. B. Hale] and "Mattie" [M. D. Hale] to GEH, June 4, 1889, HHL.

12. GEH to H. A. Rowland, May 21, June 1, 1886, MEL; GEH to [HMG], [June 6, 1889], HHL.

13. GEH to [HMG], [June 9, June 16, 1889], HHL.

14. GEH to [HMG], [June 30, 1889], HHL.

15. ESH to JAB, June 18, 1889, SLO; GEH to [HMG], Aug. 11, 1889, HHL.

16. GEH to [HMG], [Aug 4, 1889], HHL.

17. CAY to GEH, Aug. 9, Aug. 19, 1889, YOA; GEH to [HMG], [Aug 16, 1889], HHL.

18. JAB to ESH, Sept. 19, Oct. 18, 1889, Aug. 13, 1891, SLO.

19. GEH, "Photography of the Solar Prominences," B. S. Thesis, MIT, 1890, reprinted in Helen Wright, Joan N. Warnow, and Charles Weiner (eds.), *The Legacy of George Ellery Hale* (Cambridge: MIT Press, 1972).

20. GEH to [HMG], July 9, 1890, HHL.

21. GEH, *Biographical Notes*, HPM.

22. GEH to [HMG], July 30, 1890, HHL.

23. GEH to [HMG], [July 23, 1889], HHL.

24. E. A. Tanner to GEH, July 13, 1889, YOA.

25. L. I. Blake to GEH, Aug. 7, 1890, YOA.

26. GEH to H. A. Rowland, July 3, 1890, MEL.

27. CAY to GEH, July 31, 1890, YOA.

28. GEH to [HMG], Aug. 10, Sept. 19, 1890, HHL; GEH, *PASP*, 3, 30, 1890.

29. GEH to [HMG], Oct. 1, 1890, HHL.

30. GEH to [HMG], Dec. 14, 1890, Jan. 26, 1891, HHL; E. D. Eaton to GEH, Nov. 14, Nov. 22, Dec. 24, 1890, E. A. Tanner to GEH, Dec. 2, Dec. 8, 1890, Jan. 17, 1891, YOA.

31. JEK to GEH, Jan. 29, 1891, YOA; GEH, *SM*, 10, 23, 1891.

32. GEH, *SM*, 10, 257, 1891.

33. CAY to GEH, Feb. 9, May 13, May 30, 1891, YOA; CAY, *SM*, 10, 312, 1891; GEH, *SM*, 10, 321, 1891; GEH, *PASP*, 3, 300, 1891.

34. ESH to GEH, Jan. 7, 1891, JEK to GEH, Feb. 25, 1891, YOA; GEH to [HMG], May 25, June 9, 1891, HHL.

35. E. D. Eaton to GEH, July 29, 1891, YOA.

36. W. A. Gibson to W. E. Hale, July 15, 1891, HPM.

37. GEH to [HMG], Aug. 21, Oct. 8, 1891, HHL.

38. GEH, *AA*, 11, 407, 811, 1892.

39. WSA, *PASP*, 52, 165, 1940.

40. E. D. Eaton to GEH, Mar. 10, May 9, 1892, C. A. Bacon to GEH, Apr. 16, 1892, YOA; GEH to HMG, May 6, 1892, HHL.

41. Richard J. Storr, *Harper's University: The Beginnings* (Chicago: University of Chicago Press, 1966).

42. WRH to G. A. Douglass, Feb. 11, 1891, WRH to GEH, Feb. 19, Apr. 28, 1891, YOA; GEH to WRH, Apr. 30, 1891, RL.

43. WRH to GEH, May 13, May 19, 1891, YOA.

44. GEH to WRH, May 30, 1891, WRH to GEH, June 5, 1891, YOA.

45. GEH to WRH, Apr. 29, 1892, RL.

46. A. Hall to WRH, June 13, 1891, RL.

47. GEH to WRH, June 30, July 22, 1892, W. E. Hale to WRH, July 1, 1892, WRH to GEH, July 27, 1892, RL.

48. H.S. Miller, *ASP Leaflet* No. 479, 1969.

49. GEH to WRH, Sept. 23, 1892, RL.

50. WRH to F. T. Gates, [~May 1892], RL.

51. T. W. Goodspeed to F. T. Gates, Oct. 7, 1892, WRH to F. T. Gates, Oct. 10, 1892, RL.

52. GEH to E. D. Eaton, Dec. 27, 1892, MWO.

53. GEH to [HMG], May 6, 1892, HHL.

54. GEH, *PASP*, 3, 79, 1891.

55. GEH, *PASP*, 3, 81, 151, 261, 1891.

56. R. W. Pike, *PASP*, 3, 300, 1891.

57. GEH, *PASP*, 4, 49, 1892.

58. GEH to JEK, Oct. 12, Nov. 19, 1891, Sept. 13, 1892, HL.

59. *AA*, 11, 635, 1892.

60. *AA*, 12, 78, 640, 743, 1893. D. E. Osterbrock, *Chicago History*, 9, 178, 1980, gives a full description of the meeting and many references to contemporary descriptions of it.

61. *Chicago Tribune*, Aug. 24, 1893, *Chicago Daily Inter Ocean*, Aug. 24, 1893.

62. W. W. Payne, *SM*, 6, 250, 1887; O. C. Wendell, *Ap. J.*, 6, 1897; D. J. Warner, *Alvan Clark & Sons, Artists in Optics* (Washington: Smithsonian Institution Press, 1968).

63. C. S. Hastings to GEH, Oct. 26, 1892, YOA.

64. A. G. Clark, *AA*, 12, 673, 1893.

65. GEH, *PASP*, 3, 79, 1891.

66. H. H. Belfield to GEH, Oct. 29, 1891, YOA.

67. GWR to GEH, Jan. 14, Nov. 21, 1892, YOA.

68. GEH to L. F. Culver, Jan. 2, 1893, MWO.

69. GWR, *Smithsonian Contributions to Knowledge*, 34, Part 2, 1904.

70. A. A. Common, *Memoirs R.A.S.*, 46, 173, 1881.

71. GWR, *Smithsonian Contributions to Knowledge*, 34, Part 2, 1904.

72. GEH, *Ap. J.*, 5, 119, 1897.

73. H. Grubb, *Proc. Royal Dublin Society*, 1, 1, 1877.

74. D. Gill to R. S. Floyd, Sept. 18, 1876, SLO.

75. GEH to W. Huggins, Dec. 28, [189]2, MWO.

76. SWB, *SM*, 4, 193, 1885.

77. GEH to [HMG], Sept. 3, 1893, HHL; GEH to JEK, Nov. 25, 1893, Mar. 11, Apr. 26, 1894, HL.

78. D. E. Osterbrock, *Wisconsin Magazine of History*, 68, 108, 1985, gives references to many letters dealing with the choice of the Williams Bay site.

79. GEH to HMG, Sept. 22, 1894, HHL.

80. GWR to GEH, [~July 1, 1895], YOA.

81. GEH to GWR, Aug. 11, [189]5, MWO.

82. E. Petit to GWR, July 22, 1895, YOA; GEH to GWR, July 25, [189]5, MWO.

83. E. Petit to GWR, Aug. 14, Sept. 3, Sept. 26, 1895, YOA.

84. GWR to GEH, Sept. 5, [18]95, YOA.

85. GEH to GWR, Mar. 13, [189]6, MWO.

86. GWR to GEH, June 3, 1896, YOA; GEH to Compagnie des Glaces & Verres, June 5, [189]6, GEH to E. Petit, June 30, [189]6, MWO.

87. GEH to J. W. Powell, Feb. 16, [189]3, GEH to J. Hance, Feb. 16, [189]3, GEH to [HS]P, Feb. 20, Feb. 28, [189]3, GEH to JAB, Mar. 3, [189]3, MWO.

88. GEH, *AA*, 13, 662, 1894.

89. GEH, *Ap. J.*, 1, 318, 1895; GEH to JEK, Mar. 21, Aug. 2, Sept. 20, 1895, HL.

90. GEH to GWR, Sept. 20, Sept. 25, [189]5, MWO.

91. GEH to GWR, Sept. 12, Oct. 1, [189]5, MWO.

92. E. Petit to GWR, Feb. 13, 1896, YOA.

93. GWR to GEH, Mar. 25, May 25, 1896, YOA; GEH, *Ap. J.*, 5, 254, 1897.

94. GEH, *Ap. J.*, 12, 372, 1900. This paper summarizes the 1895 and 1896 attempts, as well as those in succeeding years, through the 1900 eclipse described in the next chapter.

95. GWR, *Ap. J.*, 14, 217, 1901.

96. GWR to GEH, Oct. 9, [18]96, YOA.

97. GEH to W. E. Hale, Nov. 17, 1897; GEH to WRH, Aug. 2, 1902, YOA. The second letter states that Ritchey's salary was $1,400 in early 1902, making $1,200 a good estimate for 1896.

98. GWR to GEH, June 19, July 11, Aug. 4, 1896, YOA.

99. GWR to GEH, Dec. 27, 1896, YOA; *Chicago City Director[ies]*, Chicago City Directory Co., 1889–1900. GWR is listed each year through 1896, but not thereafter.

100. WSA to WMG, Apr. 14, 1914, HPM; WGR passport application, Aug. 6, 1925, State Dept. Archives.

CHAPTER 3 YERKES OBSERVATORY: 1897–1904

1. GEH, *Ap. J.*, 5, 254, 1897.

2. GEH, *Ap. J.*, 5, 310, 1897.

3. GWR, *Ap. J.*, 5, 143, 1897.

4. GEH to JEK, Nov. 24, 1896, Jan. 20, 1897, YOA; Helen Wright, *Explorer of the Universe: A Biography of George Ellery Hale* (New York: E. P. Dutton & Co., 1966). All the facts about the early history of Yerkes Observatory not otherwise referenced are from this excellent book.

5. GEH to "My dear Father and Mother," [W. E. and M. B. Hale], [Feb. 14, 1897], HPM.

6. A. G. Clark to GEH, Feb. 17, 1897; GEH to C. T. Yerkes, May 24, 1897; GEH to A. G. Clark, May 27, 1897; GEH to Warner & Swasey, May 27, 1897; GEH to C. A. Bacon, May 27, 1897; YOA.

7. GEH to WRH, May 20, 1897, GEH to H. A. Rust, May 25, 1897, YOA.

8. GEH, Report of director of Yerkes Observatory for May 1897, [~June 7, 1897]; GEH to CAY, June 1, 1897; GEH to W. Huggins, June 12, 1897; GEH to Warner & Swasey, July 3, 1897, YOA. GEH, *Ap. J.*, 6, 37, 1897.

9. See D. E. Osterbrock, *JHA*, 15, 81, 1984, and G. Verschuur, *Interstellar Matters* (New York: Springer Verlag, 1988) for more about Barnard's life and career.

10. GEH to JEK, Oct. 4, 1894, Mar. 5, Mar. 7, Mar. 19, 1896, YOA. See also reference 2.

11. GEH to "My dear Mother" [M. B. Hale], July 28, 1897, HPM.

12. GEH, *Ap. J.*, 6, 353, 1897; *Ap. J.*, 10, 211, 1899; see also D. E. Osterbrock, *Wisconsin Magazine of History*, 68, 108, 1985, which describes the Yerkes Observatory dedication and the founding of the AAS, with many references to contemporary sources.

13. GEH, Report of the director of Yerkes Observatory for three months ending October 31, 1897, [~Nov. 15, 1897], RL; F. Ellerman to GEH, Dec. 28,

1897, Jan. 14, 1898, Dec. 30, 1899, Jan. 4, 1900, YOA.

14. D. E. Osterbrock, *James E. Keeler, Pioneer American Astrophysicist: And the Early Development of American Astrophysics* (Cambridge: Cambridge University Press, 1984). See especially pp. 257–70, and the many references to original source material listed for these pages.

15. EBF, *An Astronomer's Life* (Boston and New York: Houghton Mifflin Co., 1933); FS, *PASP*, 47, 175, 1935; OS, *BMNAS*, 19, 25, 1938; Reference 14, pp. 272–73, and references given there.

16. A. H. Joy, *BMNAS*, 31, 1, 1958; WSA, *AJ*, 20, 133, 1899; WSA, *Science*, 106, 196, 1947.

17. WSA, "Winter Quarter, 1899, Class-work Grade"; WSA to EBF, July 2, July 23, Aug. 26, 1900; WSA to GEH, [~Jan. 15, 1900], Dec. 31, 1900, YOA.

18. A. A. Michelson to WRH, May 16, 1895; T. C. Mendenhall to WRH, Jan. 31, 1898; F. L. O. Wadsworth to WRH, May 28, June 23, July 11, July 27, 1898; F. L. O. Wadsworth to T. W. Goodspeed, June 23, 1898, all at RL. GEH to WRH, Feb. 8, 1898; WRH to GEH, Feb. 21, 1898; both at YOA. JAB to GEH, Mar. 16, Mar. 25, 1899, HPM.

19. [GEH] to WRH, Nov. 11, 1898, HPM.

20. [GEH] to WRH, Nov. 11, 1898, HPM.

21. GWR, *Ap. J.*, 12, 352, 1900.

22. GWR, *Harper's Monthly*, 107, 411, 1903.

23. D. E. Osterbrock, *JHA*, 15, 81, 151, 1984.

24. GEH to SN, Mar. 12, 1901, LC; references 20 and 21.

25. Reference 20; GWR, *Harper's Monthly*, 109, 508, 1904.

26. GEH to WWC Sept. 14, Oct. 29, 1900, SLO; WWC to GEH, Oct. 19, 1900, YOA.

27. JEK to GEH, Feb. 21, 1899; WWC to GEH, Aug. 16, 1899; both at YOA.

28. GEH to WRH, Mar. 2, May 5, 1900, RL.

29. GEH, *Ap. J.*, 12, 372, 1900.

30. GEH to SPL, Feb. 25, 1900, YOA; GEH to H. Crew, June 9, 1900, NUA.

31. GEH to JEK, June 22, 1900, YOA; GEH, *Ap. J.* 12, 80, 1900; EBF, *Science*, 12, 177, 1900; EEB, *Science*, 12, 178, 1900.

32. GWR, *Ap. J.*, 12, 352, 1900.

33. GEH to C. R. Cross, Oct. 2, 1900, Apr. 17, 1901; GEH to ECP, Aug. 15, 1902; GEH to A. Ricco, Aug. 11, 1902; GEH to R. W. Wood, Nov. 20, 1902; GEH to [I.F.] Blackstone, Feb. 16, 1903; all at YOA. GWR to SPL, Jan. 3, 1903, SIA.

34. GEH to H. A. Rust, July 16, 1898; GEH to T. W. Goodspeed, July 16, 1898; GEH requisition, July 16, 1898; all at YOA.

35. GWR, *Ap. J.*, 14, 217, 1901.

36. J. Ritchey, *Pattern Making: Practical Treatise for the Pattern Maker on*

Wood Working and Wood Turning, Tools and Equipment, Construction of Simple and Complicated Patterns, Modern Molding Machines and Molding Practice, revised W. W. Monroe, C. W. Beese and P. R. Hall (Chicago: American Technical Society, 1940).

37. *Armour Institute of Technology Yearbook[s]*, 1894–95 through 1905–6; *Williams' Cincinnati City Director[ies]*, 1886–1892; *Bennett & Co. Evansville City Director[ies]*, 1890–95; *Chicago Annual Director[ies] of the City*, 1885 through 1906.

38. L. Braymer to R. E. Keim, Dec. 14, 1960, author's personal collection.

39. GEH to JEK, May 10, June 12, 1899, YOA; GEH to WWC, Aug. 5, 1901, SLO.

40. GWR, *Ap. J.*, 14, 167, 1901; GWR, *Ap. J.*, 14, 293, 1901.

41. GWR, *Ap. J.*, 15, 129, 1902.

42. J. E. Felten, *Supernova 1987A in the Large Magellanic Cloud*, ed. M. Kafatos and A. G. Michalitsianos (Cambridge: Cambridge University Press, 1988), p. 232. This paper gives many references to previous papers on the "light-echo" effect.

43. *San Francisco Chronicle*, Nov. 20, 1901.

44. WWC to GEH, Nov. 11, 1901, SLO.

45. WWC to GEH, Dec. 6, 1901, GEH to WWC, [~Dec. 12, 1901], SLO.

46. GWR, *Ap. J.*, 16, 178, 1902.

47. See Osterbrock, *James Keeler*, pp. 297–320, 325–26, and references given there. Also D. E. Osterbrock, *Astronomy Quarterly*, 5, 87, 1985.

48. GEH to G. C. Comstock, Sept. 3, 1901; GEH to SN, Oct. 30, Nov. 5, Nov. 12, 1901, Jan. 15, 1902; all at LC. *Bull. AAS*, 1, 143, 1909, GWR. *Bull. AAS*, 1, 176, 1909.

49. GWR to CGA, Aug. 28, 1902; CGA to SPL, Oct. 10, 1902; [R. Rathburn] to GWR, Oct. 18, 1902; all at SIA. GWR to ECP, Oct. 25, 1902, HCO.

50. References 21 and 24.

51. GWR to GEH, Dec. 22, Dec. 29, [18]99, Jan. 20, Feb. 9, 1900, YOA, are examples.

52. GWR to [C.] Furness, Jan. 23, 1901, Vassar College Library.

53. GEH to WRH, Jan. 17, 1901, RL; GEH to WRH, Aug. 2, 1902, YOA.

54. WWC to GWR, Oct. 19, Nov. 17 [19]00, Jan. 7, [19]01, GWR to WWC, Sept. 27, Nov. 26, 1900, Jan. 12, 1901, SLO; WWC to GEH, Jan. 23, 190[2], YOA.

55. GWR to A. Belopolsky, July 29, Sept. 5, 1902; GEH to Board of Trustees, July 28, 1902; L. P. Lewis to GEH, Oct. 17, 1902; GEH to WRH, Oct. 24, Dec. 16, 1902; all at YOA.

56. SPL to GEH, Apr. 11, May 2, May 19, 1902; GEH to SPL, Apr. 28, May 26, 1902; all at YOA. F. W. Hodge to GWR, Jan. 22, 1902; SPL to GWR,

Nov. 21, 1902, Jan. 10, Jan. 20, 1903; GWR to SPL, Jan. 17, 1903; all at SIA.

57. SPL to GWR, Mar. 31, Apr. 9, 1903; GWR to SPL, Mar. 10, Apr. 6, Apr. 21, May 22, July 28, 1903; F. W. Hodge to GWR, June 3, July 23, 1903; all at SIA.

58. N. S. Shaler, *Smithsonian Cont. to Knowledge*, 34 (part), 1, 1903.

59. GEH to SPL, May 6, 1902, Oct. 24, 1902; SPL to GEH, Oct. 17, 1902; all at YOA. SPL to GWR, Oct. 17, Oct. 21, 1902, Mar. 16, Oct. 3, Oct. 19, 1903, Jan. 26, 1904; GWR to SPL, Oct. 25, 1902, Feb. 14, June 6, Oct. 14, Dec. 22, 1903, Jan. 8, Feb. 23, Mar. 5, Apr. 4, Apr. 15, May 6, May 31, July 31, 1904; A. H. Clark to SPL, Jan. 14, 1904; GWR to A. H. Clark, July 31, [19]04; F. W. Hodge to GWR, Oct. 30, 1903, Sept. 12, 1904; all at SIA.

60. GWR, *Smithsonian Cont. to Knowledge*, 34 (part), 1, 1904.

61. GWR, *Ap. J.*, 19, 53, 1904.

62. EBF to GWR, Jan. 8, Jan. 25, Feb. 3, Feb. 16, [Feb. 20], Feb. 23, 1904, HPM. GWR to EBF, [~Feb. 1], Feb. 22, 1904, YOA.

63. GWR to EBF, [~Jan. 10], Jan. 14, 1904, YOA; GWR to GEH, Jan. 3, Mar. 21, Apr. 6, Apr. 7, 1904, HPM.

64. GWR, *Pub. Yerkes Obs.*, 3, 387, 1904; GWR to EBF, Jan. 12, 1904, YOA; GWR to GEH, Mar. 7, 1904, HPM.

65. GWR to [F.] Dyson, Apr. 2, 1903, RAS; GWR to [A. R.] Hinks, Apr. 2, 1903, pub. in *MNRAS*, 63, 395, 1903; W. H. Wesley, *MNRAS*, 64, 237, 1904.

66. GWR to W. Huggins, July 31, 1904, RAS.

67. GWR to GEH, Mar. 11, 1904; GEH to WRH, May 23, 1904, all at HPM.

68. HSP to GEH, Nov. 11, 1901, Apr. 21, 1902, HPM.

69. *Transactions of the International Union for Cooperation in Solar Research*, 1 (Manchester: Manchester University Press, 1906); WSA, *PASP*, 61, 5, 1949.

70. GEH to JAB, Jan. 29, 1896, MWO.

71. GWR to GEH, Nov. 8, Dec. 21, [18]98, Jan. 3, Jan. 20, May 10, [18]99, YOA.

72. GEH to JEK, Mar. 23, 1899, YOA.

73. F. L. O. Wadsworth, *Ap. J.*, 5, 132, 1897.

74. GWR, *Ap. J.*, 19, 53, 1904.

75. GEH to EBF, Mar. 11, 1898, YOA.

76. SWB to GEH, Nov. 17, 1898, HPM; GEH to [HM]G, Nov. 23, 1898, HHL.

77. GEH to WRH, Sept. 19, 1898, YOA.

78. GWR to LJ, Mar. 14, 1899, SGA.

79. JEK to GEH, Apr. 19, 1899; GEH to JEK, May 10, 1899; both at YOA.

80. GEH to N. B. Ream, Jan. 19, 1899; GEH to WRH, May 9, 1899; both at YOA.

81. GEH, "A Great Reflecting Telescope," [~Mar. 1, 1902], YOA.

82. GEH to WRH, Apr. 1, 1902, RL.

83. GEH to ECP, Aug. 15, 1902; GEH to L. Boss, Aug. 28, 1902; both at MWO. WWC to GEH, Sept. 13, 1902, SLO; GEH to WRH, Sept. 4, 1902, YOA.

84. GEH to WWC, Oct. 4, 1902; GEH to WRH, Oct. 15, 1902; GEH to CDW, Oct. 20, Oct. 30, 1902; all at YOA.

85. WWC to GEH, Nov. 26, 1902, YOA.

86. WWC to GEH, Dec. 6, 1902, YOA.

87. GEH to WRH, Mar. 6, 1903, RL.

88. D. Brouwer, *BMNAS*, 24, 105, 1945.

89. GEH to [HMG], Jan. 22, 1903, HHL.

90. WJH, *Report of Committee on Southern and Solar Observatories* (Washington: Carnegie Institution, 1903), p. 71.

91. GEH to [HMG], Dec. 24, 1903, HHL; GEH to SBB, Dec. 11, 1903, YOA.

92. GWR to GEH, Dec. 27, 1902, Jan. 8, Feb. 6, 1903; GEH to GWR, Jan. 7, Jan. 10, 1903; GEH to J. S. Ames, Oct. 13, 1903; GEH to A. Agassiz, Nov. 28, 1903; GEH to WSA, Dec. 12, 1903; all at YOA. GEH to WWC, Jan. 20, Feb. 12, 1904, SLO.

93. GEH to WRH, Feb. 13, 1904, RL; GEH to [HM]G, Feb. 28, 1904, HHL.

94. K. T. Compton, *BMNAS*, 21, 143, 1941; ET to GEH, Apr. 17, May 4, 1903, HPM.

95. GEH to GWR, Mar. 5, Mar. 31, 1904, GEH; Memorandum on GWR, Mar. 20, [1904]; GWR to GEH, Mar. 28, [1904], HPM.

96. GEH to EBF, Mar. 7, 1904, HPM; GEH to WWC, Mar. 23, 1904, SLO.

97. GEH, "Expedition for Solar Research," [~May 1, 1904], RL; GEH to WWC, May 19, May 26, 1904, SLO.

98. GEH to EBF, May 21, 1904, HPM; GEH to ET, May 23, 1904, APS; GEH to [HMG], May 26, 1904, HHL.

99. GWR to ET, May 31, June 10, [19]04, APS; EBF to GEH, May 31, June 2, 1904, HPM.

100. ET to GEH, June 17, 1902, Mar. 12, 1904; GEH to ET, Mar. 5, 1904, HPM.

101. GWR to ET, June 20, June 28, 1904, APS.

102. ET to GWR, July 5, July 23, 1904, HPM.

103. GEH to H. P. Judson, Oct. 27, 1902, Oct. 28, 1903, YOA.

104. WRH to GEH, Feb. 25, 1904; GEH to WRH, Mar. 7, May 20, 1904; GEH to EBF, May 26, 1904; all at HPM. EBF to WRH, June 4, 1904; GEH to WRH, June 29, 1904, both at RL.

105. GEH to [HM]G, Aug. 4, 1904, HHL.

106. [GEH], *PA*, 12, 503, 1904.
107. G[EH] to [HMG], Oct. 14, 1904, HHL.
108. GWR, *Ap. J.*, 19, 53, 1904; GWR to ET, Nov. 22, [19]04, APS.
109. GWR to ET, [Dec 2, 1904], APS; GWR to EEB, Dec. 21, [19]04, VU.
110. G[EH] to [HMG], Dec. 2, 1904, HHL.
111. GEH to HMG, Dec. 21, 1904, HHL.
112. GWR to ET, Jan. 13, 1905, APS.
113. EBF to GEH, Dec. 27, 1904, HPM.
114. WWC to GEH, Dec. 26, 1904, HPM.

CHAPTER 4 MOUNT WILSON SUCCESS: 1905–1908

1. GEH to RSW, Jan. 4, 1905; GEH to WRH, Mar. 6, 1905; both at HPM. Memorandum, "G. W. Ritchey salary payments," Oct. 15, 1919, CIW.

2. WRH to GEH, June 2, 1904, Jan. 6, 1905, HPM; WRH to "My dear Father" [S. A. Harper], June 13, 1904; WRH to [E. B.] Hurlbert, Aug. 6, 1904; WRH to GEH, Feb. 17, 1905; all at RL.

3. GEH to EBF, Nov. 21, 1904, Jan. 17, Jan. 31, 1905; EBF to GEH, Nov. 22, 1904, Jan. 5, Feb. 4, Feb. 11, 1905; S. A. S[imon], "Commentary," [Mar. 5, 1905]; WRH to GEH, Feb. 11, 1905; all at HPM. EBF to WRH, Jan. 31, Feb. 10, Feb. 11, 1905, RL; GEH to HMG, Feb. 22, 1905, HHL.

4. GEH to EBF, Jan. 7, Feb. 15, Feb. 17, Feb. 18, Feb. 20, Feb. 25, Mar. 2, Mar. 11, 1905; EBF to GEH, Feb. 24, Mar. 2, 1905; WRH to GEH, Dec. 28, 1904, Jan. 16, Feb. 4, Mar. 10, 1905; GEH to WRH, Jan. 7, Jan. 23, Jan. 31, Mar. 10, 1905; all at HPM. WRH to GEH, Mar. 17, 1905, RL.

5. EBF to WRH, Dec. 1, 1904, RL; GEH to EBF, Jan. 14, May 16, June 18, 1905; EBF to GEH, Jan. 14, 1905; both at HPM.

6. EBF to GEH, Mar. 16, 1905; GEH to EBF, Mar. 21, 1905; both at HPM.

7. GWR to EBF, Jan. 19, Jan. 23, Feb. 1, 1905; EBF to GWR Jan. 27, Feb. 3, Feb. 8, 1905, Feb. 7, 1907; all at YOA.

8. EBF to GEH, Jan. 18, 1905; GEH to EBF, Jan. 24, 1905; both at HPM.

9. GEH to WWC, Feb. 26, 1905, SLO; GWR to ET, Jan. 21, Jan. 25, Feb. 3, Feb. 20, Feb. 24, 1905, APS; GWR to EBF, Feb. 23, 1905, YOA.

10. EBF to GEH, Jan. 24, Mar. 6, Mar. 15, Mar. 17, Mar. 19, Mar. 21, Apr. 1, Apr. 20, May 8, 1905; GEH to EBF, Feb. 26, Mar. 11, Apr. 7, Apr. 12, Apr. 26, 1905; T. W. Goodspeed to EBF, Mar. 20, 1905; all at HPM. GWR to EBF, Mar. 9, 1905; GWR to SBB, Mar. 10, 1905; EBF to GWR, Mar. 15, 1905; all at YOA. GEH to H. P. Judson, Feb. 24, 1905, RL.

11. GEH, *PASP*, 17, 41, 1905.

12. GEH, *Cont. MWSO*, 1, 29, 1905; GEH, *CIW Yearbook*, 3, 155, 1905.

13. WSA to GEH, [~May 1, 1903]; WSA to EBF, Oct. 16, Nov. 13, Dec. 18, 1904, Feb. 3, Apr. 9, 1905; all at YOA. WSA to GEH, Jan. 28, Feb. 12, 1904, HPM.

14. GEH and WSA, *Ap. J.*, 11, 1906; *Ap. J.*, 23, 400, 1906; WSA, *Ap. J.*, 24, 69, 1906; GEH, WSA and H. G. Gale, *Ap. J.*, 24, 185, 1906.

15. GEH to EBF, Dec. 28, 1904, Nov. 5, Nov. 21, Dec. 21, Dec. 26, 1905, Jan. 2, May 8, 1906, EBF to GEH, May 24, Oct. 26, Nov. 15, Nov. 27, Dec. 26, Feb. 6, 1906; W. Heckman to EBF, Apr. 25, 1906; all at HPM.

16. GWR to GEH, Mar. 17, Mar. 22, [19]05; EBF to GEH, May 12, 1905; GEH to EBF, May 15, May 22, 1905; [GWR] to LJ, May 24, 1905; all at HPM. GWR to ET, Apr. 13, June 12, 1905, APS; GEH, Rep. Dir. MWSO, *CIW Yearbook*, 456, 1905.

17. GWR to "Dear Father" [J. Ritchey], June 28, 1905, HPM; *Thurston's Director[ies] of Greater Pasadena*, 1905–1919.

18. WSA, *PASP*, 50, 119, 1938.

19. GWR, *PASP*, 17, 186, 1905. EBF to GWR, Sept. 14, 1904, Aug. 8, 1905; [GWR] to EBF, Aug. 23, 1905; all at HPM.

20. GWR to WWC, June 5, 1905; WWC to GWR, June 7, 1905; GWR to R. H. Tucker, Sept. 15, 1905; [R. H. Tucker] to GWR, Sept. 18, 1905; all at SLO.

21. WWC to GEH, May 19, July 11, Aug. 13, Aug. 30, 1902, SLO.

22. WWC to GEH, Nov. 26, [Dec 5], Dec. 6, 1902, YOA; GWR to WWC, [Dec 17, 1902]; WWC to GWR, Dec. 31, 1902; both at SLO.

23. WWC to GEH, Mar. 1, 1904, HPM.

24. WWC to GWR, Dec. 7, 1904, Jan. 9, 1905; GWR to WWC, Jan. 5, Jan. 12, 1905; all at SLO.

25. GEH, *Ap. J.*, 23, 6, 1906; GEH and FE, *Ap. J.*, 23, 54, 1906; GEH, *Ap. J.*, 24, 61, 1906.

26. Transactions of the International Union for Co-operation in Solar Research, 1, 58, 1906.

27. GEH to HMG, Nov. 28, [1905], HHL.

28. Helen Wright, *Explorer of the Universe: A Biography of George Ellery Hale* (New York: E. P. Dutton & Co., 1966), p. 198.

29. GEH to EBF, Jan. 3, 1906; EBF to GEH, Jan. 26, 1906; both at HPM.

30. GWR to EEB, Apr. 9, 1906, VU.

31. GEH, Rep. Dir. MWSO, *CIW Yearbook*, 5, 160, 1906; GEH to EBF, Apr. 30, June 5, 1906, HPM.

32. GWR to Union Iron Works, [~June 7], July 30, Aug. 7, 1906; Union Iron Works to GWR, July 31, 1906 (all telegrams); GEH to EBF, Sept. 1, 1906; all at HPM.

33. PL to VMS, Oct. 18, 1905; PL to W. A. Cogshall, Mar. 2, 1906; PL to GWR, Mar. 2, 1906; PL to C. O. Lampland, Mar. 16, 1906; J. B. McDowell to VMS, Mar. 22, 1906; all at LOA.

34. PL to GWR, Mar. 28, 1906; PL to VMS, Mar. 28, 1906; both at LOA.

35. PL to GWR, Apr. 13, 1906, LOA.

36. PL to VMS, Apr. 20, Apr. 24, May 9, 1906; PL to GWR, May 9, 1906, all at LOA.

37. GWR to PL, May 22, 1906 (telegram), HPM; Lowell Observatory guest book, May 27–28, 1906.

38. PL to GWR, May 30, 1906, LOA.

39. PL to GWR, June 6, 1906; PL to VMS, June 26, 1906; both at LOA. GWR to [EC]P, May 11, 1906, HCO.

40. GWR to GEH, June 26, 1906, HPM; GEH to WWC, June 29, July 10, 1906, SLO.

41. GEH to GWR, July 15, 1906, HPM.

42. WWC to JDH, Dec. 26, 1903; WWC to GEH, Dec. 26, 1903; GEH to WWC, Jan. 20, Feb. 12, 1904; all at HPM. GEH to WWC, Mar. 7, 1904, SLO; *Los Angeles Times*, May 25, 1911.

43. GWR to [H. R.] Baynton, Jan. 6, 1905; H. R. Baynton to GWR, Jan. 9, 1905; [GWR] to JDH, Sept. 19, 1905; all at HPM.

44. [GWR] to JDH, July 27, 1906, HPM.

45. [JDH] to GEH, Sept. 14, 1906, HPM.

46. GWR to ET, Sept. 17, 1906, APS. [GWR] to LJ, Apr. 3, Apr. 26, May 25, June 14, 1906; [GWR] to JDH, Oct. 22, Nov. 10, 1906; JDH to GWR, Nov. 8, 1906; all at HPM.

47. GEH, *Ap. J.*, 24, 214, 1906; EBF to GEH, Sept. 20, 1906; GEH to EBF, Sept. 25, 1906; all at HPM.

48. GEH to WWC, Sept. 29, 1906, SLO.

49. GEH to JDH, Dec. 13, 1906, Feb. 8, 1907; JDH to GEH, Dec. 17, Dec. 31, 1906; A. Carnegie to JDH, Dec. 18, 1906; all at HPM.

50. GEH to SN, Mar. 10, Mar 21, Oct. 25, 1906, LC.

51. GWR to GEH, Oct. 22, Nov. 20 (telegram), 1906; GWR to W. S. Atkinson, Dec. 11, 1906; all at HPM. GWR to ET, May 6, May 28, 1907, APS.

52. A. B. Porter to EBF, Mar. 30, 1907, YOA. FGP to GEH, Aug. 3, 1907, HPM. [WWC] to [A. B. Porter], May 24, Sept. 3, 1907, SLO.

53. WSA to [FG]P, May 2, May 17, Oct. 7, Nov. 11, 1907, Jan. 7, 1908; FGP to GEH, Mar. 4, Apr. 6, 1908; [GEH] to FGP, Mar. 21, 1908; all at HPM.

54. GWR to GEH, May 21, June 17, 1907; GEH to EBF, July 3, Sept. 21, 1907, HPM. Rep. Dir. MWSO, *CIW Yearbook*, 6, 134, 1907.

55. GEH and WSA, *Ap. J.*, 25, 75, 1907; GEH and WSA, *Ap. J.*, 25, 300, 1907; WSA, *Ap. J.*, 26, 203, 1907; EBF to WSA, May 13, May 24, June 17, 1907; WSA to EBF, May 7, Sept. 23, Nov. 23, Dec. 3, 1907; all at YOA.

56. GEH to HMG, Feb. 24, [1906], Nov. 27, 1907, both at HHL. GEH to WWC, Apr. 4, 1907, HPM. GEH to H. P. Judson, Oct. 11, 1907, RL.

57. GEH to EBF, Jan. 10, 1908, HPM.

58. GWR to EEB, Feb. 27, [19]08, VU. GWR to ET, May 11, [19]08, APS.

59. GWR to GEH, Apr. 11, 1908, HPM. GWR to ET, Apr. 27, [19]08, APS.

60. GEH to GWR, Apr. 1, 1908, HPM.

61. WSA to GEH, June 30, 1908, HPM. R. J. Wallace to [WS]A, Dec. 28, [1908], MWO.

62. GEH to EBF, May 19, 1908, HPM.

63. GEH, *Ap. J.*, 28, 100, 1908.

64. GEH, *Ap. J.*, 25, 68, 1907; GEH, *Ap. J.*, 27, 204, 1908.

65. GEH to HMG, July 7, 1908, HHL. GEH, *Ap. J.*, 28, 100, 1908; GEH, *Ap. J.*, 28, 315, 1908.

66. [E. E. Wilhoit] to [GE]H, Aug. 11, 1908, HPM.

67. GWR to ET, June 15, [19]08, APS.

68. GWR to ET, Sept. 18, [19]08, APS.

69. GWR to ET, Oct. 2, 1908, APS; GEH to HMG, Oct. 13, 1908, HHL.

70. GEH to W. Huggins, Dec. 21, 1908; GEH to WWC, Dec. 23, 1908; GEH to EBF, Dec. 23, 1908; all at HPM.

71. GWR, *Ap. J.*, 29, 198, 1909.

72. GEH to HMG, Jan. 7, 1909, HHL; GWR to [E.E.] Wilhoit, Jan. 16, 1909, HPM; GWR to GEH, Jan. 17, 1909, MWO.

CHAPTER 5 MOUNT WILSON PROBLEMS: 1909–1914

1. [GWR] to LJ, Apr. 3, June 14, 1906, HPM. GWR to LJ, Sept. 15, 1906, SGA.

2. [GEH] to LJ, Apr. 4, 1907; LJ to GEH, Apr. 13, 1907; LD to GEH, June 29, 1907; all at HPM.

3. H. de Coquereaumont to [LD], Oct. 17, 1907, SGA; LJ to GWR, Oct. 25, 1907, HPM.

4. [WSA] to JDH, Jan. 18, 1907; [GWR] to JDH, May 17, Oct. 17, Oct. 29, 1907; all at HPM.

5. [GWR] to JDH, Oct. 30, 1907, Feb. 10, Apr. 16, May 19, June 3, June 6, 1908; [GWR] to LJ, June 22, Sept. 17, Oct. 29, 1907, Feb. 6, Apr. 10, 1908, all at HPM.

6. [GWR] to JDH, July 24, 1908, HPM.

7. H. de Coquereaumont to [LD], Aug. 12, 1908; M. Surelle to [LD], Oct. 14, 1908, both at SGA.

8. LJ to GWR, June 8, Aug. 19, Aug. 27, Sept. 2, Nov. 11, 1908; [GWR] to LJ, Nov. 24, 1908; [GWR] to JDH, Oct. 6, Nov. 25, Dec. 11, 1908; all at HPM.

9 [GWR] to LJ, Jan. 5, 1909, HPM.

10. [GWR] to LJ, Oct. 31, 1907, Aug. 11, 1908, both at HPM.

11. Reference 9; GWR to E. E. Wilhoit, Jan. 16, 1909; both at HPM.

12. GWR to [L]D, Feb. 3, Feb. 10, 1909, SGA. GWR to GEH, Feb. 1, Feb. 5, Feb. 12, Feb. 21, 1909; GEH to [WS]A, Dec. 13, 1910; all at HPM. GWR to SBB, Feb. 22, [19]09; SBB to GWR, Mar. 9, 1909; both at YOA.

13. GWR to GEH, Mar. 9, Mar. 12, 1909; GEH to JDH, Feb. 16, Feb. 22, Mar. 16, 1909; all at HPM. GWR to [L]D, Mar. 10, 1909, SGA.

14. GWR to [W. H.] Wesley [~Feb. 10, 1909], RAS; GWR to [L]D, Apr. 2, 1909, SGA; *Observatory*, 32, 187, 1909; R.A.S. Club dinner lists, (661) Apr. 7, 1909 (privately printed).

15. [GEH] to [ECH], Apr. 20, 1909, HPM; GWR to [L]D, Apr. 2, Apr. 21, June 5, 1909, SGA; GWR to ET, May 6, 1909, APS.

16. GEH to W. Huggins, Feb. 24, 1909, HPM.

17. [GEH] to [ECH], Apr. 7, Apr. 12, Apr. 25, Apr. 29, July 13, 1909, HPM.

18. [GEH] to [ECH], May 10, June 10, June 13, 1909; H. Camberou to GEH, May 26, 1909; LD to GEH, June 8, 1909; all at HPM.

19. GWR to [L]D, June 22, 1909, SGA; LJ to Solar Observatory, Aug. 10, 1909, HPM.

20. [GEH] to WSA, Jan. 20, 1909; GEH to WWC, Jan. 22, 1909; both at HPM. GEH to SN, Jan. 19, 1909, LC.

21. GWR to GEH, Feb. 21, [19]09; GEH to GWR, Mar. 26, 1909; both at HPM.

22. GEH to WSA, Mar. 26, Aug. 4, 1909, HPM.

23. GWR to [L]D, Sept. 7, [19]09, SGA; GWR to [EE]B, Sept. 7, [19]09, VU.

24. GEH to WWC, Sept. 21, Oct. 6, Oct. 11, Oct. 14, Oct. 23, 1909; WWC to GEH, Oct. 1, Oct. 9, 1909; all at SLO. WWC to GEH, Oct. 20, 1909, HPM.

25. GWR to [L]D, Oct. 20, Nov. 17, Dec. 23 [19]09, SGA; [GWR] to [L]D, Oct. 12, 1909; LD to GWR, Oct. 27, 1909; both at HPM.

26. GEH to EBF, Nov. 1, 1909; GEH to D. Gill, Nov. 6, 1909; both at HPM.

27. [GWR] to JDH, Feb. 16, 1910; [GWR] to LJ, Feb. 14, May 31, 1910; LJ to GWR, Feb. 16 (telegram), Feb. 18, May 20, 1910; all at HPM.

28. GEH to GWR, Apr. 11, 1910; GWR to GEH, Apr. 16 [19]10 (telegram); both at HPM.

29. GEH to HMG, Apr. 30, 1910, HHL.

30. "Agreement" [between Mount Wilson Solar Observatory, by GEH, and GWR], May 31, 1910, HPM.

31. WSA to J. L. Wirt, July 20, [1910] (telegram); J. L. Wirt to WSA, July 21, 1910 (telegram); Memorandum, GWR salary payments, Oct. 15, 1919; all at CIW. GWR advertisement, [~May 31, 1910], HPM.

32. F. Le Guet Tully, *Henri Chrétien, des Etoiles au Cinemascope* (Nice: Cercle Scientifique et Technique Henry Chrétien, 1987).

33. HC to G. Bassot, Mar. 8, Apr. 1, May 1, 1910, CHC.

34. D. B. Pickering, *PA*, 40, 595, 1932; Helen Wright, *Explorer of the Universe: A Biography of George Ellery Hale* (New York: E. P. Dutton & Co., 1966), p. 331.

35. HC to G. Bassot, May 20, June [~15], 1910, CHC.

36. GWR to D. Gill, Apr. 30, May 21, 1910, RGS; GWR to EEB, May 13, June 3, 1910, VU.

37. GWR, *MNRAS*, 70, 632, 1910.

38. GWR, *Ap. J.*, 32, 26, 1910.

39. GWR, *Harper's Monthly*, 121, 740, 1910.

40. HC, Annual report, Feb. 10, [1911], CHC; HC, *Revue d'Optique*, 1, 13, 49, 1922; GWR, *JRAS Canada*, 22, 159, 1928.

41. GWR to [L]D, May 16, 1910; GEH to [L]D, July 9, 1910; both at SGA. [GEH] to LJ, June 10, 1910; LJ to Mount Wilson Solar Observatory, June 12, July 27, Aug. 25, Oct. 10, 1910; [WSA] to LJ, Sept. 25, 1910; all at HPM.

42. *Transactions of the International Union for Co-operation in Solar Research*, 3, 1911. This volume is the source for the remainder of the discussion of this meeting, unless otherwise referenced.

43. GWR to ET, July 18, 1910, APS.

44. [WSA] to HHT, July 1, 1910; GEH to [WS]A, July 16, 1910; J. C. Kapteyn to GEH, July 18, 1910; [WSA] to EBF, July 22, 1910; [WSA] to [GE]H, July 22, 1910; HHT to [GE]H, July 30, 1910; HPM; FHS to EBF, Aug. 5, 1910; all at YOA.

45. [GEH] to [ECH], Aug. 6, 1910, HPM.

46. [WSA] to GEH, Sept. 14, 1910, HPM.

47. GWR to [EE]B, Aug. 15, [19]10, VU; GWR to ET, Sept. 7, Sept. 12, 1910, APS; GWR to D. Gill, Sept. 17, 1910, RGS; S. D. T[ownley], *PASP*, 22, 199, 1910.

48. GWR to ET, Sept. 27, 1910, APS; GWR to HC, Sept. 23, Sept. 24, Oct. 14, Nov. 13, 1910, CHC.

49. [WSA] to JDH, Sept. 14, Sept. 19, Oct. 18, Oct. 31, 1910; JDH to WSA, Sept. 16, 1910; GEH to JDH, Sept. 22, 1910; JDH to GEH, Sept. 26, Oct. 6, 1910;

EEB to JDH, Sept. 22, 1910; all at HPM. GEH to WWC, Sept. 22, 1910; [WWC] to GEH, Sept. 29, 1910; both at SLO.

50. GWR to ET, Sept. 27, Dec. 10, 1910, APS; GWR to HC, Oct. 12, Dec. 10, Dec. 22, 1910; Feb. 10, 1911; all at CHC. GWR to EEB, Dec. 8, 1910, VU.

51. GWR to ET, Oct. 18, Nov. 7, 1910, APS; GWR to D. Gill, Nov. 27, 1910, RGS.

52. GWR to [HN]R, Nov. 9, 1910, PUA.

53. GEH to JDH, Nov. 6, 1910; [WSA] to JDH, Nov. 14, 1910; JDH to WSA, Dec. 12, 1910; JDH to GEH, Dec. 15, 1910; J. C. Kapteyn to GEH, Nov. 9, 1910; all at HPM.

54. GEH to JDH, Dec. 12, 1910, JDH to GEH, Dec. 16, 1910, GEH to [WS]A, Dec. 13, 1910, WSA to GEH, Dec. 28, 1910, Jan. 6, 1911, [WSA] to JDH, Dec. 22, 1910, Jan. 3, 1911, HPM.

55. GEH to WSA, Jan. 12, Jan. 23 (telegram), Jan. 25, 1911.

56. GEH to [H. F.] Newall, Oct. 27, Nov. 2, Dec. 1, Dec. 21, 1910; ECH to WSA, Dec. 24, 1910; all at HPM. GEH to [W. H.] Wesley, Jan. 4, Jan. 13, 1911, RAS; D. Gill to GWR, Dec. 15, 1910, RGS; GEH to HMG, Nov. 9, Dec. 17, 1910, and Jan. 1, Jan. 11, 1911; all at HHL.

57. GEH to [W. H.] Wesley, Jan. 21, 1911; ECH to [W. H.] Wesley, Feb. 27, 1911; both at RAS. GEH to HMG, Feb. 15, Mar. 26, 1911, HHL.

58. GEH to [WS]A, Jan. 23, 1911, HPM.

59. WSA to RSW, Jan. 13, 1911, CIW; [WSA] to [GW]R, Feb. 20, 1911, MWO; GWR to [WS]A, [~Feb. 21, 1911], HPM.

60. GWR to HC, Feb. 10, 1911, CHC; GWR to ET, Feb. 3, Mar. 11, 1911, APS; GWR to EEB, Jan. 11, Feb. 23, Apr. 14, May 8, 1911, VU; *PA*, 19, 386, 1911.

61. [WSA] to GWR, May 16, May 27, Aug. 30, 1911, HPM; [LG]R to RSW, July 27, 1911, CIW; GWR to [RS]W, July 31, 1911; RSW to [WS]A, Aug. 15, 1911, both at MWO.

62. WSA to GEH, Mar. 9, 1911; GEH to [WS]A, Mar. 24, Aug. 30, 1911, [WSA] to EBF, Apr. 20, May 4, 1911; [GEH] to EBF, Nov. 7, Dec. 9, 1911; all at HPM. GEH, Rep. Dir. MWSO, *CIW Yearbook*, 11, 172, 1912.

63. [WSA] to D. Gill, Mar. 24, May 3, 1911; GEH to D. Gill, Apr. 1, 1911; D. Gill to GEH, Apr. 12, 1911; W. B. Hale to WSA, Apr. 10, 1911; all at HPM. [WSA] to [L]D, May 1, 1911; ECH to [L]D, May 22, 1911, SGA.

64. ECH to [W. H.] Wesley, Mar. 11, Apr. 23, Oct. 3, 1911; GEH to [W. H.] Wesley, June 14, 1911, all at RAS. [GEH] to [ECH], July 17, 1911, HPM; GEH to HMG, June 12, June 27, 1911, HHL.

65. GEH to ECH, July 29, July 30, Aug. 5, Aug. 11, Aug. 13, Aug. 15, Aug. 24, Aug. 27, Sept. 9, [Sep 22], 1911; J. G. Gehring to GEH, Sept. 25, 1911; all at HPM. GEH to HMG, July 30, Aug. 13, Sept. 7, 1911, HHL.

66. WSA to GEH, Sept. 14, Sept. 28 (telegram), 1911; GEH to WSA, [~Oct. 1, 1911] (telegram); [WSA] to D. H. Burnham, Oct. 18, 1911; all at HPM.

67. RAS to GWR, May 13, Oct. 12, 1911; GWR to RAS, July 8, 1911; all at ROE.

68. GEH to GWR, Oct. 28, 1911, HPM; GWR to EEB, Dec. 22, 1910; Feb. 6, May 24, July 12, 1911; all at VU.

69. J. H. McBride to GEH, July 11, [19]11; [GEH] to HHT, Dec. 2, 1911; both at HPM. GEH to HMG, Dec. 18, 1911, HHL.

70. R. E. Smith to [C.] Furness, Jan. 18, 1912, Vassar College Library.

71. GWR to ET, Jan. 20, July 8, 1912, APS; GWR to HC, May 15, July 20, Aug. 7, Aug. 16, 1912, CHC.

72. GWR to RAS, May 19, July 8, 1912; RAS to GWR, Jan. 18, 1913; all at ROE. RAS, *Phil. Trans. Roy. Soc. Lon. A*, 213, 27, 1913.

73. GWR to HC, Oct. 2, 1912, HC to GWR, [~Oct. 20, 1912], CHC.

74. ECP to GWR, Aug. 14, 1912, HCO; GWR to ET, Oct. 27, Nov. 17, 1912, Jan. 9, 1913, APS; GWR to HC, Dec. 19, 1912, CHC; *Rumford Fund of the American Academy of Arts and Sciences* (Boston: American Academy, 1950), p. 27.

75. GEH to ECH, Jan. 25, Feb. 5, Feb. 17, Feb. 20, Nov. 1, Nov. 5, 1912, HPM; GEH to HMG, Mar. 13, Mar. 27, Nov. 10, 1912, Jan. 31, 1913, HHL; GEH, *Ap. J.*, 38, 27, 1913.

76. WSA to GEH, Apr. 30, Aug. 4, 1912; GEH to HHT, July 8, 1912; [GEH] to LJ, July 8, 1912; LD to FHS, June 18, 1912; FHS to GEH, Aug. 1, [1912]; [GEH] to [L]D, Aug. 27, 1912; [GEH] to LJ, Oct. 5, Nov. 25, 1912; GEH to D. Gill, Nov. 6, 1912; all at HPM.

77. GEH to D. Gill, Dec. 26, 1912; [WSA] to [L]D, Mar. 13, 1913; both at HPM. GEH to [L]D Mar. 4, 1913, SGA; Rep. Dir. MWSO, *CIW Yearbook*, 12, 195, 1913.

78. GEH to ECH, Nov. 13, 1912, HPM.

79. [GEH] to HHT, Feb. 19, 1913, HPM. Whether or not Ritchey actually suffered epileptic attacks is not a matter of record. He himself reported "fainting spells" to R. S. Woodward (reference 61), and Dr. J. H. McBride, who breached the confidential physician-patient relationship to reveal his "diagnosis" to Hale, knew that by this time his much more important patient hated the optician. Catherine Ritchey Miller, Ritchey's niece, has stated that there is absolutely no history of epilepsy in the family, as she learned from closely questioning her father, who was an M.D. (C. R. Miller to author, Feb. 28, 1986).

80. GEH to WSA, Mar. 11, 1913; GEH to A. S. King, Apr. 9, 1913; GEH to ECH, June 6, June 10, 1913; HHT to [WS]A, June 11, 1913; GEH to WSA, July 1, 1913; all at HPM. GEH to [W. H.] Wesley, May 7, May 18 (telegram), 1913, RAS.

81. WSA to GEH, May 19, 1913; [WSA] to WMG, Oct. 11, Nov. 29, 1913; both at HPM.

82. GWR to ET, July 12, Aug. 5, Sept. 10, Oct. 2, Oct. 14, Nov. 3, Dec. 5, Dec. 30, 1913, APS; GWR to RAS, Feb. 12, Apr. 24, Sept. 30, 1913, ROE.

83. HHT to GEH, June 18, 1912; [WSA] to [HH]T, July 14, Aug. 18, Dec. 24, 1913; all at HPM. Copy of report from WSA to HHT on "Ritchey's 30" mirror for Helwan," Jan. 20, 1914, ROE.

84. GWR to ET, Jan. 15, 1914, APS; RAS to GWR, Mar. 10, 1913, May 29, 1914, ROE.

85. WWC, RGA and, C. D. Perrine to GEH, Jan. 7, 1905; GEH, GWR, and F. Ellerman to WWC, C. D. Perrine, and RGA, Jan. 16, 1905; WWC to GEH, Oct. 21, 1913; all at SLO.

86. GEH to WWC, Nov. 1, 1913; GEH to WWC, Mar. 6, 1914; both at SLO.

87. GWR to WWC, Oct. 18, 1910, Nov. 23, 1913; WWC to GWR, Oct. 22, 1910; Nov. 24, 1913; all at SLO.

88. [WWC], "A Brief History of Astronomy in California," (manuscript), [1914], SLO.

89. [GEH] to WWC, May 27, 1914; WWC to GEH, May 29, 1914; both at HPM.

90. WWC to GEH, May 18, May 22, May 29, 1914; WWC to Z. S. Eldredge, May 29, 1914; all at SLO. GEH to WWC, June 1, 1914; WWC to GEH, June 4, Nov. 9, 1914; both at HPM. Two *additional* letters from GEH to WWC, also dated May 27 and June 1, 1914, are referred to in this correspondence, but no copies still exist. Probably they were more vituperative and were labeled to be destroyed after reading.

91. WWC, "A Brief History of Astronomy in California," in Z. S. Eldredge (ed.), *History of California*, 5, 231 (New York: Century History Co., 1915).

CHAPTER 6 MOUNT WILSON FAILURE: 1914–1919

1. GWR to ET, Aug. 12, 1914, APS; GEH, Rep. Dir. MWSO, *CIW Yearbook*, 13, 241, 1914.

2. GEH to [HMG], June 24, 1914, HHL; GEH to ECH, June 28, [1914], HPM.

3. WSA to RGA, Aug. 7, 1914; GEH to WWC, Oct. 13, Nov. 5, 1914; all at SLO.

4. GEH, Rep. Dir. MWSO, *CIW Yearbook*, 14, 251, 1915.

5. L. Holcomb, *PA*, 22, 471, 1914; S. C. Hunter, *PA*, 23, 341, 1915.

6. GEH to WWC Nov. 5, 1914; Feb. 1, 1915; SLO.

7. GEH to [HMG], Mar. 8–14, Apr. 11, Dec. 12, 1915, Feb. 27, 1916; GEH

to R. C. Maclaurin, June 19, June 20, 1914; both at HHL.

8. GEH to RGA, Dec. 6, 1915, SLO; CGA to GWR, Dec. 7, 1915; GWR to CGA, Dec. 21, 1915; both at SIA.

9. WSA to WWC, Apr. 16, 1914, SLO; [WSA] to GEH, Apr. 29, 1916, HPM; GWR to RSW, May 1, 1916, CIW; [Optical Shop] "Log[s] of 100" mirror," 1–7, Observatories of the CIW, Pasadena.

10. [WSA] to GEH, July 26, Sept. 8, 1916, HPM; GWR to ET, Oct. 13, 1916, APS; GEH, Rep. Dir. MWSO, *CIW Yearbook*, 15, 227, 1916.

11. [A. E. Douglass] to GWR, Oct. 18, 1916, Jan. 4, 1917; GWR to A. E. Douglass, Nov. 6, Nov. 28, 1916, Jan. 15, 1917; all at A. E. Douglass Papers, Special Collections, University of Arizona Library, Tucson.

12. [WSA] to GEH, June 15, Sept. 6, Oct. 12, 1916; GEH to [WS]A, July 15, Nov. 1, Nov. 9, 1916; all at HPM.

13. [WSA] to GEH, June 7, 1916, HPM; WSA to WWC, June 17, Nov. 29, 1916; GEH to WWC, Aug. 1 (telegram), Dec. 1, 1916; all at SLO. The many letters to his wife are [GEH] to [ECH], HPM. Hale's NRC and other preparedness and war activities are fully described by Helen Wright, *Explorer of the Universe: A Biography of George Ellery Hale* (New York: E. P. Dutton & Co., 1966), especially chapter 14.

14. GWR to ET, Jan. 1, Mar. 5, 1917, APS; [WMG] to Cadwalader, Wickersham & [Taft], Oct. 14, 1919; Memorandum, GWR salary payments, Oct. 15, 1919; all at CIW.

15. [WSA] to EEB, Jan. 3, 1917, MWO; Death certificate, James Ritchey, Pasadena, Apr. 19, 1917; *Pasadena Star News*, Apr. 20, 1917; GWR to ET, May 2, 1917, APS.

16. GEH to WWC, Mar. 6, 1917, SLO; WSA to GEH, Mar. 29, May 18, May 28, May 31, 1917; GEH to WSA, Apr. 4, May 29, 1917; [GEH] to W. Crozier, June 9, 1917; E. R. Wilson to WMG, June 19, 1917; all at HPM.

17. WSA to GEH, May 18, July 15, 1917, HPM.

18. *New York Times*, July 29, 1917; *Pasadena Star News*, July 30, 1917; GWR to EEB, Aug. 15, [19]17, VU.

19. GWR, *PASP*, 29, 210, 1917.

20. GWR to ECP, Aug. 15, 1917; ECP to GWR, Aug. 23, 1917; all at HCO.

21. GWR, *PASP*, 29, 257, 1917, 30, 162, 1918.

22. H. D. Curtis, *PASP*, 29, 210, 1917 and 29, 206, 1917.

23. HS to HNR, Sept. 3, 1917, HCO; HS, *PASP*, 29, 213, 1917.

24. GWR to ET, Oct. 8, 1917, APS.

25. [WSA] to EEB, Feb. 8, 1917, MWO; FGP, *PASP*, 29, 256, 1917; GWR, *PASP*, 29, 256, 1917 and 30, 163, 1918.

26. [WSA] to EEB, Oct. 6, Oct. 29, 1917; EEB to GWR, Dec. 6, 1917; all at MWO. GEH to WWC, Oct. 19, 1917, SLO.

27. GEH to WWC, Mar. 29, Apr. 7, 1916; WWC to GEH, Apr. 6, 1916; all at SLO.

28. WWC to GEH, Nov. 14, Dec. 10 (2 letters), Dec. 28, 1917, Jan. 15, 1918 (2 letters); WWC to CGA, Jan. 15, Jan. 19, 1918; all at SLO. GEH to WWC, Jan. 7, 1918; WWC to GEH, Jan. 15, 1918; both at HPM.

29. WWC to GEH, Sept. 6, 1918; GEH to WWC, Sept. 14, 1918, Mar. 4, 1919; WWC to CGA, Sept. 26, 1918; all at SLO.

30. WWC to CGA, Oct. 4, Oct. 31, 1919, Nov. 2, 1920, Oct. 19, 1921, SLO.

31. GEH to HHT, July 25, 1917; GEH to ECH, Aug. 17, 1917; [GEH] to EBF, Oct. 8, 1917; GEH to J. Jeans, Nov. 3, 1917; all at HPM. HS to HNR, Oct. 31, 1917, HCO.

32. GEH to ECH, May 6, May 21, Nov. 5, 1918, HPM.

33. GWR to ET, Dec. 18, [19]17, Feb. 15, 1918, APS.

34. WSA to GEH, Feb. 11, 1918; GEH to WSA, Mar. 5, 1918; FGP to WSA, Mar. 24, Apr. 27, 1918; all at HPM.

35. WSA to GEH, Mar. 22, [~Mar. 31], Apr. 1, 1918; GEH to WSA, Apr. 11, 1918; all at HPM.

36. RGA, *Adolfo Stahl Lectures in Astronomy* (San Francisco: Stanford University Press, 1919), p. 246.

37. WSA to GEH, May 3, 1918, HPM.

38. GEH to HMG, June 21, July 14, 1918, HHL; GEH to WMG, Aug. 29, 1918, HPM.

39. G. R. Nichols to RSW, July 16, 1918; [RSW] to G. R. Nichols, July 17, 1918; [RSW] to G. E. Tripp, July 30, 1918; [RSW] to GEH, Aug. 13, Aug. 21, 1918; all at CIW.

40. RSW to E. S. Hughes, Aug. 20, 1918; GEH to RSW, Aug. 19, Aug. 29, Sept. 11, 1918; WSA to RSW, Sept. 11, 1918; [RSW] to WSA, Sept. 13, 1918; all at CIW.

41. GWR to W. C. North, Aug. 19, 1918; H. P. Bailey to GWR, Aug. 27, 1918; both at HPM.

42. GEH to WSA, Sept. 11, 1918 (2 letters), HPM.

43. GEH to HMG, Oct. 3, 1918, HHL; GEH to ECH, Oct. 23, 1918, HPM.

44. WSA to GEH, Sept. 7, 1918; GEH to [RS]W, Sept. 16, 1918; RSW to WSA, Sept. 17, 1918; all at HPM.

45. J. G. Scrugham to RSW, Oct. 30, 1918; [RSW] to J. G. Scrugham, Nov. 1, 1918; GWR to RSW, Feb. 6, Feb. 20, 1919; J. I. Corbett to GWR, Feb. 14, 1919; RSW to GWR, Feb. 26, 1919; all at CIW. WSA to RSW, Oct. 29,

Nov. 9, 1918; RSW to WSA, Nov. 4, Nov. 8, 1918; all at HPM.

46. [WSA] to RSW, Dec. 2, Dec. 30, 1918; WSA to GEH, Dec. 8, 1918, Feb. 13, Mar. 7, Mar. 20, Apr. 3, 1919; GEH to WSA, Apr. 19, 1919 (telegram); all at HPM.

47. GWR to ET, Jan. 15, 1919, APS.

48. GEH to HMG, Jan. 6, 1919, HHL; WSA to GEH, Feb. 2, 1919; GEH to WSA, Feb. 8, Feb. 27, Mar. 8, 1919; all at HPM.

49. GEH to WSA, Mar. 12, 1919; WSA to GEH Mar. 23, 1919; both at HPM. WSA to RSW, Mar. 14, 1919; [RSW] to WSA, Mar. 20, 1919; both at CIW.

50. [WSA] to GEH, Mar. 15, 1919; GEH to WSA, Apr. 6, 1919; both at HPM. WSA to RSW, Apr. 5, 1919.

51. GEH to WWC, May 13 (telegram), May 23, 1919, SLO.

52. WWC, *PASP*, 31, 249, 1919.

53. WWC to GEH, June 9, 1919, SLO.

54. WSA to GEH, June 29, Aug. 28 (telegram), 1919, HPM.

55. WWC to GEH, Aug. 30, 1919, SLO.

56. J. H. Moore, *PASP*, 31, 238, 1919; HS to HNR, June 25, 1919, HCO.

57. GEH to HHT, June 24, 1919; GEH to EBF, July 1, 1919; GEH to FHS, July 16, 1919; all at HPM.

58. GWR to ET, Sept. 16, 1919, APS; A. van Maanen, *Ap. J.*, 44, 331, 1916.

59. GWR to HHT, Aug. 15, 1919; GEH to EBF, Oct. 13, 1919; both at HPM. GEH to WWC, Sept. 12, 1919, SLO.

60. GEH, *PASP*, 31, 257, 1919.

61. GWR to ET, Oct. 15, 1919, Jan. 24, 1920, APS.

62. GEH to RSW, Aug. 19, 1919, CIW.

63. GEH to RSW, Aug. 24, [1919] (telegram), CIW.

64. RSW to GEH, Aug. 26, 1919, HPM.

65. [RSW] to Cadwalader, Wickersham & Taft, Oct. 3, 1919, CIW.

66. Cadwalader, Wickersham & Taft to RSW, Oct. 8, 1919, CIW.

67. [WMG] to Cadwalader, Wickersham & Taft, Oct. 14, Oct. 15, 1919, CIW.

68. Cadwalader, Wickersham & Taft to RSW, Oct. 22, 1919, CIW.

69. GEH to RSW, Oct. 6, Oct. 8, Oct. 11 (telegram), Oct. 22, 1919; RSW to GEH, Oct. 13, 1919; all at HPM.

70. RSW to GEH, Oct. 27, Oct. 31, 1919 (both telegrams); GEH to RSW, Oct. 30, 1919; all at HPM.

71. [GEH] to GWR, Oct. 31, 1919, HPM.

72. GEH to RSW, Nov. 5, 1919; RSW to GEH, Nov. 12, 1919; GEH to GWR, Nov. 5, Nov. 20, Nov. 24, 1919; GWR statement, Nov. 22, 1919; GWR to GEH, Nov. 24, 1919; all at HPM.

73. GWR to GEH, Nov. 29, 1919; GEH to RSW, Dec. 2, 1919; GEH to GWR, Jan. 19, 1920; all at HPM.

74. GEH to HMG, Oct. 14, 1919, HHL; GEH to HHT, Nov. 3, 1919, HPM.

75. AHJ to D. J. Mills, June 24, 1964, MWO.

CHAPTER 7 AZUSA: 1919–1924

1. GWR to ET, Jan. 24, 1920, Jan. 30, 1921, APS.

2. GWR to C. R. Cross, Apr. 8, 1919, American Academy of Arts and Sciences.

3. GWR to ET, [~May 1, 1922], APS.

4. HC, *Revue d'Optique*, 1, 13, 49, 1922.

5. HC to K. Schwarzschild, Feb. 28, 1913, AIP.

6. GEH, Rep. Dir. MWSO, *CIW Yearbook*, 11, 172, 1912.

7. HC, *Calcul des Combinaisons Optiques*, 4th edition, Libraire de Bac, Paris, [1958], p. 384.

8. GWR to ET, July 17, Aug. 13, 1923, APS.

9. GEH to HMG, Mar. 31, 1920, HHL; GEH to ECH, [Apr 11], Apr. 22, 1920; GEH to WSA, Apr. 17, 1920; both at HPM. GEH to JCM, Apr. 7, 1920, LC.

10. GEH to HMG, Aug. 29, Oct. 10, Nov. 11, 1920, HHL; GEH to JCM, Oct. 22, Nov. 4, Nov. 27, 1920; JCM to GEH, Oct. 27, Nov. 11, Dec. 3, 1920; all at HHL.

11. GEH to JCM, Sept. 6, Dec. 16, 1920; JCM to GEH, Dec. 3, Dec. 23, 1920; all at LC. GEH to HMG, Dec. 18, 1920, HHL; GEH, Rep. Dir. MWO, *CIW Yearbook*, 19, 209, 1920.

12. GEH to HMG, Feb. 20, Sept. 22, 1921, HHL; GEH to JCM, Jan. 7, Feb. 10, May 7, May 29, June 5, Aug. 27, 1921, HHL.

13. FGP to GEH, Aug. 17, Sept. 21, 1921, HPM; FGP, *Ap. J.*, 46, 24, 1917, *Ap. J.*, 51, 276, 1920; GEH, *PASP*, 32, 112, 1920.

14. GEH to JCM, Dec. 23, 1921, Jan. 5, Feb. 9, Mar. 13, Mar. 17, 1922; JCM to GEH, Mar. 18, Mar. 25, 1922; all at LC. GEH to HMG, Feb. 2, Feb. 5, Mar. 28, 1922, HHL.

15. GEH to JCM, May 16, 1922; JCM to GEH, May 23, 1922; both at LC. JCM to GEH, Apr. 22, 1922, HPM.

16. GEH to JCM, June 30, July 3, Aug. 8, Sept. 4, Nov. 17, 1922; Jan. 5, Feb. 16, Mar. 2, May 10, July 17, 1923; all at LC. GEH to HMG, June 29, 1922, HHL.

17. GEH to JCM, Oct. 8, Dec. 3, 1922, Mar. 2, Mar. 6, 1923, LC; GEH to HMG, Sept. 7, 1922, HHL.

18. GEH to JCM, Mar. 29, 1923, HPM.

19. GEH to HMG, June 12, 1922, June 1, 1923, HHL; GEH to JCM, Sept. 17, 1922, Apr. 3, June 8, 1923; JCM to GEH, May 4 (cablegram), May 7, 1923; all at LC. GEH to ECH, Apr. 18, 1922; Observatory Staff to GEH, July 2, [1923] (cablegram); JCM to GEH, Dec. 22, 1923; all at HPM.

20. GEH to HMG, Nov. 3, 1922, HHL; GEH to JCM, Nov. 9, 1922, LC.

21. GEH to JCM, Apr. 29, May 1, Aug. 24, 1923, LC; GEH to HMG, Sept. 4, 1923, HHL.

22. GEH to JCM, Oct. 31, Nov. 25, 1923; JCM to GEH, Nov. 10, 1923; all at LC.

23. GEH to JCM, Dec. 3, 1923, Jan. 21, Feb. 10, 1924, LC; GEH to EBF, Oct. 11, Nov. 22, Dec. 27, 1923, YOA.

24. B. Baillaud to GEH, Mar. 30, July 18, 1920; GEH to B. Baillaud, May 21, Nov. 20, 1920; all at HPM.

25. *L'Astronomie*, 38, 210, 1924; *L'Astronomie*, 39, 371, 1925; *L'Astronomie*, 53, 76, 1939; *L'Illustration*, Apr. 24, 1926; *Cincinnati Post*, Sept. 24, 1938; *Cincinnati Times-Star*, Sept. 24, 1938; *Cincinnati Enquirer*, Sept. 25, 1938.

26. A. Danjon, *Courte Histoire de l'Observatoire de Haute Provence*, OHP 1965; C. Fehrenbach, *Histoire et Avenir de l'OHP*, ed. A. A. Chalabaev et M. J. Vin, OHP 1987, p. 1. All the descriptions of the French telescope project to the end of the chapter are based on these two sources, except where other references are given.

27. A. Danjon, "Avant-Projet d'Organisation d'un Observatoire d'Astronomie Physique," July 15, 1923, CHC.

28. A. Dina to HC, July 23, Aug. 19, 1923; HC to A. Dina, July 27, Aug. 23, 1923; all at CHC.

29. A. Dina to HC, Aug. 30, 1923; HC to [A. Dina], Sept. 9, 1923; M. W. Dina to [M.] Chrétien, Oct. 8, 1923; all at CHC.

30. A. Danjon to HC, Oct. 14, 1923; HC to A. Danjon, Oct. 17, 1923; both at CHC.

31. A. Dina, Sept. 26, 1923, "Notice sur les télescopes" (manuscript); [G.-A.] Ferrié to HC, Oct. 9, 1923; HC to [G.-A.] Ferrié, Oct. 29, 1923; HC [~Oct. 29, 1923], "Re sujet de la note de M. Dina Sur les télescopes" (manuscript); all at CHC.

32. C. Fehrenbach, *L'Astronomie*, 81, 328, 1967.

33. C. Fehrenbach, *Des Hommes, des Télescopes, des Étoiles* (Paris: CNRS, 1990), pp. 68–72.

34. GWR to ET, [~Mar. 20, 1924], APS.

CHAPTER 8 PARIS: 1924–1930

1. GWR to ET, [~Mar. 20], Apr. 2, 1924; ET to GWR, Mar. 25, 1924; both at APS.

2. *L'Astronomie*, 38, 274, 1924; *CR*, 179, 1482, 1924; *New York Times*, Oct. 27, 1924.

3. GWR to ET, June 12, 1924, APS.

4. GWR to HC, May 18, May 26, 1924, CHC.

5. [G.-A.] Ferrié to HC, May 17, 1924, CHC.

6. A. Danjon, *Courte Histoire de l'Observatoire de Haute Provence*, OHP 1965; C. Fehrenbach, *L'Astronomie*, 81, 328, 1967; C. Fehrenbach, *Histoire et Avenir de l'OHP*, ed. A. A. Chalabev et M. J. Vin, OHP 1987, p. 1; C. Fehrenbach, *Des Hommes, des Télescopes, des Étoiles* (Paris: CNRS, 1990). All the descriptions of the French project in this chapter not otherwise referenced are based on these four sources. The patent is "Brevet d'Invention," Dec. 11, 1924, RBC.

7. AT to WR, July 14, 1926, RAC.

8. GWR to HC, May 23, 1924, CHC; GWR to ET, June 17, 1924, APS.

9. GWR to ET, July 1, July 9, 1924, APS.

10. GWR to ET, Sept. 30, 1924, APS.

11. "Father" [GWR] to W[GR], Jan. 24, 1925, RBC; GWR to ET, May 16, 1925, APS; GWR, *CR*, 181, 208, 1925.

12. GWR, *CR*, 185, 758, 1927.

13. GWR, *L'Astronomie*, 41, 529, 1927.

14. GWR to W. H. Evans, Feb. 28 [19]25, NO.

15. [G.-A.] Ferrié to HC Feb. 25, 1925, CHC.

16. LGR to ET, May 1, 1925, APS.

17. GWR to HC, Mar. 3, Nov. 17, 1925; A. Couder to HC, Oct. 19, 1925; all at CHC.

18. A. Danjon, "Sur les Projects Relatifs au Grand Télescope" (manuscript), Dec. 17, 1925, CHC.

19. GWR to HC, Dec. 22, 1925, Jan. 7, Jan. 26, 1926, CHC.

20. A. Dina to HC, Mar. 19, 1926; GWR to HC, Mar. 23, 1926; both at CHC. GWR to [L]D, Mar. 25, 1926; GWR, "Liste des different travaux exécutés . . ." (manuscript), Mar. 27, 1926; both at SGA.

21. AT, "Memorandum of conversation with General Ferrié . . ." (manuscript), Sept. 14, 1926, RAC.

22. "Extrait des Minutes de Greffe du Tribunal Civil de la Seine Seant au Palais de Justice a Paris," Mar. 7, 1928, CHC.

23. J. Baillaud to HC, July 17, Aug. 2, 1926; H. Deslandres to HC, July 22, 1926; A. de la Baume Pluvinel to HC, Aug. 2, Aug. 7, 1926; all at CHC.

24. HC to F. A. Bouelle, July 20, 1926, CHC.

25. GWR to HC, Apr. 7, July 15, July 22, Aug. 18, 1926, CHC.

26. GWR to HC, Aug. 26, Aug. 31, 1926; GWR, blueprint, Aug. 25, 1926; all at CHC.

27. AT to WR, Sept. 2, Oct. 21, 1926; AT, Memorandum of visit to Paris Observatory, Sept. 17, 1926; J. Baillaud to J. D. Rockefeller, Jr., Oct. 7, 1926; AT, Memorandum of conversation with Fabry, Oct. 12, 1926; E. Picard to AT, Oct. 22, 1926; all at RAC.

28. GWR to C. A. Chant, Aug. 9, Oct. 19, 1926, DDO.

29. EBF to GWR, Oct. 16, 1914; C. C. Crump to EBF, Apr. 12, [~June 26], Sept. 16, Sept. 27, 1926; EBF to C. C. Crump, Sept. 21, 1925, June 7, Sept. 21, Sept. 22, Oct. 25, 1926; OS to GWR, Oct. 23, 1933; all at YOA. EBF to HYB, Feb. 25, 1926, UT.

30. GWR to HC, Aug. 19, Sept. 14, Sept. 21, [Sept. 28], 1926, CHC.

31. GWR to HC, Oct. 31, 1926, CHC.

32. AT, Memorandum of conversation with E. Picard, Dec. 29, 1926, RAC; [B.] Baillaud, *Rapport Annuel sur l'Etat de l'Observatoire de Paris*, 1926.

33. A. Couder, "Rapport sur la Fabrication et l'Utile d'un Mirior de Télescope de 80 cm de Diametre" (manuscript), Apr. 6, 1927, SGA; GWR to HC, Apr. 8, 1927, CHC.

34. GWR, Shop notebook, [1926–1932], RBC; GWR to HC, June 29, 1927, CHC; [H.] Deslandres, *Rapport Annuele sur l'Etat de l'Observatoire de Paris*, 1927.

35. GWR and HC, *CR*, 185, 265, 1927, reprinted also in *L'Astronomie*, 41, 541, 1927.

36. H. B. Hinton to HC, Sept. 15, 1927; W. K. Castleton to C. Fabry, Sept. 29, 1927; W. K. Castleton to HC, Oct. 4, Oct. 11, 1927; W. W. Crotch to HC, Oct. 28, Nov. 4, 1927; all at CHC.

37. HC to GWR, [~Feb. 22, 1927]; GWR to HC, Feb. 25, June 9, June 13, 1927; all at CHC.

38. GWR, *CR*, 185, 1024, 1927.

39. GWR, *CR*, 185, 640, 1927.

40. GWR per LGR to HC, July 19, 1927; GWR to [A.] de Gramont, June 22, [1930]; both at CHC. GWR to ET, Nov. 16, 1927, APS.

41. VN to HC, Aug. 6, 1927, CHC; Z. Kopal, *Of Stars and Men: Reminiscences of an Astronomer* (Bristol: Adam Hilger, 1986).

42. D. B. Pickering, *PA*, 35, 258, 1927.

43. D. B. Pickering, *PA*, 35, 550, 1927.

44. *PA*, 35, 301, 1927; D. B. Pickering, *PA*, 36, 135, 1928.

45. GWR to HC, Oct. 22, 1927; Mar. 12, Mar. 30, Apr. 13, 1928; all at CHC.

46. GWR, *L'Astronomie*, 42, 27, 1928; *L'Astronomie*, 42, 60, 1928; *L'Astronomie*, 42, 121, 1928.

47. GWR, *L'Astronomie*, 42, 169, 1928.

48. GWR, *L'Astronomie*, 42, 225, 1928; *L'Astronomie*, 42, 281, 1928.

49. GWR, *JRAS Canada*, 22, 159, 1928.

50. GWR, *JRAS Canada*, 22, 207, 1928; *JRAS* Canada, 22, 303, 1928.

51. GWR, *JRAS Canada*, 22, 359, 1928.

52. GWR, *JRAS Canada*, 23, 15, 1929; *JRAS Canada*, 23, 167, 1929.

53. GWR, Shop notebook, [1926–1932], RBC.

54. GWR, *L'Astronomie*, 42, 169, 1928.

55. HC, *CR*, 185, 1125, 1927.

56. GWR, *JRAS Canada*, 23, 15, 1929; *JRAS Canada*, 23, 167, 1929.

57. GWR to RAS, Nov. 6, 1927; RAS to J. Jackson, Nov. 17, 1927; both at RAS.

58. GWR to T. E. R. Phillips, Nov. 29, 1927, Jan. 12, Feb. 23, Apr. 16, 1928, RAS.

59. *Observatory*, 51, 151, 1928; *Observatory*, 51, 177, 1928; *Science*, 67, 602, 1928; *L'Astronomie*, 42, 411, 1928; Royal Astronomical Society Club dinner lists, (815) May 11, 1928 (privately printed).

60. GWR, *Trans. Opt. Soc.*, 29, 197, 1928.

61. E. Touchet, *L'Astronomie*, 42, 333, 1928; *L'Astronomie*, 42, 350, 1928; "Transparencies and Negatives of G. W. Ritchey's Photographs" (manuscript), [~May 1, 1928]; GWR to HC, May 26, [1928]; Invitation to exhibition, June 14, 1928; all at CHC.

62. [A.] de Gramont to HC, May 21, 1928, CHC.

63. *L'Astronomie*, 42, 258, 1928.

64. *New York Times*, July 23, Sept. 30, 1928.

65. HS to WSA, Oct. 11, 1926, MWO.

66. HS to GWR, Oct. 28, 1926; GWR to HS, Nov. 27, Dec. 27, 1926; all at HCO.

67. HS to GWR, Dec. 14, Dec. 21, 1926, Jan. 8, 1927; GWR to HS, Feb. 25, 1927; all at HCO.

68. HS to GWR, Mar. 1, Sept. 19, Nov. 3, 1927; GWR to HS, Oct. 5, Oct. 22, 1927; all at HCO.

69. GWR to ET, Nov. 16, 1927, APS; GWR to HS, Nov. 28, 1927, HCO; EBF to C. C. Crump, Feb. 4, 1928, YOA.

70. HS to JCM, Sept. 23, 1927; JCM to HS, Sept. 26, 1927; both at CIW.

71. HSP to JCM, Dec. 19, 1927; [WMG] memo "concerning Dr. Pritchett's

letter of December 19, 1927," Dec. 21, 1927; JCM to HSP, Jan. 11, 1928; all at CIW.

72. GWR to [JC]M, Feb. 11, 1928, CIW.

73. JCM to WSA, Feb. 17, 1928; WMG to WSA, Feb. 25, 1928; both at MWO. WMG to GWR, Feb. 25, 1928, CIW.

74. WSA to JCM, Mar. 3, 1928, MWO.

75. WMG to WSA, Mar. 8, 1928; JCM to WSA, Apr. 11, 1928; both at MWO.

76. GWR, *L'Astronomie*, 42, 362, 1928.

77. GWR to JCM, June 22, 1928, CIW; GWR to Director of National Park Service, July 5, 1928, LC; GWR to HC, June 26, [1928], CHC.

78. GWR to JCM, July 8, 1928; JCM to GWR, July 20, July 24, 1928; all at CIW. A. E. Demaray to JCM, July 18, 1928; JCM to A. E. Demaray, July 24, 1928; JCM to M. R. Tillotson, July 25, 1928; FEW, "Yavapai Point Station report 16–29 July 1928" (manuscript), July 19, 1928; all at LC.

79. C. B. Ford diary extracts, Oct. 18, 1928–Jan. 19, 1929; C. B. Ford to author, Dec. 3, 1984, Jan. 25, Feb. 2, Feb. 12, 1985; C. B. Ford, *Some Stars, Some Music: The Memoirs of Clinton B. Ford* (Cambridge: A.A.V.S.O., 1986).

80. E. M. Antoniadi, *Journal of the British Astronomical Association*, 39, 326, 1929.

81. GWR, *L'Evolution de l'Astrophotographie et les Grands Télescopes de l'Avenir* ([Paris]: Sociéte Astronomique de France, 1929).

82. E. Touchet, *L'Astronomie*, 43, 547, 1929; RAS, *Nature*, 125, 169, 1930.

83. GWR, "Points on the Use on the Sky of the Ritchey-Chrétien Telescope and Other Matters" (manuscript), [~Nov. 17, 1927]; GWR, "Mr. Chrétien" (handwritten memorandum), [~Apr. 1, 1928]; all at CHC.

84. [GWR], *Revue d'Optique*, 8, 76, 1929.

85. [A.] de Gramont to HC, June 20, 1930, CHC; GWR to [A.] de Gramont, June 22, [1930], Archives, Academy of Sciences, Paris; GWR to JR, Jan. 26, [1931]; GWR to JFH, Jan. 27, 1931; both at NA.

86. Photographs of "Field of faint stars in the Milky Way in Cygnus," "Central stars of the Pleiades," and "Photo. at Vallière with the 50-cm Ritchey-Chrétien reflector," [1930], RBC.

87. GWR, *CR*, 191, 22, 1930. In this paper the focal ratio of the telescope is stated to be f/6.25, corresponding to Ritchey's own personal 0.5-meter Ritchey-Chrétien optics, but this is undoubtedly an error. The primary mirror of the f/6.25 optics was not pierced, while pictures of the telescope mounted at Vallière shows that it was pierced, as the f/6.8 which Ritchey had made, and which Ritchey and Chrétien had described and exhibited, was. The scale

of the photographs, as given in terms of the diameter of the field in minutes of arc and millimeters in this paper, corresponds to the f/6.8 reflector.

88. GWR to C. A. Chant, Feb. 26, 1930, DDO; GWR to HC, [~Mar. 20], Apr. 3, 1930, CHC.

89. VN to HC, Mar. 24, 1930, CHC.

90. GWR, *L'Illustration*, Feb. 20, 1930, p. 256.

91. GWR to HC, Dec. 10, 1930, Jan. 29, 1931, CHC.

CHAPTER 9 PASADENA AND PALOMAR: 1919–1931

1. FGP, *PASP*, 31, 276, 1919.

2. [FGP], *PASP*, 31, 294, 302, 1919.

3. GEH, *PA*, 27, 563, 1919.

4. P. W. Merrill, *PASP*, 31, 305, 1919; WSA and AJH, *PASP*, 31, 307, 308, 1919.

5. GEH, *PA*, 27, 635, 1919.

6. GEH, *PA*, 28, 599, 1920.

7. WSA and FHS, *PASP*, 33, 31, 1921.

8. GEH to WWC, Sept. 21, 1919; WWC to GEH, Oct. 1, 1919; W.H. Wright to GEH, Feb. 20, 1920; all at SLO. GEH to EBF, Jan. 9, 1920, HPM.

9. F. L. Gianetti to JCM, Oct. 18, Oct. 27, 1920; JCM to [GE]H, Mar. 26, 1921; all at LC.

10. WCW to Lick Observatory, Feb. 10, [19]21; A. G. Marshall to WCW, Mar. 2, 1921; WCW to A. G. Marshall, Mar. 9, May 2, June 1, June 22, Sept. 26, [19]21; all at SLO.

11. [AHJ] to WJH, Aug. 15, 1921; WJH to AHJ, Aug. 24, Aug. 30, 1921; all at MWO.

12. WCW to A. G. Marshall, Oct. 10, [19]21; [WWC] to WCW, Oct. 18, Oct. 31, 1921; WCW to WWC, Oct. 25, [19]21; all at SLO.

13. WCW to AHJ, June 28, July 28, [19]21; [AHJ] to WCW, July 7, 1921; all at MWO.

14. [AHJ] to WJH, Aug. 31, 1921, MWO.

15. WCW to AHJ, Oct. 11, [19]21; [AHJ] to WCW, Oct. 21, Nov. 1, 1921; [AHJ] to WJH, Oct. 22, 1921; all at MWO.

16. GEH to WCW, Oct. 27, 1921, HPM; [AHJ] to WCW, Nov. 1, 1921, MWO.

17. WCW to AHJ, Jan. 12, [19]22; WCW "To . . . Superintendents . . . of the Public High Schools of . . . Michigan and . . . Ontario," Jan. 1922; [GEH] to [AHJ], [~Jan. 17, 1922]; [AHJ] to WCW, Jan. 18, 1922; [AHJ] to WJH, Jan. 27, 1922; all at MWO.

18. WCW to GEH, Apr. 21, 1922; GEH to WCW, May 4, 1922; both at HPM.

19. [M. Zeigen] to [WCW], Jan. 8, [19]25; EBF to WCW, Feb. 17, 1925; H. D. Curtis to WCW, Apr. 3, 1925; W. C. Rufus to WCW, June 30, 1925; J. E. Mellish to WCW, July 6, 1925; H. B. Weber to [GW]R, June 16, 1930; all at BHL.

20. P. H. Abelson, *BMNAS*, 47, 27, 1975.

21. [GEH] to [FG]P, Apr. 5, 1922; FGP to [GE]H, July 6, 1922; May 23, 1923; all at HPM.

22. WSA to HSP, July 18, 1923, MWO; GEH to JCM, Apr. 15, June 7, 1924; L. F. Barker to GEH, May 19, 1924; GEH to E. D. Burton, May 26, 1924; all at LC.

23. G[EH] to [HMG], Aug. 23, [1924]; Jan. 26, Oct. 29, 1925, [Feb. 28], June 13, Sept. 8, Oct. 22, Dec. 17, 1926; all at HHL.

24. G[EH] to [HMG], Oct. 22, Nov. 10, 1925, Jan. 31, May 3, 1926, Feb. 18, June 11, 1927, HHL.

25. JCM, *CIW Yearbook*, 24, 1, 1925; FEW, *CIW Yearbook*, 26, 383,1927.

26. FEW to FER, Feb. 8, 1926; EBF to FEW, Feb. 11, 1926; both at YOA.

27. H. D. Babcock to [GE]H, May 24, June 21, July 24, 1925, HPM; CGA to WSA, Mar. 20, 1926, MWO; H. W. Babcock, interview, Apr. 12, 1984.

28. FGP, *PASP*, 38, 194, 1926.

29. HNR, *Scientific American*, 135, 174, 1926.

30. OS to EBF, Sept. 13, 1926, YOA.

31. GEH, *The New Heavens* (New York: Charles Scribner's Sons, 1922); GEH, *The Depths of the Universe* (New York: Charles Scribner's Sons, 1924); GEH, *Beyond the Milky Way* (New York: Charles Scribner's Sons, 1926).

32. H. J. Thorkelson to WSA, Nov. 1, 1926; WSA to HSP, Mar. 30, 1927; both at MWO.

33. [GEH] to [ECH], Apr. 26, Apr. 28, May 12, 1927; W. B. Terhune to GEH, Apr. 22, 1927; A. F. Riggs to GEH, May 31, June 8, 1927; GEH to A. F. Riggs, June 8, 1927; all at HPM. WSA to GEH, May 9, 1927, MWO. G[EH] to [HMG], Oct. 13, Dec. 12, 1927, HHL.

34. GEH, *Harper's Magazine*, 156, 639, 1928.

35. GEH to [WS]A, Mar. 15, 1928, MWO.

36. [WSA] to GEH, Mar. 30, 1928, MWO.

37. [GEH] to [ECH], Apr. 10, Apr. 12, Apr. 17, Apr. 21, 1928, HPM. GEH to WSA, Apr. 12, 1928; WSA to GEH, Apr. 13, 1928 (both telegrams); both at MWO. GEH to WWC, July 18, 1928, SLO.

38. JCM to GEH, Apr. 3, 1928, HPM.

39. [GEH] to [ECH] May 3, May 8, May 11, May 14, 1928; JCM to WR, May 3, 1928; JCM to GEH, May 7, 1928; all at HPM. GEH to WSA, May 4 (telegram), May 6, 1928, MWO.

40. GEH to [WS]A, Apr. 7, Apr. 28, 1932, MWO.

41. H[MG] to [GEH], Apr. 18, [1928], HPM; G[EH] to [HMG], June 2, 1928, HHL.

42. [GEH] to [J. J. Carty], June 12, 1928; G[EH] to [HMG], June 15, 1928; [WR], "200-inch Reflecting Telescope to be included in formal request from the Trustees of the California Institute of Technology," [~June 6, 1928]; [GEH], "Astrophysical Observatory of the California Institute of Technology," [~June 12, 1928]; all at HHL.

43. I. S. Bowen, *BMNAS*, 36, 1, 1962; JCM to WSA, Aug. 19, 1928; JCM to WMG, Aug. 19, 1928; all at LC.

44. GEH to [HH]T, Apr. 27, Apr. 28, 1929, HPM. These are successive drafts of a letter that Hale never sent (as he explained in GEH to HHT, Dec. 17, 1929, HPM). In them he stated that they confirmed a long conversation he had had with Turner a few days previously, in London.

45. GEH to [JA]A, June 29, June 30, 1928, HPM.

46. WSA, JAA, F[G]P, FHS to [GE]H, July 3, 1928, CIT.

47. C. E. St. John, "Report of Investigations Suggested by Dr. Hales Cablegram of July 8, 1928" (manuscript), Aug. 30, 1928, CIT.

48. JAA to [GE]H, July 24, 1928, HPM; [WSA] to FER, July 28, Aug. 3 (telegram), 1928; FER to WSA, July 31, (telegram and letter), Aug. 6, Oct. 3, 1928; all at MWO. FER to EBF, Aug. 14, Aug. 28, [19]28, YOA.

49. JAA to [GE]H, Sept. 5, 1928, HPM.

50. FGP, *PASP*, 40, 11, 1928.

51. HS to GWR, July 25, 1928; GWR to HS, Aug. 6, 1928; both at HCO.

52. GEH to JAA, July 19, 1928; JAA to [GE]H, Aug. 24, 1928; both at HPM. GEH to [RG]A, Sept. 17, 1928, SLO.

53. JAA to [GE]H, [Sept.] 26, Sept. 28 (telegram), [1928], HPM; WSA to FER, Oct. 31, 1928, MWO; G[EH] to [HMG], Nov. 2, 1928, HHL.

54. O. Gingerich, *JHA*, 21, 77, 1990.

55. [GEH] to [ECH], Oct. 6, 1928, HPM; WSA to JCM, Oct. 24, 1928, LC.

56. W. Kaempffert to "Director, Mount Wilson Observatory," Sept. 5, 1928; WSA to W. Kaempffert, Sept. 5, 1928 (both telegrams); both at MWO.

57. GEH to C. F. Marvin, July 12, Aug. 4, Aug. 21, Oct. 3 (telegram), 1928; C. F. Marvin to GEH, July 18, Aug. 10, 1928; JAA to C. F. Marvin, July 31, 1928; all at HPM.

58. E. Hubble to JCM, Oct. 5, 1928; JCM to E. Hubble, Oct. 8, 1928; JCM to

M. R. Tillotson, Oct. 8, 1928; all at LC. E. Hubble to VMS, Oct. 5, Nov. 2, 1928; VMS to [W. A.] Cogshall, Dec. 11, 1928; VMS to R. L. Putnam, Jan. 12, 1929; all at LOA.

59. GEH to [JA]A, Dec. 7, 1928, July 30, 1930, CIT.

60. R. W. Porter, *PA*, 26, 147, 1918.

61. AGI, *Amateur Telescope Making: Book One*, p. 59 (New York: Scientific American, 1962); B. C. Willard, *Russell W. Porter: Arctic Explorer, Artist, Telescope Maker* (Freeport: Bond Wheelwright Co., 1976).

62. GEH to RGA, Nov. 4, 1929, SLO.

63. GEH, *Harper's Magazine*, 159, 720, 1929.

64. ET to FGP, May 28, 1926; ET to GEH, Mar. 19, Mar. 23, 1928; G. Swope to H. M. Robinson, July 26, 1928 (telegram); all at HPM. GEH to ET, March 26, 1928; ET to GEH, Mar. 28, 1928; both at APS.

65. ALE, "Fused Quartz Mirror 300 Inches Dia." (memorandum), Mar. 30, 1928; GEH to ET, July 11, July 24, 1928; ET to GEH, July 16, 1928; JAA to GEH, Sept 21, 1928; A. L. Willis to JAA, Nov. 20, 1928; all at HPM.

66. JAA to ET, Dec. 28, 1928; ET to GEH, Mar. 7, Apr. 10, May 9 (telegram), June 4, June 19, 1929; GEH to ET, Jan. 22, May 10 (telegram), 1929; all at HPM.

67. ET to GEH, Nov. 8, 1929, Feb. 15, 1930; GEH to ET, Nov. 14, 1929; ALE to JAA, Mar. 4, 1930; GEH to JAA, Mar. 20, 1930; GEH to HHT, Jan. 24, 1930; all at HPM.

68. GEH to ET, Dec. 20, 1930, Feb. 6, 1931, APS; ET to GEH, Jan. 20, Feb. 17, 1931, HPM; GEH to HMG, Oct. 22, 1930, Jan. 15, 1931, HHL.

69. [GEH] to [JA]A, Mar. 21, 1931; GEH to ET, Mar. 5, Apr. 28, 1931; ALE to GEH, Mar. 23, 1931; all at HPM.

70. ET to GEH, May 5, May 22, June 11, July 28, 1931; GEH to ET, May 8, May 27, May 29 (telegram), June 5, July 20, 1931; ALE to GEH, May 22, July 14 (telegram), 1931; all at HPM.

71. GEH to ALE, Sept. 10, 1931; ALE to GEH, Sept. 18, 1931; both at HPM.

CHAPTER 10 MIAMI AND WASHINGTON: 1930–1936

1. G. H. Lutz, *Sky and Telescope*, Nov. 1943, no. 25, p. 18.

2. GWR to HC, Apr. 3, 1930, CHC; GWR, "Application for passport," Apr. 11, 1930, Paris, Department of State Archives, Washington; G. Corbet to GWR, Apr. 13, 1930, RBC.

3. EBF to WCW, Sept. 1, 1927, YOA; *Detroit News*, May 4, June 29, 1930; *PA*, 38, 450, 1930.

4. *Daily Science News Bulletin*, Oct. 9, 1930, HPM; *Los Angeles Times*,

Nov. 9, 1930, *New York Times*, Nov. 9, 1930; C. A. C[hant], *JRAS Canada*, 24, 199, 1930.

5. *Miami Herald*, Dec. 28, 1930, Jan. 1, Jan. 4, Jan. 7, 1931; Detroit Times, Dec. 31, 1930; *New York Times*, Jan. 1, 1931.

6. On the history of the U. S. Naval Observatory, see S. J. Dick and L. E. Doggett (eds.), *Sky With Ocean Joined: Proceedings of the Sesquicentennial Symposia of the U. S. Naval Observatory, December 5 and 8, 1980*. U. S. Naval Observatory, 1983, and further references given in it.

7. H. Plotkin, *Proceedings of the American Philosophical Society*, 122, 385, 1978.

8. GEH to FS, Jan. 12, Feb. 20, 1924; FS to GEH, Jan. 23, 1924; all at YUA. WWC to J. H. MacLafferty, Jan. 14, 1924, BL.

9. Navy Department press release, June 5, 1943, NO.

10. W. D. Uptegraff to D. A. Reed, Nov. 3, 1926; W. D. Uptegraff to S. W. Dempsey, Nov. 3, 1926; W. D. Uptegraff to E. T. Pollock, Dec. 8, 1926, NA.

11. E. T. Pollock to CSF, Jan. 3, 1928, NA.

12. A. Hall to A. Clark & Sons, Nov. 1, 1927; C.A.R. Lundin to A. Hall, Nov. 21, 1927; both at NA.

13. A. Hall to H. D. Curtis, Nov. 1, 1927; F. B. Littell to FS, Nov. 4, 1927; A. Hall to EBF, Nov. 10, 1927; CSF to H. D. Curtis, Mar. 16, 1928; CSF to HNR, Mar. 16, 1928; all at NA.

14. H. D. Curtis to A. Hall, Nov. 4, 1927; EBF to A. Hall, Dec. 14, 1927; HNR to CSF, Mar. 20, [~Mar. 27], 1928; [F]S to [F. B.] Littell, Nov. 8, 1927; FS to CSF, Jan. 9, Jan. 16, 1928; CSF to FS, Mar. 16, Apr. 27, 1928; all at NA. FS to CSF, Apr. 17, 1928, YUA.

15. "Minutes of Conference of Professors Frost, Russell and Shapley with Superintendent of Naval Observatory," Apr. 27, 1928; CSF to H. D. Curtis, Apr. 28, 1928; both at NA.

16. "$160,000 Modernization of Naval Observatory" [~Jan. 1929], (large manuscript sheet), NO.

17. J. S. Hall, *Astronomy Quarterly*, 5, 227, 1987. This paper is an excellent account of Ritchey's years at the Naval Observatory and gives an exceptionally good portrait of Robertson.

18. JR to SN, Dec. 27, 1905, Apr. 26, 1906; SN to JR, Feb. 1, 1906; all at LC. JR, *AJ*, 35, 190, 1924.

19. JR to [CSF], Oct. 25, 1928, NA; JR to [H]C, Mar. 13, 1929, CHC.

20. Hearing before Subcommittee of House Committee on Appropriations Bill for 1930, *Congressional Record, 70th Congress, 2nd Session,* 801, 1930; *Philadelphia Inquirer*, Feb. 4, 1929.

21. WWC to C. D. Wilbur, Feb. 6, 1929; WWC to FS, Feb. 6, 1929; FS to

WWC, Feb. 13, Mar. 19, 1929; WWC to C. F. Adams, Apr. 4, 1929; all at YUA. HS to C. F. Adams, Mar. 16, Aug. 15, 1929, HCO. [WSA] to GEH, Sept. 11, 1929; [WSA] to G. K. Burgess, Sept. 14, 1929; [WSA] to W. S. Eichelberger, Sept. 16, 1929; all at MWO. GEH to A. L. Barrows, Sept. 13, 1929; A. L. Barrows to [C. F. Adams], Sept. 14, 1929; E. W. Brown to A. L. Barrows, Sept. 17, 1929; all at NAS.

22. GEH to FS, Oct. 23, 1929; FS to FER, Jan. 4, 1930; FER to FS, Jan. 1930; all at YUA. W. H. Wright to JFH, Sept. 28, 1935, Jan. 20, 1936, SLO.

23. C. F. Adams to HS, Aug. 21, 1929, HCO; A. L. Barrows to WSA, Sept. 17, 1929, NAS; FS to GEH, Nov. 11, 1929; FS to WWC, Nov. 11, 1929; both at YUA.

24. *Hearings before Committee on Naval Affairs of the House of Representatives, 1929–30, 71st Congress,* Feb. 27, Mar. 25, Apr. 16, 1920; *Congressional Record,* 71st Session, 2nd Session, 1930, HR 9370, "A bill to provide for the modernization of the U.S. Naval Observatory," Apr. 6, May 10, May 21, June 3, June 5, June 11, 1930.

25. JFH to [V. K.] Coman, [F. S.] Littell, et al., Nov. 12, 1930, NA.

26. JFH to GWR, Jan. 22, 1931, NA.

27. GWR to JR, Jan. 26, 1931, NA.

28. GWR to JFH, Jan. 28, 1931, NA.

29. GWR to B. F. Ashe, Jan. 25, 1931, University of Miami Archives; GWR to HC, Jan. 29, 1931, CHC; WMB, "Notes," Feb. 6–Dec. 28, 1931, NO. Many of the dates cited in this chapter come from this record, drawn up by Browne from his daily rough notes, soon after the events occurred.

30. JFH to [CSF], Feb. 5, Feb. 11, 1931; CSF to [JFH], Feb. 9, Feb. 16, 1931; all at NA.

31. G. H. Peters to J. W. Fecker, Oct. 22, 1930; J. W. Fecker to [JFH], Nov. 15, 1930; Glaswerk Jena to [JFH], Feb. 28, 1931; J. Vaurie to JFH, Mar. 18, 1931; all at NA.

32. H. E. Burton to [JFH], Feb. 20, 1931; C. B. Watts to [JFH], Feb. 20, 1931; JFH, "Questionnaire for All Astronomers," Feb. 25, 1931; all at NA.

33. H. D. Curtis to CSF, Mar. 19, Apr. 20; HNR to CSF, Mar. 27, 1928; CSF to FS, May 28, 1928; all at YOA. JFH to E. Lohmann, July 30, 1930, NA.

34. JFH to Secretary of the Navy, Feb. 17, 1931; Judge Advocate General to [JFH], Feb. 21, Mar. 18, 1931; JFH to Judge Advocate General, Feb. 24, Mar. 20, 1931; JFH to F. B. Upham, May 31, 1931; Secretary of the Navy to [JFH], Mar. 28, 1931; all at NA.

35. JFH to P. Fox, Feb. 20, 1931; JFH to GWR, Mar. 20, 1931; JFH to G. H. Lutz, Mar. 20, 1931; G. H. Lutz to JFH, Mar. 24, 1931; all at NA. GWR to WMB, Mar. 24, [1931], NO.

36. GWR to B. F. Ashe, May 31, 1931, University of Miami Archives; GWR to [A.] de Gramont, June 12, 1931, Archives, Academy of Sciences, Paris.

37. "Contract for Reflecting Telescope Amount $76,000," June 5, 1931; JFH to Judge Advocate General, June 5, 1931; H. E. Knauss to [JFH], June 10, 1931; all at NA.

38. JFH to Judge Advocate General, June 24, 1931; D. F. Sellers to [JFH], July 10, 1931; "Voucher favor of GWR," July 30, 1931; all at NA.

39. CSF to Directors Yerkes Observatory, . . . May 29, 1928, WMB; "General Plan of Astrographic Work," Feb. 19, 1931; both at NA. *The Tester*, U. S. Naval Air Test Center, Patuxent River, Maryland, Mar. 8, 1963.

40. [GWR], Ledger, May 25, 1931–March 3, 1933, RBC; [WMB], "Construction of 40-inch Reflector. Personnel" [~Oct. 1934]. Copies of this record, and of the similar [WMB] "Construction of 40-inch Reflector. Optical Work," [~Oct. 1934], based on his daily rough notes, were kindly given to me by W. Malcolm Browne. They are referred to as "Personnel" and "Optical Work" in the remainder of this chapter.

41. "Notes"; JFH to Potomac Electric Power Co, July 1, 1931; all at NA.

42. *New York Times*, June 16, 1931; *Roanoke Times*, July 20, 1931; W. H. Cassell to Naval Observatory, July 22, 1931; JFH to W. H. Cassell, July 28, 1931; all at NA.

43. *Washington Post*, Oct. 25, 1931; JFH to WSA, Oct. 28, 1931; JFH to Editor, *Washington Post*, Oct. 28, 1931; J. L. Coontz to JFH, Nov. 2, 1931; WSA to JFH, Nov. 2, 1931; JFH to J. L. Coontz, Nov. 3, 1931; all at NA.

44. "Notes," "Optical Work," Ledger.

45. VN to HC, Aug. 20, Oct. 18, Oct. 29, Nov. 30, Dec. 10 [~Dec. 15], Dec. 20, 1931; [HC] to GWR [~Aug. 15, 1931]; all at CHC.

46. VN, *Czechoslovak Journal of Mathematics and Physics*, 67, 142, 1938; F. Nusl to V. V. Heinrich, [~Jan. 1938], Archives of Charles University, Prague.

47. GWR, Shop Notebook, [1926–1932], RBC.

48. GWR to JFH, July 13, 1932; [JFH] to Chief Bureau of Navigation, July 15, 1932 (2 letters); C. F. Adams to GWR, Aug. 2, 1932; JFH to GWR, Aug. 6, 1932; JFH, "Memorandum regarding delay," Sept. 7, 1933; GWR to Comptroller General, June 2, 1934; JFH to Comptroller General, Dec. 5, 1934; all at NA. "Optical Work," Ledger.

49. JFH to GWR, Oct. 20, Nov. 2, 1932; V. K. Coman to [JFH], Nov. 1, 1932; JFH to G. H. Lutz, Jan. 25, Feb. 27, 1933; JFH to H. Spencer Jones, June 13, 1935; all at NA.

50. V. K. Coman to [JFH], Mar. 31, May 31, 1933; JFH to GWR, May 4, 1933; JFH to Judge Advocate General, May 11, May 16, 1933; JFH to Chief Bureau of Navigation, May 19, 1933; F. B. Upham to Judge Advocate General,

May 27, 1933; Judge Advocate General to Chief Bureau of Navigation, June 3, 1933; JFH to Secretary of the Navy, Sept. 15, 1933; Judge Advocate General to [JFH], Nov. 7, 1933; all at NA. Ledger.

51. "Optical Work."

52. JFH to Astronomical Council, Apr. 27, 1934; JFH to Chief Bureau of Navigation, May 4, 1934; Astronomical Council to JFH, June 16, 1934; M. B. Mott to W. P. Habel, Sept. 14, 1934; all at NA. Annual Report of the Naval Observatory, Report of Superintendent, July 1, 1934, NO.

53. GWR to Secretary of the Navy, May 15, 1934; JFH to Chief Bureau of Navigation, May 16, 1934; Chief Bureau of Navigation to Judge Advocate General, May 17, 1934; Secretary of the Navy to [JFH], May 26, 1934; all at NA.

54. C. B. Ford to GWR, June 5, 1934; GWR to C. B. Ford, June 26, 1934; Clinton B. Ford Personal Files.

55. JFH to W. D. Bergman, Sept. 12, 1933; JFH to W. D. Leahy Oct. 4, 1930, 1933; E. S. Henkel to [W. D. Leahy], Oct. 9, 1933; W. D. Leahy to [JFH], Nov. 7, 1933; JFH to Secretary of the Navy, Nov. 13, 1933; JFH to Chief of Naval Operations, May 2, 1934; all at NA.

56. JFH to Administrative Committee, May 21, 1934, NA. WMB to [H. E.] Burton, Aug. 19, 1934; WMB to [JF]H, Apr. 22, 1935; both at NO.

57. WMB Interview with J. S. Hall and A. A. Hoag, Dec. 7, 1982; JFH to [B. B.] Snowdon, Apr. 8, 1933; JFH to A. B. Dickerson, May 19, 1934; JFH to V. B. Sease, July 2, 1934; A. B. Dickerson to JFH, July 12, 1934; C. E. K. Mees to JFH, July 27, 1934; all at NA.

58. JFH to Secretary of the Navy, Apr. 1, 1932; G. A. McKay to chief of Naval Operations, Apr. 13, June 2, 1932; JFH to Chief of Naval Operations, Apr. 22, 1932; L. H. Reichelderfer to JFH, May 24, 1934; J. C. Gotwals to JFH, May 26, 1932; all at NA.

59. JFH to F. H. Rogers, July 9, 1934, NA.

60. "Optical Work"; JFH to Secretary of the Navy, July 9, 1934; GWR to [JFH], Oct. 24, 1934; Chief Bureau of Navigation to [JFH], Nov. [1], 1934; JFH to Chief of Naval Operations, Nov. 1, 1934; all at NA.

61. GWR to HC, Aug. 27, 1931, Aug. 29, 1932, CHC; GWR to [A.] de Gramont, June 15, 1932, Archives, Academy of Sciences, Paris; JR to JFH, Oct. 17, Oct. 18, 1933, NA; HS to GWR, Aug. 27, 1934, HCO.

62. Reports of Administrative Council, July 12, Nov. 21, [1933], NO.

63. WWC to FS, Dec. 31, 1932, Jan. 3, Feb. 9, Feb. 22, Mar. 21, 1933; WWC to Anon, Mar. 10, 1933; all at YUA. The last letter was to a person close to President Roosevelt. Campbell cut his name out of this copy, which he sent to Schlesinger.

64. JFH to [CSF], Apr. 17, 1931; JFH to R. McLean, May 4, 1931; JFH to C. E. Adams, Nov. 16, 1931; all at NA.

65. GEH to [RG]A, Nov. 4, 1929, SLO; WSA to FER, May 12, 1930, MWO.

66. FER to C. Fabry, Mar. 18, 1931, CHC.

67. FER to WSA, Sept. 15, Oct. 9, 1934; [WSA] to FER, Oct. 4, 1934; all at MWO. FER, *PASP*, 46, 339, 1934; *JOSA*, 25, 42, 1935.

68. [WSA] to FGP, Oct. 12, 1934; FGP to [WS]A, Oct. 21, 1934; both at MWO. FGP, *JOSA*, 25, 156, 1935; Ralph Haupt interviews, Feb. 22, July 9, 1986.

69. [WSA] to FER, Oct. 16, 1934; FGP to WSA, Oct. 31, Nov. 5, Nov. 12, Nov. 22, 1934; FEW to [WS]A, Nov. 6, 1934; all at MWO. FER to RGA, Oct. 22, Nov. 3, [19]34; [RGA] to [FE]R, Oct. 27, 1934; all at SLO.

70. J. D. Phenix to HYB, Feb. 1, 1932, UT; HYB to OS, Feb. 10, Feb. 17, 1933; OS to HYB, Feb. 15, Feb. 21, 1933; OS to H. J. Lutcher Stark, [Feb. 21], [Mar. 6] (telegram), 1933; all at YOA.

71. OS to WSA, June 26, 1933; OS to J. S. Plaskett, June 26, 1933; WSA to OS, June 28, 1933; J. S. Plaskett to WSA, June 29, 1933; OS to R. M. Hutchins, July 26, 1933; R. M. Hutchins to OS, Aug. 12, 1933; all at YOA. See also D. S. Evans and D. J. Mulholland, *Big and Bright: A History of McDonald Observatory* (Austin: University of Texas Press, 1986).

72. GWR to [JFH], Oct. 1, 1934; JFH to General Accounting Office, Oct. 1, 1934; GWR, Sworn statement, [Dec 5, 1934]; JFH to GWR, June 6, 1935; all at NA. S 1036, "Bill authorizing adjustment of the claim of Dr. George W. Ritchey," *Congressional Record*, 74th Congress First Session, Jan. 3–Aug 26, 1935.

73. WMB, "Proposed Program of Observing for the 40-Inch Reflector," Feb. 2, [19]32; JR, "Memorandum on the Use of the 40-Inch Reflector," [Feb. 3, 1932]; WMB to [JFH], July 30, [19]34; all at NO.

74. 40-inch Journal #2, 1934 November to 1940 May 28, NO (afterward referred to as "Journal" in this chapter).

75. *Washington Post*, Jan. 6, 1935; JFH to Editor, *Washington Post*, Jan. 12, 1935; both at NA. JFH, *U. S. Naval Institute Proceedings*, 61, 9, 1935.

76. JFH to [H. E.] Burton, Apr. 3, 1935; WMB to [JF]H, Apr. 22, 1935; WMB, "40-inch Ritchey-Chrétien Reflector, Annual Report for Fiscal Year 1935," June 5, 1935, Journal; all at NO. JFH to Chief, Bureau of Navigation, May 7, 1935, NA. *Washington Herald*, May 5, 1935; WMB, Interview with J. S. Hall and A. A. Hoag, Dec. 7, 1982; Interviews Jan. 14, Jan. 16, June 26, 1985. These interviews are partial sources for many other episodes described in this chapter, but especially this one.

77. JFH to GWR, June 1, 1935; JFH to E. P. Burrell, June 3, 1935; both at NA. Journal, NO.

78. HC to GWR, [~Oct. 20, 1912], CHC; FER to RGA, Nov. 13, 1934, SLO;

JFH to Agfa Ansco Corp. et al., June 18, 1934; JFH to F. O. Strowger, July 5, 1934; JFH to S. L. Boothroyd, Mar. 23, 1935; all at NA. WMB, "Annual Report . . . 1935," June 5, 1935, Journal, NO.

79. [WMB], "Repairs and Improvements Requested for 40-Inch Reflector," [~July 1, 1935]; JFH to Administrative Committee, Oct. 24, 1935; [Administrative Committee] to [JFH], Nov. 7, 1935; J. E. Willis, "The 40-inch Ritchey-Chrétien Telescope of the U. S. Naval Observatory" manuscript, July 18, 1938; all at NO. AGI to GWR, Nov. 27, 1935, SIA.

80. WMB, *AJ*, 46, 145, 1937.

81. A. S. Hickey to GWR, July 3, 1935; JFH to GWR, Sept. 9, Nov. 15, 1935, Feb. 19, 1936; GWR to [JF]H, Sept. 11, Nov. 16, 1935; JFH to G. H. Lutz, Feb. 12, 1940; all at NA.

82. GWR to AGI, Feb. 19, May 10, 1936, SIA; GWR to C. P. Olivier, Feb. 15, 1936, APS; J. E. Willis to JFH, Feb. 26, 1940, NA.

83. HC to GWR, Apr. 13, Oct. 2, 1933; HC to [J]R, June 3, 1935; GWR to HC, Feb. 20, 1936; all at CHC.

84. GWR to W. H. Wright, Aug. 24, Sept. 8, 1935; W. H. Wright to GWR, Aug. 28, 1935; all at SLO. GWR to [C. O.] Lampland, Mar. 3, 1935; C. O. Lampland to GWR, Oct. 14, Nov. 4, Nov. 25, 1935; GWR to VMS, Oct. 25, Nov. 20, 1935; all at LOA. GWR to K. Williams, July 30, Aug. 29, Sept. 12, 1935, RAS.

85. OS to GWR, Oct. 23, Nov. 16, 1933, Feb. 26, 1936, YOA.

86. GWR to K. Williams, [~May 1, 1938], RAS; H. G. Patrick to G. E. Pendray, Aug. 12, 1938, NA.

CHAPTER 11 SOUTHERN CALIFORNIA: 1932–1945

1. *Azusa City Directory*. Baldwin Park, Coulson Directory Co., 1940.

2. Paul Nakada interview, May 2, 1986.

3. Nakada interview; GWR, Shop Notebook, [1926–1932], RBC; GWR to VMS, Oct. 25, 1935, LOA.

4. GWR to K. Williams [~May 1, 1938], RAS; [GWR], draft chapters and articles, mostly undated; some of the last notes dated are Apr. 10, [19]42, Jan. 25, Feb. 25, 1945, RBC.

5. [GEH] to [ECH], Apr. 26, May 3, May 12, May 14, May 27, June 1, 1932, HPM.

6. G[EH] to [HMG], Dec. 13, 1932, HHL; K. Williams to [RG]A, Sept. 29, 1933, SLO.

7. HSP to JCM, Aug. 22, [1934]; JCM to HSP, Aug. 25, 1934; both at LC. GEH to W. H. Wright, Oct. 12, 1934, SLO.

8. Helen Wright, *Explorer of the Universe: A Biography of George Ellery Hale* (New York: E. P. Dutton & Co., 1966). Many of the events not otherwise referenced in this chapter are based on this book, pp. 407–29.

9. G[EH] to [HMG], Jan. 14, 1932, May 15, Sept. 29, Oct. 26, 1933, Jan. 21, 1934, HHL.

10. GEH, *Amateur Telescope Making Advanced, Book One*, ed. AGI (New York: Scientific American, 1937), p. 180.

11. ECH to WSA, Mar. 19, 1934 (telegram), MWO; [GEH] to [ECH], Mar. 16, Mar. 18, 1934; GEH to "President and Executive Committee of the CIW," Mar. 24, 1934; both at HPM. JCM to HSP, Apr. 7, June 11, 1934, LC. The bookkeeper's name has been disguised, but in all other respects this event is described as it occurred.

12. JCM to WSA, Feb. 2, Feb. 18, 1935; [J. Herbert] to JCM, Mar. 2, 1935; all at MWO.

13. JCM, "Walter S. Adams" (memo to file), Nov. 6, 1933, LC; GEH to E. Root and HSP, Apr. 6, 1934; G[EH] to [HMG], Apr. 19, June 17, 1934; all at HHL.

14. E. Hubble to WSA, Apr. 24, 1934; WSA to E. Hubble, May 7, 1934; both at MWO.

15. GEH to RGA, Apr. 11, 1935; GEH to W. H. Wright, Oct. 30, Nov. 29, 1935; all at SLO. G[EH] to [HMG], Apr. 7, 1936, HHL.

16. GEH, *Ap. J.*, 82, 111, 1935; GEH, *Nature*, 127, 221, 1936.

17. G[EH] to [HMG], Apr. 14, 1936, HHL.

18. *Pasadena Star News*, Feb. 7, 1938; *Los Angeles Times*, Feb. 8, Feb. 9, 1938.

19. WSA, *PASP*, 50, 119, 1938; G. Stromberg, *PA*, 46, 357, 1938; JAA, *JOSA*, 29, 306, 1939.

20. WSA, *Ap. J.*, 87, 369, 1938; H. D. Babcock, *PASP*, 50, 156, 1938; T. Dunham, *MNRAS*, 99, 322, 1939.

21. WSA, *BMNAS*, 21, 181, 1940; reference 8.

22. JAA, *PASP*, 51, 24, 1939.

23. R. S. Richardson, *PASP*, 60, 215, 1948.

24. *Azusa Herald and Pomotropic*, Nov. 8, 1945; *PASP*, 57, 319, 1945.

25. *PA*, 54, 51, 1946; D. Hoffleit, *Sky and Telescope*, 5, 53:11, March 1946.

26. F. J. Hargreaves, *MNRAS*, 107, 36, 1947.

27. Navy Department press release, Feb. 9, 1950, NO.

28. Annual Report of the Naval Observatory, Report of Superintendent, July 1, 1938; Annual Report of the Naval Observatory, July 1, 1939; Annual Report of the Naval Observatory, July 1, 1940; J. E. Willis, "The 40-inch Ritchey-Chrétien Telescope of the U. S. Naval Observatory" (manuscript),

July 18, 1938; all at NO. J. E. Willis and H. R. J. Grosch, *AJ*, 50, 14, 1942.

29. JFH to [H. E.] Burton, May 17, 1939; H. E. Burton to [JFH], May 19, June 3, 1939; JFH to W. Davis, June 5, 1939; all at NA.

30. A. H. Carpenter to [AG]I, Dec. 31, 1945; B. King to AGI, Jan. 7, 194[6]; J. Stokley to AGI, Jan. 7, 1946; JFH to AGI, Jan. 11, Feb. 15, 1946; all at SIA. Memo, "Concerning the statement that no astronomer would think of getting a Ritchey (Ritchey-Chrétien) telescope," [~Feb. 10, 1946], NO.

31. J. S. Hall and A. H. Mikesell, *Pub. U.S. Naval Obs.*, 17, 1, 1950.

32. J. S. Hall to VMS, Oct. 18, 1948; J. S. Hall to F. E. Roach, Mar. 25, 1949; all from J. S. Hall personal files.

33. JR to Y. [Chrétien], Dec. 19, 1950; *Congressional Record* [~Dec. 1, 1950]; *Washington Post* [~Dec. 1, 1950]; all at CHC.

34. J. S. Hall to J. Irwin, Oct. 20, 1953, J. S. Hall personal files.

35. J. S. Hall, *Astronomy Quarterly*, 5, 227, 1987. This excellent article is the source for many other of the statements in this chapter not otherwise referenced.

36. J. S. Hall and A. A. Hoag, *Sky and Telescope*, 16, 4, 1956.

37. *Sky and Telescope*, 23, 4, 1962; J. S. Hall, *Bull. AAS,* 14, 834, 1982; A. Meinel to author, Oct. 6, 1990.

38. RAS, *Phil. Trans. Roy. Soc. Lon. A*, 213, 27, 1913.

39. R. Angel, *QJRAS*, 31, 141, 1990.

40. C. A. Coulman, *Annual Review of Astronomy and Astrophysics*, 23, 19, 1985; H. W. Babcock, *Science*, 249, 253, 1990.

41. AHJ, *BMNAS*, 31, 1, 1958.

Bibliography

BOOKS

Fehrenbach, Charles. *Des Hommes, des Télescopes, des Étoiles*. Paris: Editions du Centre National de la Recherche Scientifique, 1990.

Kargon, Robert H. *The Rise of Robert A. Millikan: Portrait of a Life in American Science*. Ithaca, New York: Cornell University Press, 1982.

Kevles, Daniel T. *The Physicists: The History of a Scientific Community in Modern America*. New York: Knopf, 1978.

King, Henry C. *The History of the Telescope*. High Wycombe, England: Charles Griffin & Co., 1955.

Miller, Howard S. *Dollars for Research: Science and Its Nineteenth Century Patrons*. Seattle: University of Washington Press, 1970.

Osterbrock, Donald E. *James E. Keeler: Pioneer American Astrophysicist and the Early Development of American Astrophysics*. Cambridge, Massachusetts: Cambridge University Press, 1984.

Pendray, G. Edward. *Men, Mirrors and Stars* (revised edition). New York: Funk & Wagnalls, 1939.

Smith, Robert W. *The Expanding Universe: Astronomy's Great Debate, 1900–1931*. Cambridge, Massachusetts: Cambridge University Press, 1982.

Willard, Berton C. *Russell W. Porter: Arctic Explorer, Artist, Telescope Maker.* Freeport, Maine: Bond Wheelwright Co., 1976.

Woodbury, David O. *The Glass Giant of Palomar* (revised edition). New York: Dodd, Mead & Co., 1970.

Wright, Helen. *Explorer of the Universe: A Biography of George Ellery Hale.* New York: E. P. Dutton & Co., 1966.

_________. *Palomar: The World's Largest Telescope.* New York: Macmillan, 1952.

PAPERS

Adams, Walter S. "George Ellery Hale 1868–1938," *BMNAS*, 21, 181–241, 1939.

Hall, John S. "Biographical Note: Henri Chrétien and his Magnificent Contribution to the Design of Telescopes," *Bull. AAS*, 14, 834–836, 1982.

_________. "The Ritchey-Chrétien Reflecting Telescope: Half a Century from Conception to Acceptance," *Astronomy Quarterly*, 5, 227–251, 1987.

Hoskin, M. A. "Ritchey, Curtis, and the Discovery of Novae in Spiral Nebulae," *JHA*, 7, 45–53, 1976.

Hoskin, M. A. "The 'Great Debate': What Really Happened," *JHA*, 7, 169–182, 1976.

Joy, Alfred H. "Walter Sydney Adams 1876–1956," *BMNAS*, 31, 1–31, 1958.

Mills, Deborah J. "George Willis Ritchey and the Development of Celestial Photography," *American Scientist*, 54, 64–93, 1966.

Osterbrock, Donald E. "America's First World Astronomy Meeting: Chicago 1893," *Chicago History*, 9, 178–85, 1980.

_________. "The Minus First Meeting of the American Astronomical Society," *Wisconsin Magazine of History*, 68, 108–18, 1985.

_________. "The Quest for More Photons: How Reflectors Supplanted Refractors as the Monster Telescopes of the Future at the End of the Last Century," *Astronomy Quarterly*, 5, 87–95, 1985.

Index

References to illustrations are printed in italics.

About the Author

DONALD E. OSTERBROCK was director of Lick Observatory, University of California, Santa Cruz, from 1973 until 1981. Since then he has continued at the university as professor of astronomy and astrophysics.

He received his Ph.D. in 1952 from the University of Chicago, and in 1986 Ohio State University awarded him an honorary D.Sc. degree. He has also served on the faculties of the California Institute of Technology and the University of Wisconsin.

He is a member of the National Academy of Sciences and a former president of the American Astronomical Society. He has published research papers on gaseous nebulae, interstellar matter, and active galactic nuclei, and is the author of *Astrophysics of Gaseous Nebulae* and *Astrophysics of Gaseous Nebulae and Active Galactic Nuclei.*

The author has also studied and written extensively on the history of astronomy, particularly on the rise of astrophysics and the big-telescope era in the United States. Other books he has written are *James E. Keeler, Pioneer American Astrophysicist: And the Early Development of American Astrophysics*, and with coauthors John R. Gustafson and W. J. Shiloh Unruh, *Eye on the Sky: Lick Observatory's First Century.*